Advances in Analytical Strategies to Study Cultural Heritage Samples, 2nd Edition

Advances in Analytical Strategies to Study Cultural Heritage Samples, 2nd Edition

Guest Editors

Maria Luisa Astolfi
Maria Pia Sammartino

MDPI

Basel • Beijing • Wuhan • Barcelona • Belgrade • Novi Sad • Cluj • Manchester

Guest Editors
Maria Luisa Astolfi
Department of Chemistry
Sapienza University of Rome
Rome
Italy

Maria Pia Sammartino
Department of Chemistry
Sapienza University of Rome
Rome
Italy

Editorial Office
MDPI AG
Grosspeteranlage 5
4052 Basel, Switzerland

This is a reprint of the Special Issue, published open access by the journal *Molecules* (ISSN 1420-3049), freely accessible at: https://www.mdpi.com/journal/molecules/special_issues/W5CEW55MGP.

For citation purposes, cite each article independently as indicated on the article page online and as indicated below:

Lastname, A.A.; Lastname, B.B. Article Title. *Journal Name* **Year**, *Volume Number*, Page Range.

ISBN 978-3-7258-6476-8 (Hbk)
ISBN 978-3-7258-6477-5 (PDF)
https://doi.org/10.3390/books978-3-7258-6477-5

Contents

Preface

This Reprint, *Advances in Analytical Strategies to Study Cultural Heritage Samples—2nd Edition*, brings together a selection of papers that exemplify the rapid evolution and interdisciplinary nature of modern heritage science. Over the past decade, significant progress in analytical chemistry, spectroscopy, imaging, and data-driven methodologies has transformed our ability to investigate cultural heritage materials while minimizing the need for invasive sampling. The contributions collected here highlight how these advances allow researchers to explore the composition, degradation mechanisms, and conservation needs of artworks and historical artifacts with increasing precision.

The purpose of this Reprint is to provide readers with a coherent overview of state-of-the-art analytical approaches applied to cultural heritage research. The articles included were originally published in the Special Issue of *Molecules* devoted to this topic and have been assembled in book form to offer a comprehensive reference for researchers, conservators, and students entering the field. By gathering these studies together, this Reprint aims to illustrate the breadth of current methodologies—from spectroscopy and chromatography to chemometrics, nanomaterials, and non-destructive imaging—and to showcase how such integrated workflows can support both fundamental understanding and practical conservation efforts.

The collection also reflects the deeply interdisciplinary dimension of heritage science, where chemists, physicists, conservators, art historians, archaeologists, and data scientists collaborate to address complex questions related to material characterization, authenticity, provenance, and long-term preservation. By presenting a wide range of case studies, this Reprint demonstrates how analytical research can respond to real problems encountered in museums, archives, archaeological sites, and conservation laboratories.

We hope this Reprint will serve as a valuable resource for the scientific and conservation communities, offering inspiration for future investigations and supporting the continued development of innovative, sustainable, and minimally invasive strategies for the study and preservation of cultural heritage.

Maria Luisa Astolfi and Maria Pia Sammartino
Guest Editors

molecules

MDPI

Editorial

Advances in Analytical Strategies to Study Cultural Heritage Samples—2nd Edition

Maria Luisa Astolfi [1,2,*] and Maria Pia Sammartino [1,†]

1 Department of Chemistry, Sapienza University of Rome, Piazzale Aldo Moro 5, 00185 Rome, Italy; mariapia.sammartino@uniroma1.it
2 Research Center for Applied Sciences to the Safeguard of Environment and Cultural Heritage (CIABC), Sapienza University of Rome, Piazzale Aldo Moro 5, 00185 Rome, Italy
* Correspondence: marialuisa.astolfi@uniroma1.it
† The author is currently retired.

1. Introduction

The study of cultural heritage samples represents a unique intersection of science, history, and art, where the careful application of analytical techniques can uncover invaluable information about materials, manufacturing processes, degradation phenomena, and conservation strategies. In recent years, advances in analytical chemistry, spectroscopy, microscopy, and multi-modal imaging have revolutionized the approach to heritage science, allowing researchers to obtain detailed compositional and structural information while minimizing sample destruction [1]. These methods are particularly crucial for irreplaceable artifacts, historical documents, wall paintings, textiles, and archaeological finds, where traditional sampling strategies may be limited or impossible.

Recent trends in heritage science have emphasized non-invasive and minimally invasive techniques, reflecting both ethical considerations and technological progress [1,2]. For instance, vibrational spectroscopies such as Fourier transform infrared (FTIR) and Raman spectroscopy, in combination with chemometric data analysis, provide precise information on organic and inorganic components without compromising the artifact's integrity [3,4]. Similarly, hyperspectral imaging and portable X-ray fluorescence (pXRF) allow in situ chemical mapping of materials in complex cultural contexts [5]. The application of mass spectrometry, including gas chromatography–mass spectrometry (GC–MS) and ambient mass spectrometry techniques, has further enabled the detailed characterization of organic binders, pigments, and volatile compounds [6–10]. Such technological integration has also been complemented by nanomaterial applications, where engineered nanoparticles and functionalized hydrogels offer innovative strategies for conservation, cleaning, and protective coatings [11,12].

The multidisciplinary nature of heritage research has encouraged collaborations among chemists, materials scientists, conservators, art historians, and archaeologists. This collaborative approach not only improves the reliability of analytical results but also ensures that scientific findings are translated into practical conservation strategies. In addition, computational modeling and data-driven approaches, including multivariate statistical analysis and machine learning algorithms, have become increasingly valuable for interpreting complex datasets and predicting long-term behavior of materials under various environmental conditions [13].

Recent studies have highlighted the importance of developing comprehensive analytical workflows that integrate multiple techniques, allowing for cross-validation of results and deeper insights into the chemical and physical transformations occurring in cultural heritage objects [14,15]. Examples include the identification of metal soap formation in oil paints, the detection of early synthetic dyes in textiles, and the non-invasive characterization of historic print techniques [16–20]. The continuous refinement of these methodologies enhances both preventive conservation strategies and forensic studies, providing robust evidence for provenance, authenticity, and degradation assessment [21–24].

Moreover, the growing awareness of environmental and anthropogenic impacts on heritage materials has reinforced the need for continuous monitoring and tailored interventions [25,26]. From airborne pollutants affecting paper and textiles to light-induced photodegradation of pigments, understanding the mechanisms of deterioration is essential for designing effective conservation protocols [27–31]. By integrating chemical, physical, and environmental data, contemporary heritage science enables informed decision-making for the preservation of artifacts across centuries.

Overall, the contributions to the second edition of the Special Issue "Advances in Analytical Strategies to Study Cultural Heritage Samples" exemplify the diversity and sophistication of current research, illustrating how modern analytical strategies can both expand scientific understanding and support practical conservation efforts.

2. An Overview of Published Articles

The second edition of the Special Issue comprises nine contributions, which collectively highlight state-of-the-art in heritage science analytical strategies. Doménech-Carbó et al. (contribution 1) represented a compelling integration of heritage conservation methodologies with forensic science approaches, addressing a highly sensitive historical context: the mass graves of the Francoism era in Spain (1939–1956). The authors employed a comprehensive multi-analytical strategy to examine personal objects recovered from these sites, including clothing, metal objects, and personal effects, aiming to both preserve the material integrity of the items and extract critical forensic information. Non-destructive techniques such as XRF, FTIR, and optical microscopy were combined with targeted sampling strategies to provide a detailed chemical and physical characterization of the artifacts. The study not only revealed the composition and degradation patterns of the materials but also offered insights into the manufacturing techniques and potential provenance of the objects. By bridging conservation science and forensic analysis, this work underscores the potential of analytical chemistry to support both historical reconstruction and ethical handling of sensitive cultural heritage. Furthermore, it highlights the importance of developing protocols that minimize sample handling while maximizing the information obtainable from fragile and invaluable items, setting a precedent for future studies in similar socio-historical contexts. This contribution exemplifies how interdisciplinary approaches can enhance the understanding of both material culture and historical events, combining methodological rigor with societal relevance.

Chiodini et al. (contribution 2) highlighted the application of modern analytical chemistry techniques to historical book conservation. The authors investigated volatile organic compounds (VOCs) emitted by two rare books, aiming to understand degradation processes and chemical interactions within paper and ink components. Using gas chromatography–mass spectrometry (GC–MS) and headspace analysis, they identified specific VOC profiles associated with oxidation, hydrolysis, and microbial activity. The work demonstrates how VOC monitoring can serve as an early diagnostic tool to detect degradation before visible damage occurs, providing crucial information for preventive conservation strategies. This non-invasive approach ensures the preservation of the integrity of the historical texts while offering actionable insights for long-term storage and environmental control.

Botti et al. (contribution 3) explored innovative nanocomposite hydrogels as cleaning and protective agents for paper-based artworks. The study focused on poly(acrylic acid)/TiO_2 formulations, which combine mechanical cleaning efficacy with photocatalytic antimicrobial properties. The hydrogels were thoroughly characterized for their rheological behavior, swelling capacity, and interaction with paper fibers. Application tests demonstrated effective removal of surface contaminants without compromising the paper structure. This research underscores the potential of nanotechnology for cultural heritage conservation, offering environmentally friendly solutions that enhance both cleaning efficiency and long-term protection.

Labate et al. (contribution 4) employed a combination of non-invasive techniques—including XRF, UV–visible diffuse reflectance spectrophotometry with optical fibres (FORS), and portable X-ray diffraction (XRD)—to investigate the chemical transformations responsible for the blackening of originally blue paint layers. The multi-analytical approach allowed the identification of pigment composition, degradation products, and environmental factors contributing to discoloration. The study provides valuable insights into both the deterioration mechanisms and suitable conservation strategies, demonstrating how integrated non-destructive analysis can support informed decision-making in artwork preservation. Further exploring paint systems, the study by Poli et al. (contribution 5) provided initial insights into metal soap formation in alkyd-based paints using FTIR spectroscopy. The authors investigated the early stages of this common degradation phenomenon, monitoring the chemical interactions between metal ions and fatty acid components. The results offer proof-of-concept data that enhance understanding of degradation pathways and underscore the importance of early detection for preventive conservation. Such findings can guide conservators in selecting suitable environmental and storage conditions to mitigate metal soap-related deterioration. Freixas-Jambert et al. (contribution 6) focused on bromoil photographic prints, which present unique conservation challenges due to their complex organic and inorganic components. Using external reflection FTIR spectroscopy as a non-invasive technique, the study successfully identified key binder and pigment constituents, providing insights into historical photographic processes and degradation mechanisms. This work underscores the value of minimally destructive analytical methods for the characterization, study, and preservation of sensitive photographic heritage.

Textile heritage was explored by Forleo et al. (contribution 7), who applied ATR-FTIR spectroscopy combined with multivariate curve resolution–alternating least squares (MCR-ALS) algorithms to profile early synthetic dyes on historic woolen samples. This approach allowed the authors to deconvolute complex overlapping spectral signals, identifying the presence of early synthetic dyes while preserving the integrity of the delicate fibers. By providing detailed chemical information on historical dyeing techniques, this study contributes to understanding the material composition, degradation pathways, and conservation needs of textile artifacts.

The insights gained from such advanced spectroscopic and chemometric analyses resonate with complementary strategies demonstrated in other contributions of the Special Issue. De Caro et al. (contribution 8) explored the functionalization of artwork packaging using Ag-doped TiO_2 and ZnO nanoparticles, providing innovative solutions for the preventive conservation and protection of cultural heritage materials. Similarly, Nádvorníková et al. (contribution 9) employed GC–MS combined with principal component analysis to differentiate lipid binders in wall paintings, illustrating how sophisticated analytical workflows can resolve complex organic mixtures and inform appropriate conservation strategies. Together, these studies exemplify the integration of non-destructive and minimally invasive analytical techniques with chemometrics and material functionalization approaches, highlighting the breadth of modern heritage science from chemical characterization to preventive conservation.

3. Conclusions

The second edition of the Special Issue highlights the continued evolution of analytical strategies in heritage science, emphasizing minimally invasive and multi-modal approaches. The studies presented demonstrate that integrating advanced spectroscopy, mass spectrometry, imaging, and nanomaterials can provide comprehensive insights into both the composition and degradation of cultural artifacts. Furthermore, these contributions underscore the importance of multidisciplinary collaboration, combining expertise in chemistry, materials science, conservation, and art history to achieve meaningful results.

The editorial emphasizes that future research should continue to develop accessible, robust, and non-destructive methods while considering sustainability and environmental impacts. As analytical technologies advance, researchers are better equipped to address long-standing conservation challenges, improve preventive care, and inform policy decisions regarding cultural heritage management. The achievements of this Special Issue exemplify the scientific and practical value of contemporary analytical strategies, and they provide a roadmap for future investigations in the field.

Funding: This research received no external funding.

Acknowledgments: I would like to thank all the authors who have chosen to report their results in this Special Issue and thank the contributions of the academic editors; all peer reviewers; and editorial team members.

Conflicts of Interest: The author declares no conflicts of interest.

List of Contributions:

1. Doménech-Carbó, M.T.; Pasíes Oviedo, T.; Roca, R.C.; Múgica Mestanza, J. Between Heritage Conservation and Forensic Science: An Analytical Study of Personal Items Found in Mass Graves of the Francoism (1939–1956) (Spain). *Molecules* **2025**, *30*, 2783. https://doi.org/10.3390/molecules30132783.
2. Chiodini, C.; Rovellini, P.; Chiodini, M.; Giacomelli, L.; Baglio, D.; the 5B IISS Torno Working Group. Modern Analytical Chemistry Meets Heritage Books: Analysis of Volatile Organic Compounds (VOCs) from Two Books Preserved at the Biblioteca Capitolare of Busto Arsizio. *Molecules* **2025**, *30*, 2447. https://doi.org/10.3390/molecules30112447.
3. Botti, S.; Bonfigli, F.; D'Amato, R.; Rodesi, J.; Santonicola, M.G. Poly(Acrylic Acid)/TiO2 Nanocomposite Hydrogels for Paper Artwork Cleaning and Protection. *Molecules* **2025**, *30*, 75. https://doi.org/10.3390/molecules30010075.
4. Labate, M.; Aceto, M.; Chiari, G.; Baiocco, S.; Operti, L.; Agostino, A. Multi-Analytical and Non-Invasive Approach for Characterising Blackened Areas of Originally Blue Paints. *Molecules* **2024**, *29*, 6043. https://doi.org/10.3390/molecules29246043.
5. Poli, T.; Haaf, M.P.; Piccirillo, A.; Costa, A.P.; Craig, R.L.; Pozzi, F. First Insights into the Formation of Metal Soaps in Alkyd-Based Paints: A Proof-of-Concept Investigation Using FTIR Spectroscopy. *Molecules* **2024**, *29*, 5840. https://doi.org/10.3390/molecules29245840.
6. Freixas-Jambert, R.; Ruiz-Recasens, C.; Nieto-Villena, A.; Oriola-Folch, M. Non-Invasive Characterisation of Bromoil Prints by External Reflection FTIR Spectroscopy. *Molecules* **2024**, *29*, 5833. https://doi.org/10.3390/molecules29245833.
7. Forleo, T.; Giannossa, L.C.; De Juan Capdevila, A.; Lagioia, G.; Mangone, A. Hats Off to Modeling! Profiling Early Synthetic Dyes on Historic Woolen Samples with ATR-FTIR Spectroscopy and Multivariate Curve Resolution–Alternating Least Square Algorithm. *Molecules* **2024**, *29*, 4651. https://doi.org/10.3390/molecules29194651.

8. de Caro, T.; Toro, R.G.; Cassone, L.; Barbaccia, F.I.; Zaratti, C.; Colasanti, I.A.; La Russa, M.F.; Macchia, A. Functionalization of Artwork Packaging Materials Utilizing Ag-Doped TiO2 and ZnO Nanoparticles. *Molecules* **2024**, *29*, 3712. https://doi.org/10.3390/molecules29153712.
9. Nádvorníková, J.; Pitthard, V.; Kurka, O.; Kučera, L.; Barták, P. Egg vs. Oil in the Cookbook of Plasters: Differentiation of Lipid Binders in Wall Paintings Using Gas Chromatography–Mass Spectrometry and Principal Component Analysis. *Molecules* **2024**, *29*, 1520. https://doi.org/10.3390/molecules29071520.

References

1. Astolfi, M.L. Advances in analytical strategies to study cultural heritage samples. *Molecules* **2023**, *28*, 6423. [CrossRef] [PubMed]
2. Ma, C.; Dou, H.; Zhao, Z.; Qiu, X.; Li, H.; Wang, X. Review of in-situ non- and micro-destructive techniques for pigment analysis in architectural heritage. *npj Herit. Sci.* **2025**, *13*, 222. [CrossRef]
3. Yogurtcu, B.; Cebi, N.; Koçer, A.T.; Erarslan, A. A review of non-destructive Raman spectroscopy and chemometric techniques in the analysis of cultural heritage. *Molecules* **2024**, *29*, 5324. [CrossRef]
4. Daher, C.; Bellot-Gurlet, L.; Le Hô, A.S.; Paris, C.; Regert, M. Advanced discriminating criteria for natural organic substances of cultural heritage interest: Spectral decomposition and multivariate analyses of FT-Raman and FT-IR signatures. *Talanta* **2013**, *115*, 540–547. [CrossRef]
5. Possenti, E.; Miliani, C.; Cotte, M.; Realini, M.; Colombo, C. SR-based µXRD–µXRF 2D mapping to study Mg-rich historical frescoes subjected to inorganic conservation treatments. *Analyst* **2025**, *150*, 1590–1604. [CrossRef]
6. Fu, Y.; Huang, Y.; Shi, S.; Zong, S.; Li, Y.; Wei, S. Identification of organic binding media in ancient architectural decoration by THM–Py–GC/MS. *J. Cult. Herit.* **2024**, *67*, 421–429. [CrossRef]
7. Tamburini, D.; Fulcher, K.; Briggs, L.; von Aderkas, N.; Pulak, C.; Stacey, R. Advances in the characterisation and identification of mastic (*Pistacia* sp.) resin in archaeological samples by GC-QToF-MS. *RSC Adv.* **2024**, *14*, 836–854. [CrossRef]
8. Ding, L.; Yang, Q.; Liu, J.; Lee, Z. Evaluating volatile organic compounds from Chinese traditional handmade paper by SPME-GC/MS. *Herit. Sci.* **2021**, *9*, 145. [CrossRef]
9. Tammekivi, E.; Vahur, S.; Vilbaste, M.; Leito, I. Quantitative GC–MS analysis of artificially aged paints with variable pigment and linseed oil ratios. *Molecules* **2021**, *26*, 2218. [CrossRef]
10. Vettorazzo, C.; Sandström, E.; Troalen, L.G.; Mackay, C.L.; Hulme, A.N. Heritage science applications of ambient mass spectrometry. *Anal. Methods* **2025**, *17*, 3357–3369. [CrossRef]
11. Mastrangelo, R.; Chelazzi, D.; Baglioni, P. New horizons on advanced nanoscale materials for cultural heritage conservation. *Nanoscale Horiz.* **2024**, *9*, 566–579. [CrossRef] [PubMed]
12. Ioan, M.; Anghel, D.F.; Anastasescu, M.; Gifu, I.C.; Alexandrescu, E.; Matei, R.I.; Petcu, C.; Stanculescu, I.; Sanda, G.A.; Bala, D.; et al. Hybrid materials based on ZnO nanoparticles and organo-modified silica coatings as eco-friendly anticorrosive protection for metallic historic artifacts. *Coatings* **2023**, *13*, 1193. [CrossRef]
13. Fiorini, L.; Conti, A.; Pellis, E.; Bonora, V.; Masiero, A.; Tucci, G. Machine learning-based monitoring for planning climate-resilient conservation of built heritage. *Drones* **2024**, *8*, 249. [CrossRef]
14. Catelli, E.; Li, Z.; Sciutto, G.; Oliveri, P.; Prati, S.; Occhipinti, M.; Tocchio, A.; Alberti, R.; Frizzi, T.; Malegori, C.; et al. Towards the non-destructive analysis of multilayered samples: A novel XRF-VNIR-SWIR hyperspectral imaging system combined with multiblock data processing. *Anal. Chim. Acta* **2023**, *1239*, 340710. [CrossRef]
15. Zhang, M.; Liu, S.; Shao, H.; Ba, Z.; Liu, J.; Albu Kaya, M.G.; Tang, K.; Han, G. Development trend in non-destructive techniques for cultural heritage: From material characterization to AI-driven diagnosis. *Heritage* **2025**, *8*, 381. [CrossRef]
16. Hermans, J.J.; Keune, K.A.; van Loon, A.N.; Stols-Witlox, M.J.; Corkery, R.W.; Iedema, P.D. The synthesis of new types of lead and zinc soaps: A source of information for the study of oil paint degradation. In Proceedings of the Building Strong Culture through Conservation: Pre-prints ICOM-CC 17th Triennial Conference, Melbourne, Australia, 15 September 2014; pp. 17–19.
17. Quintero Balbas, D.; Lanterna, G.; Cirrincione, C.; Fontana, R.; Striova, J. Non-invasive identification of textile fibres using near-infrared fibre optics reflectance spectroscopy and multivariate classification techniques. *Eur. Phys. J. Plus* **2022**, *137*, 85. [CrossRef]
18. Koochakzaei, A.; Mobasher Maghsoud, E.; Jelodarian Bidgoli, B. Non-invasive imaging and spectroscopy techniques for identifying historical pigments: A case study of Iranian manuscripts from the Qajar era. *Herit. Sci.* **2023**, *11*, 157. [CrossRef]
19. Tarilonte, E.; González-Mendia, O.; Costantini, I.; Castro, K.; Maguregui, I. A multi-analytical approach for the identification of surface whitening phenomena in contemporary oil painting and its application to metal soaps. *J. Cult. Herit.* **2025**, *74*, 195–203. [CrossRef]
20. Poli, T.; Haaf, M.P.; Piccirillo, A.; Costa, A.P.; Craig, R.L.; Pozzi, F. First insights into the formation of metal soaps in alkyd-based paints: A proof-of-concept investigation using FTIR spectroscopy. *Molecules* **2024**, *29*, 5840. [CrossRef]

21. Edwards, H.G.; Vandenabeele, P.; Colomban, P. *Raman Spectroscopy in Cultural Heritage Preservation*; Springer: Cham, Switzerland 2023.
22. Donais, M.K.; Vandenabeele, P. Portable spectroscopy for on-site and in situ archaeology studies. In *Portable Spectroscopy and Spectrometry*; Wiley: Hoboken, NJ, USA, 2021; pp. 523–544. [CrossRef]
23. Baglioni, M.; Poggi, G.; Chelazzi, D.; Baglioni, P. Advanced materials in cultural heritage conservation. *Molecules* **2021**, *26*, 3967 [CrossRef]
24. Magdy, M. Analytical techniques for the preservation of cultural heritage: Frontiers in knowledge and application. *Crit. Rev. Anal. Chem.* **2022**, *52*, 1171–1196. [CrossRef]
25. Sammartino, M.P.; Grendene, A.; Astolfi, M.L.; Marcheggiani, S.; Mancini, L.; Vitali, M.; Antonucci, A.; Baldassarri, P.; Della Giovampaola, I.; Visco, G. Ancient spring waters still emerging and accessible in the Roman Forum area: Chemical–physical and microbiological characterization. *Open Chem.* **2023**, *21*, 20230366. [CrossRef]
26. Colantonio, C.; Baldassarri, P.; Avino, P.; Astolfi, M.L.; Visco, G. Visual and physical degradation of the black and white mosaic of a Roman Domus under Palazzo Valentini in Rome: A preliminary study. *Molecules* **2022**, *27*, 7765. [CrossRef] [PubMed]
27. del Hoyo-Meléndez, J.M. Physico-chemical characterisation and light stability of dyes and pigments found in cultural heritage objects: Insights from microfading testing for assessing light fastness. *Color. Technol.* **2025**, *141*, 265–290. [CrossRef]
28. Paolin, E.; Strlič, M. Volatile organic compounds (VOCs) in heritage environments and their analysis: A review. *Appl. Sci.* **2024**, *14*, 4620. [CrossRef]
29. Uring, P.; Chabas, A.; Alfaro, S.C. Textile ageing due to atmospheric gases and particles in indoor cultural heritage. *Environ. Sci. Pollut. Res.* **2021**, *28*, 66340–66354. [CrossRef]
30. Harth, A. The study of pigments in cultural heritage: A review using machine learning. *Heritage* **2024**, *7*, 3664–3695. [CrossRef]
31. Coletti, C.; Antonelli, F.; Germinario, L.; Maritan, L.; Piovesan, R.; Tesser, E.; Mazzoli, C. Investigating stone materials from some European cultural heritage sites for predicting future decay. *Rend. Lincei Sci. Fis. Nat.* **2025**, *36*, 103–127. [CrossRef]

molecules

MDPI

Article

Between Heritage Conservation and Forensic Science: An Analytical Study of Personal Items Found in Mass Graves of the Francoism (1939–1956) (Spain)

María Teresa Doménech-Carbó [1,*], Trinidad Pasíes Oviedo [2], Ramón Canal Roca [2] and Janire Múgica Mestanza [1,2]

[1] Instituto de Restauración del Patrimonio, Universitat Politècnica de València, Camí de Vera 14, 46022 València, Spain; janire.mugica@gmail.com
[2] Museu de Prehistòria de València, Carrer de la Corona, 36, 46003 València, Spain; trini.pasies@dival.es (T.P.O.); ramon.canal@dival.es (R.C.R.)
* Correspondence: tdomenec@crbc.upv.es

Abstract: This article describes the case of the personal items found in common graves dated between 1939 and 1956 after the Spanish Civil War (1936–1939), located in Paterna's cemetery (Spain). It was important in this study to know the state of the conservation of the objects and to obtain clues about their origin and use just as in a forensic study. This would allow the moral restitution of the historical memory of the victims of the war conflict. The multi-technique strategy has included light and electron microscopy, infrared spectroscopy and X-ray diffraction. Materials of the early 20th century used in pencil sharpeners, glasses, cutlery, lighters, rings, and buttons or medications contained in small bottles and boxes have been identified and have enabled the lives of their owners to be reconstructed during their imprisonment and execution. All these objects exhibited a thin layer of adipocere, a well-known compound in forensic science formed during the decomposition of human and animal corpses. Interestingly, rare corrosion processes have been identified in two of the objects analyzed, which are linked to their proximity to the decomposing corpses of the deceased. Copper sulfides and/or sulfates have been identified in the lighter, and scholzite, a zinc and calcium phosphate, has been identified in the glasses.

Keywords: forensic science; conservation of heritage; FTIR spectroscopy; FESEM–EDX; light microscopy; XRD; adipocere; scholzite

1. Introduction

The conservation of cultural heritage usually returns to a distant past and involves archaeological remains and works of art. However, researchers dedicated to the examination and analysis of cultural heritage must sometimes resort to forensic science to obtain clues about the causes and mechanisms that generate the alterations identified in the objects and their origin and use. This is the case for the clothes and personal objects exhumed in several mass graves dated between 1939 and 1956, located in the cemetery of Paterna (Valencia, Spain) [1] (Figure 1).

At the end of the Spanish Civil War (1936–1939), many civilians were retaliated against by Francisco Franco's government. Many of them were imprisoned, and after a summary trial, they were executed and buried in mass graves without having been afforded the guarantees of a fair trial. In 2012, a systematic and scientific exhumation and identification works began on the remains of people buried in mass graves by an interdisciplinary team formed by archaeologists, anthropologists, historians, restorers, chemists, biologists, forensic scientists, etc. These works were carried out within the legal framework provided by a series of laws and decrees approved in the Spanish parliament and the Valencian government to restore the historical memory of the victims of the war conflict. These laws and decrees regulate how to proceed with the exhumed human remains and their belongings [2–5].

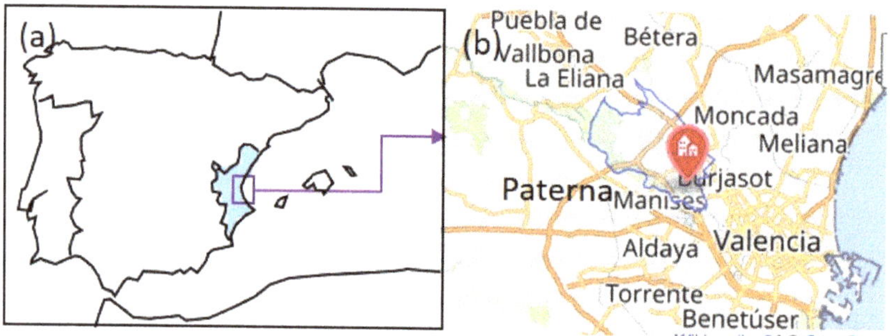

Figure 1. (**a**) Map of Spain; in blue is marked the Valencian Community; (**b**) detail of the map showing the area covered by the Paterna municipality and the town (red point). Adapted from URL: https://www.google.es/maps (accessed on 1 June 2025).

The execution of 2237 victims by the defenders of the Franco regime began just four days after the end of the civil war in 1939, and it did not end until seventeen years later, in 1956. The executions took place in Paterna, a town near Valencia (Spain). Once processed, the prisoners were transferred from the prison to the cemetery and were shot at the Terrer, a wall located 500 m from the cemetery. The victims were deposited in the so-called common or mass graves of Paterna. These were up to one hundred square pits that measured 2 and 2.5 m on each side and 1 to 6 m deep (Figure 2a). The bodies were directly thrown into the graves without any care, so they appeared in piles. The condemned men and women, who were taken out of prison in groups to be shot, constituted a *saca* (sack). The different sacas stacked in the pit can be well distinguished because each was covered with a layer of quicklime. The entire process passed quickly in such a way that the deceased were dressed in the street clothes they were wearing at the moment of the execution. The executors did not bother to take them the street clothes and belongings they were carrying at that time in their pockets. The Valencian Museum of Ethnology (l'ETNO), which custodies the collection, and the Prehistory Museum of Valencia, in collaboration with the Universitat Politècnica de València, undertook the task of restoring the clothes and personal objects of the deceased identified through forensic procedures to creating an exhibition project [6–12] to pay tribute to the victims. On 5 July 2023, the temporary exhibition "Archaeology of memory. The Paterna graves" was inaugurated by the Prehistory Museum of Valencia (Figure 2b). This temporary exhibition was part of the "The Graves of Francoism: Archaeology, Anthropology and Memory" project co-organized with L'ETNO and the Department of Democratic Memory of the Valencia Provincial Council [13–15].

This research was challenging for two reasons. Firstly, due to the great variety of both organic and inorganic materials. Additionally, the personal items of Paterna's mass graves constitute a peculiar case study due to the particular state of preservation in which they were found, given the aggressive conditions of the burial site. The layers of quicklime between each saca of stacked bodies must have significantly increased the pH of the soil, at least for a time, until the moisture and the rains infiltrated into the ground and reduced the alkalizing effect of the lime that was transformed into calcium carbonate. Another unique aspect of this case study was the presence of a gray–whitish and waxy substance that permeated all objects, especially clothing and other belongings made of organic materials such as rope or wood. The analyses carried out in the project confirmed the hypothesis on the substance's identity. It was adipocere formed on the corpses by saponification of the fats during the decomposition process of the deceased, which diffused to the clothes and objects in the pockets. In forensic science, the formation of adipoceres is an alteration mechanism that the buried bodies of animals and humans undergo frequently [16–25]. The adipoceres make it challenging to identify the original components of the objects as they interfere during the analyses. Therefore, the transfer of knowledge from forensic science to archaeology and heritage conservation regarding sample collection, appropriate analytical workflows, and standardized analytical protocols could be beneficial [22].

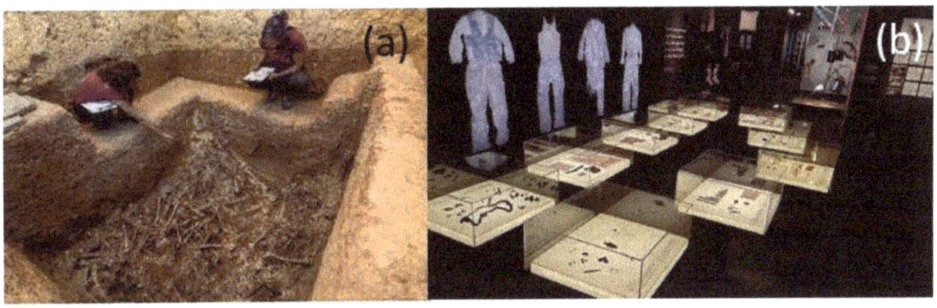

Figure 2. (**a**) The archaeologist Eulàlia Diaz (left) and the anthropologist Gemma López (right), making field work in the mass grave 112 of the Cemetery of Paterna. Courtesy by the photography author: Eloy Ariza—Asociación Científica ArqueoAntro; (**b**) the temporary exhibition held in the Museu de Prehistoria de Valencia: Courtesy by the Archivo SIP del Museu de Prehistòria de València.

The large number of samples and the great variety of materials that have constituted the objective of this restoration project have required the application of different instrumental techniques following a multi-technique strategy. Gas chromatography–mass spectrometry (GC–MS) has been widely used for analyzing organic materials because of its high sensitivity and ability to separate and identify the chemical compounds of a polymeric or complex mixture of materials. Nevertheless, this technique is time-consuming as it requires pretreatments of solid samples such as hydrolysis and derivatization. As an alternative, Fourier Transform Infrared Spectroscopy (FTIR) has seen increasing application in heritage science, being comparatively fast and easy to use as none-to-minimal sample preparation is required [26]. These characteristics are essential when many samples with various compositions must be identified. For this reason, this is one of the preferred instrumental techniques used in forensic practice [27]. Optical and scanning electron microscopy techniques (OM, SEM, and field emission FESEM) are also valuable for providing morphological data and elemental composition coupled with an X-ray microanalysis (EDX) system. The minimal amount and the easy and simple preparation of samples required have converted these instrumental techniques into excellent tools for routine analysis in the heritage and forensic sectors. X-ray diffraction (XRD) has also been used for studying the soil composition of the graves.

The analytical chemical study carried out to characterize the personal objects recovered from the Paterna graves is presented in this contribution. Through a multi-technique strategy, it has been possible to answer three of the main questions that the team of historians and restorers raised: From the perspective of the conservation of cultural heritage, what materials were the objects made from? Moreover, what alterations do the objects present? From the perspective of forensic sciences, what is the meaning and function of the belongings for their owners?

Figure 3 shows a flowchart illustrating the meta-structure of the experimental procedure implemented. The analytical study included the previously indicated instrumental techniques OM, FESEM–EDX, FTIR, and XRD. The multi-technique strategy was based on three criteria that allowed for the selection of analytical techniques and their sequencing in the experimental procedure. The first criterion refers to the minimum requirements that instrumental techniques must meet to be applied to cultural goods (minimum sample size and high sensitivity) and their capabilities that can add value to the analytical process, such as the versatility to analyze materials of all types present in the same sample, performance as a multi-component technique, and capacity to simultaneously provide data of different types (morphological, compositional, and structural). These capabilities determine the qualitative information and quantitative data to be obtained, which constitute the second criterion. These data are necessary to achieve the objectives proposed in the research. The objectives respond to the issues raised on both levels: the knowledge of the personal items in the strict field of the conservation of cultural heritage and the archaeology/history, the latter falling within the stricter scope of forensic science. The third criterion refers to the sample pretreatment requirements of each instrumental technique. While all are non-destructive, the preparation procedure varies from direct sample examination in OM and FESEM–EDX to milling in FTIR and XRD to the more time-consuming cross-section preparation. Analysis time has been another factor to consider, leading to XRD being placed last in the operational sequence and restricted to a small number of soil samples. It is worth noting that most of the conclusions reached about the composition, state of preservation, and functionality of the personal belongings were only possible by the synergies generated by the combined use of the instrumental techniques selected in an appropriate sequence (see Table S1 provided as Supplementary Electronic Material).

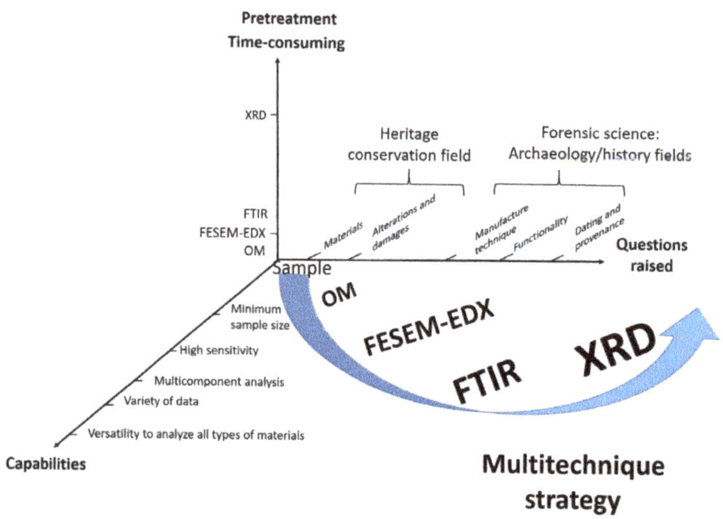

Figure 3. Flowchart illustrating the metastructure of the experimental procedure implemented.

2. Results

2.1. The Collection of Personal Items of the Mass Graves of Paterna

The collection of objects presented in this paper is a representative sample of the set of belongings analyzed. It is composed of 16 items that are shown in Figures 4–7. A detailed description of the items is summarized in Table S1.

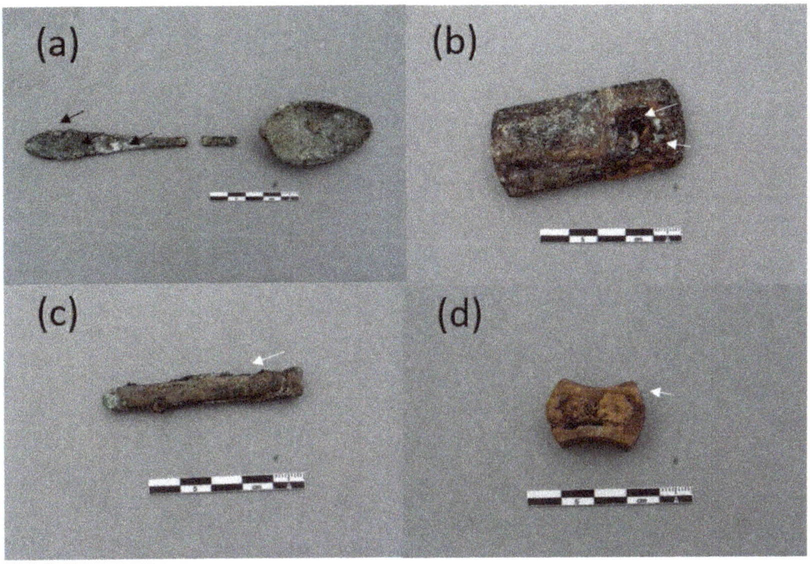

Figure 4. Objects and sampling points. (**a**) Spoon ETN006, deceased 43, mass grave 115; (**b**) cigarette lighter ETN009, deceased 131, mass grave 115; (**c**) pencil extender ETN010, deceased 144, mass grave 115; (**d**) pencil sharpener ETN012, deceased 107, mass grave 115. Photographs by courtesy of the Archivo SIP del Museu de Prehistòria de València. Arrows highlight the sampling points in the artefacts. Marks in the scalebars correspond to 1 cm.

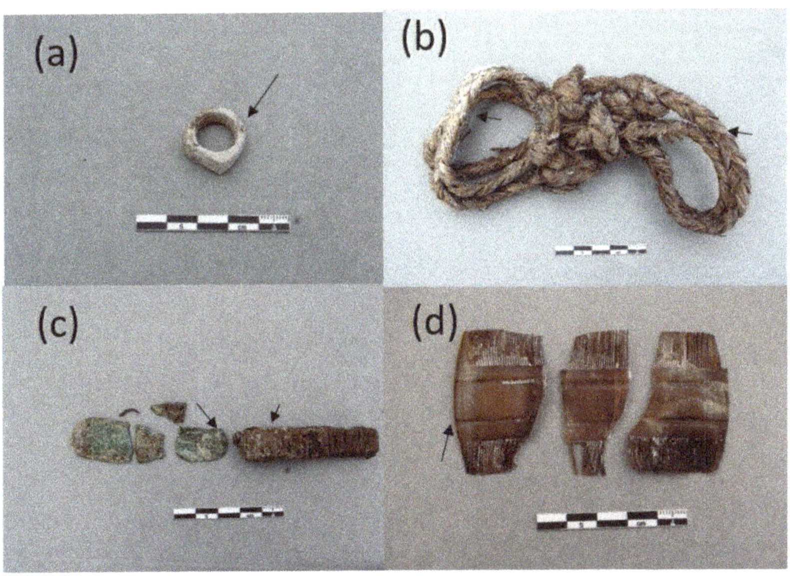

Figure 5. Objects and sampling points. (**a**) Stone ring ETN053, deceased 28, mass grave 94; (**b**) rope handcuffs ETN045, deceased 92 identified by DNA analysis as Mr. José Peiró I Grau, mass grave 112; (**c**) knife ETN048, deceased 15, mass grave 100; (**d**) nit comb ETN060, deceased 66 identified by DNA analysis as Mr. Juan Bautista Solanes i González, mass grave 112. Photographs by courtesy of the Archivo SIP del Museu de Prehistòria de València. Arrows highlight the sampling points in the artefacts. Marks in the scalebars correspond to 1 cm.

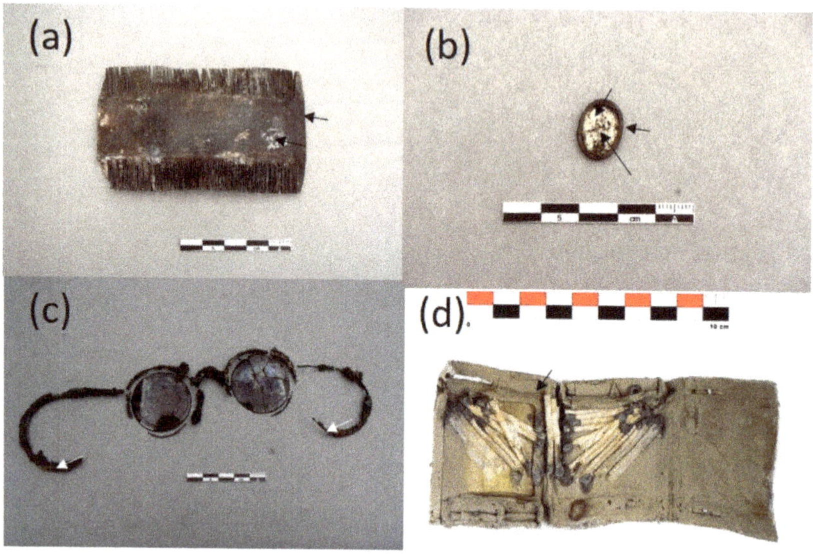

Figure 6. Objects and sampling points. (**a**) Nit comb ETN058, deceased 95, mass grave 112; (**b**) medal ETN067, deceased 68, mass grave 128; (**c**) spectacles frame ETN073, deceased 83, mass grave 111; photographs (**a**–**c**) by courtesy of Archivo SIP del Museu de Prehistòria de València; (**d**) cigarette paper in a matchbox F111-141, deceased 141, identified by DNA analysis as Mr. Vicente Orti, mass grave 111. Courtesy by the photography author: Eloy Ariza—Asociación Científica ArqueoAntro. Arrows highlight the sampling points in the artefacts. Marks in the scalebars correspond to 1 cm.

Figure 7. Objects and sampling points. (**a**) Organic remains inside a glass bottle with the inscription "Naranjil Purgante ideal" LLUE001, deceased 14, identified by DNA analysis as Mr. Manuel Lluesma i Masiá, cemetery niche 645; (**b**) button MCP001, deceased 19, mass grave 114; (**c**) devotional medallion with rests of a rope drawstring and textile fibers from clothes adhered MCP029, deceased 176, mass grave 114; (**d**) metal ring ETN033, deceased 73, mass grave 112. Photographs by courtesy of the Archivo SIP del Museu de Prehistòria de València. Arrows highlight the sampling points in the artefacts. Marks in the scalebars correspond to 1 cm.

2.2. Identification of Materials

2.2.1. Elemental Composition of Metallic Objects

Several personal items of metallic nature were identified by its elemental composition, which was determined by using FESEM–EDX. Table 1 summarizes the results obtained.

Table 1. Elemental composition (weight %) of the metallic items analyzed.

Ref.	Elemental Composition (wt%)							
	Cu [a]	Zn	Ni	Pb	Sn	Fe [a]	Al [a]	Ag [a]
ETN006	50.5 ± 3	19.1 ± 3	8.3 ± 0.5	n.d.	n.d.	n.d.	n.d.	n.d.
ETN009-1	62.7 ± 0.9	37.3 ± 0.9	n.d.	5 ± 1 [b]	15.9 ± 0.5 [b]	n.d.	n.d.	n.d.
ETN009-2	n.d.	n.d.	n.d.	n.d.	n.d.	72.1 ± 0.7	n.d.	n.d.
ETN010-1	71.7 ± 0.5	4.6 ± 0.5	n.d.	1.51 [b]	n.d.	n.d.	n.d.	4.5 [b]
ETN010-2	n.d.	n.d.	n.d.	n.d.	n.d.	77.7 ± 0.9	n.d.	n.d.
ETN033-1	7.2 ± 0.4	n.d.	n.d.	n.d.	n.d.	n.d.	n.d.	84.7 ± 0.5
ETN033-2	6.6 ± 0.5	n.d.	n.d.	n.d.	n.d.	2.7 ± 0.3	n.d.	28.3 ± 0.6
ETN048	53.6 ± 0.9	11.8 ± 0.9	7.9 ± 0.6	n.d.	n.d.	n.d.	n.d.	n.d.
ETN067	90.9 ± 0.7	9.1 ± 0.4	n.d.	1.6 [b]	n.d.	n.d.	n.d.	n.d.
MCP029	n.d.	n.d.	n.d.	n.d.	n.d.	n.d.	52.9 ± 0.2 [a]	n.d.

n.d.: not detected; [a] oxygen wt% complete to 100% the sample elemental composition; [b] value corresponding to a spot analysis.

In the samples from the spoon ETN006 and the knife ETN048, copper, zinc, and nickel were identified as the main components. A quaternary brass alloy was identified in the sample ETN009-1 excised from the upper layer of the cigarette lighter case. The major elements in the brass are copper and zinc, accompanied by smaller amounts of lead and tin found in some spot analyses. Iron was associated with the inner layers of the case and other lighter pieces, such as the hinges, that could be made of steel (sample ETN009-2). Copper, zinc, and silver were the elements identified in the sample ETN010 from the pencil extender. These metals are associated with a binary brass alloy with silver plating. Aluminum was identified in the sample MCP029 from a devotional from a medallion, and silver–copper alloy was used in the plating (ETN033-1) of a ring with iron body (ETN033-2). The X-ray spectra acquired in the sample ETN067 from the medal metal mount exhibited emission lines of copper, zinc, and lead associated with a ternary brass. In the X-ray spectrum of a second sample excised in the central part of the medal, intense sulfur and barium emission lines were identified associated with the white pigment barite. Traces of cobalt suggest that a blue cobalt pigment was also used in the medal decoration.

2.2.2. Mineralogical and Micromorphological Characterization of Inorganic Materials

Some of the materials that compose the objects were identified directly by optical and electron microscopy examination. That morphological identification was further confirmed from their elemental and molecular analysis using FESEM–EDX and FTIR spectroscopy. For example, the secondary electron images obtained in a small sample of the item MCP001, a shirt button, are shown in Figure 8. The button's nacre (mother-of-pearl) material is easily recognized by its microscopical morphology. That is an organic–inorganic composite material of biogenic origin produced by some mollusks as an inner shell layer. The nanoscopic and polygonal platelets that form the laminas of the nacre "brick-and-mortar" structure are seen in Figure 8a. Figure 8b shows the mineral bridges and the micro-roughness of the surface platelets. Figure 8c shows the IR absorption spectrum acquired in this sample. In the inset, the chemical composition obtained by FESEM–EDX is shown, which suggests the calcium carbonate composition of this biomineral. The position of the IR bands confirms that the nacre is mainly composed of the mineralogical species aragonite (CO_3 asymmetric stretching ν_3 at 1419 cm^{-1}; CO_3 symmetric stretching ν_1 at 1082cm^{-1}; CO_3 asymmetric deformation ν_2 at 862cm^{-1}; and CO_3 symmetric deformation ν_4 at 712 and 699cm^{-1}) [28].

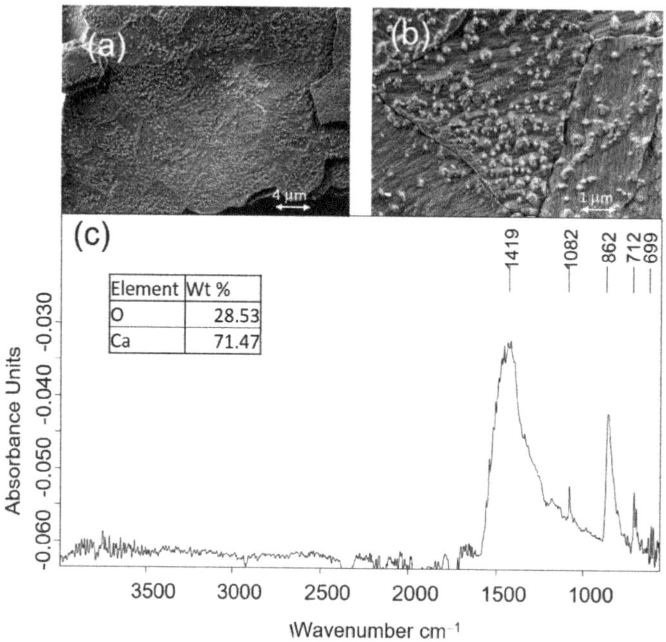

Figure 8. (**a**) Secondary electron images of the sample MCP001; (**b**) detail of the aragonite platelets and mineral bridges; (**c**) IR absorption spectrum. The inset shows the elemental composition obtained using FESEM–EDX.

Other materials of an inorganic nature that compose several personal items were identified from the IR spectra acquired in the samples excised from these items. Table 2 shows the leading IR absorption bands of analytical interest and their assignation to chemical compounds that constitute the ring ETN053. The IR absorption spectrum is shown in Figure 9a. Limestone, mainly composed of calcite with minor amounts of clays, quartz, and organic matter of lipid characteristics, is the material that composes the ring. In Figure 9b, several overlapped bands appear in the 1600–600 cm^{-1} region. Remarkably, the band at 1574 cm^{-1} is ascribed to calcium carboxylate complexes [29].

Table 2. IR bands identified in the sample ETN053 and assignment to chemical groups and compounds [28–30].

Assignment and IR Band Position (cm^{-1})	Material Identified
$2\nu_2 + \nu_4$ CO_3^{2-} stretching (2509), $\nu_1 + \nu_4$ CO_3^{2-} stretching (1797), ν_3 CO_3^{2-} asymmetric stretching (1398), ν_2 CO_3^{2-} asymmetric deformation (874), ν_4 CO_3^{2-} symmetric deformation (712)	Limestone
ν Si-O symmetric stretching (1113^{sh}), ν Si-O-Si antisymmetric in-plane stretching 1027), δ (Al_2OH) bending (915, 937), Si-O vibrations (748^{sh}, 694)	Dioctahedral 1:1 clay
ν Si-O-Si stretching (1155^{sh}, 1086^{sh}), Si-O symmetric stretching (799, 777), Si-O of SiO_4 symmetric bending (694)	Quartz
ν CH_2, CH_3 asymmetric and symmetric stretching (2918, 2851), ν C=O stretching (1730); calcium carboxylate stretching (1574)	Lipid matter

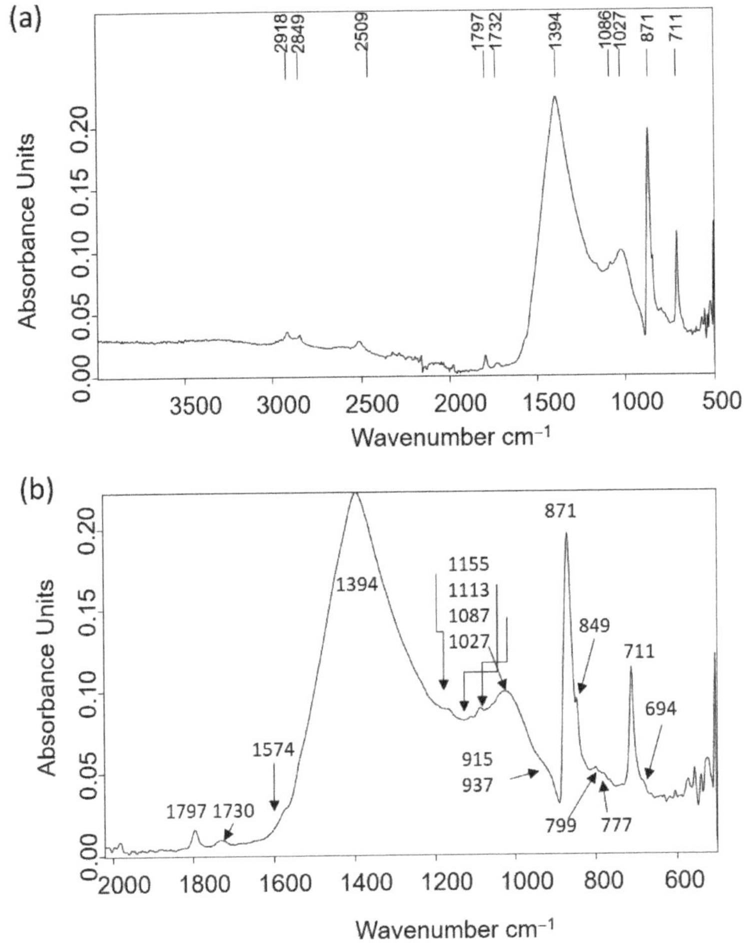

Figure 9. (**a**) IR absorption spectrum of sample ETN053 excised from a limestone ring; (**b**) detail of the IR spectrum in the 600–2000 cm^{-1} region.

2.2.3. Molecular Structure and Micromorphological Characterization of Organic Materials

The natural fibers found in the personal items ETN048 and MCP029 could be identified by their characteristic micromorphology observed in the microphotographs and secondary electron images shown in Figure 10. The fibers that formed the yarns wound around the tag of the knife ETN048 presented ribbons with convolution and a kidney shape in the cross-section characteristic of cotton (Figure 10a,b). The average fiber diameter measured is 21 µm. An IR absorption spectrum acquired in this sample (Figure S1) allowed us to recognize a proteinaceous substance, which is associated with some binding substance used for adhering the fibers to the metallic tang of the knife. This material was characterized by IR bands ascribed to amide A (NH- stretching) at 3264 cm^{-1}; amide B (NH- stretching) at 3072 cm^{-1}; CH_2 and CH_3 stretching vibrations at 2956, 2922, and 2852 cm^{-1}; amide I band at 1625 cm^{-1}; amide II band at 1539 cm^{-1}; amide III band and bending vibrations of the CH bonds of methyl and methylene groups at 1451 and 1399–90 cm^{-1}; and stretching vibrations of triglyceride ester linkages at 1232, 1159, and 1033 cm^{-1}. Cotton fibers formed the cord drawstring attached to the devotional medallion MCP029 (Figure 10c). This image also shows a thin crust of minerals and corrosion products covering the fibers. Figure 10d shows in detail the fibril micromorphology typical of cotton fibers and the high degree of deterioration due to the aggressive burial environment. Figure 10e shows an image of the esparto fibers that compose the rope handcuffs ETN045 characterized by the bundles of cellulosic fibrils (Figure 10f).

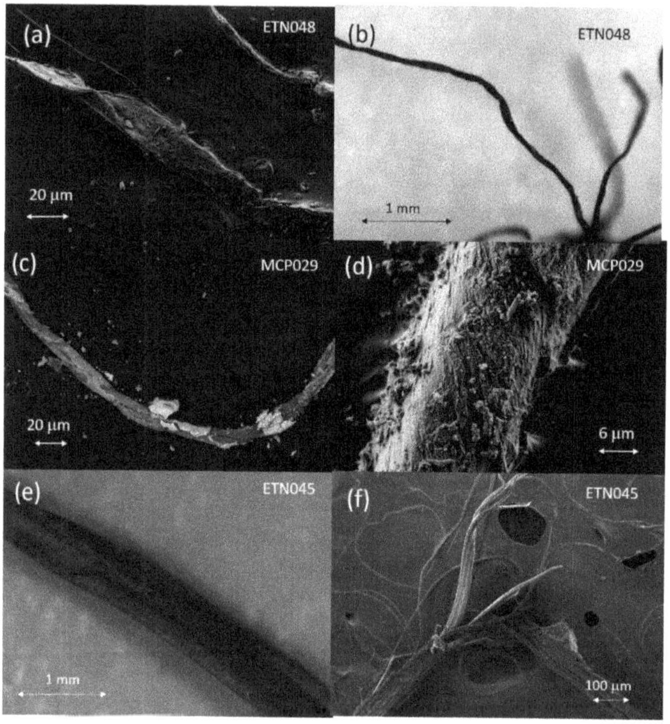

Figure 10. (**a**) Secondary electron images of fibers that formed the thread wound around the handle of the knife ETN048; (**b**) microphotograph taken in sample ETN048; (**c**) backscattered electron images of cotton fibers of the cord drawstring attached to the devotional medallion MCP029; (**d**) detail of the micromorphology of the MCP029 fibers; (**e**) microphotograph taken in rope handcuffs fibers of the sample ETN045; and (**f**) secondary electron image of fibrils of the sample ETN045.

The materials that compose the pencil sharpener ETN012 and the rope handcuffs ETN045 were identified as wood and vegetal fibers of esparto, respectively, using FTIR spectroscopy. The IR spectrum of sample ETN012 (Figure 11a) exhibited bands characteristic of wood ascribed to vibrations of H-bonded to OH-groups intermolecular in cellulose at 3262 cm^{-1}, C-H stretch in methyl and methylene groups at 2923 cm^{-1}, CH_2 asymmetric stretching and methoxyl C-H stretching at 2849 cm^{-1}, adsorbed water at 1638 cm^{-1}, aromatic skeletal vibrations plus C=O stretch at 1593 cm^{-1}, overlapped HOC bending, aromatic skeletal vibration combined with C-H in plane deformation and CH_2 scissoring at 1431 cm^{-1}, overlapped aliphatic C-H stretch in CH_3, symmetric C-H bending from the methoxy group and OH in plane deformation vibration at 1364 cm^{-1}, CH_2 rocking vibration at 1321 cm^{-1}, G ring plus C=O stretch at 1258 cm^{-1}, OH in plane deformation at 1210 cm^{-1}, aromatic C-H in plane deformation typical for G units, whereby G condensed >G etherified, at 1143 cm^{-1}, overlapped asymmetric in-phase ring stretching, aromatic in-plane deformation plus C-O deformation in primary alcohols plus C=O stretch and C_{alkyl}-O ether vibration in guaiacol at 1023 cm^{-1}, vibrations of glucomannan at 814 cm^{-1}, and rocking vibration of cellulose at 745 cm^{-1} [31,32].

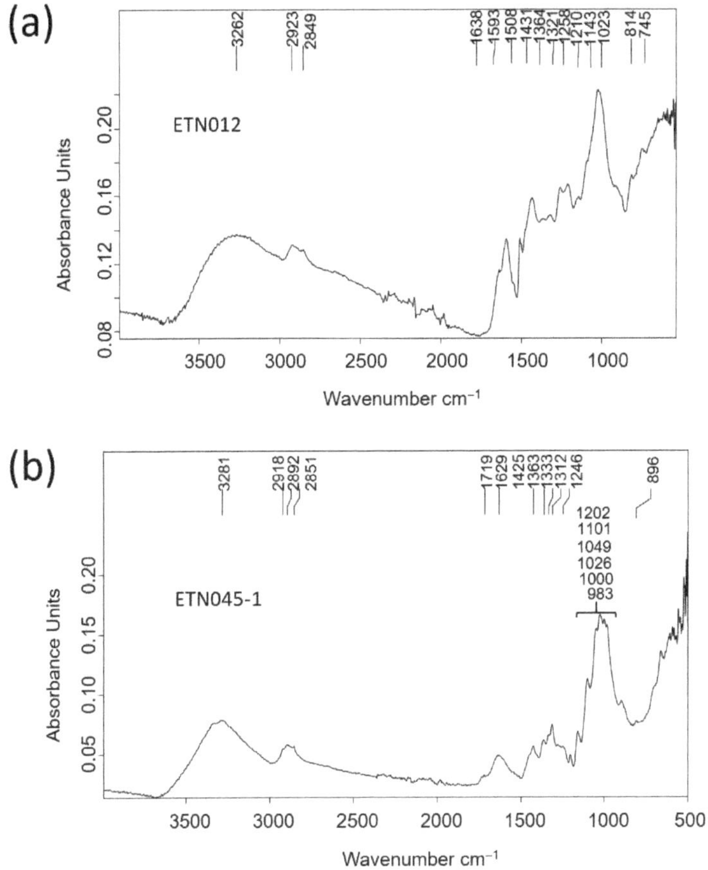

Figure 11. IR absorption spectra acquired in samples excised of the personal items: (**a**) sharpener ETN012; (**b**) Esparto grass rope handcuffs ETN045-1.

The rope handcuff fibers of the sample ETN045 were identified as esparto by their characteristic micromorphology observed in the optical microphotographs and secondary electron images obtained by OM and FESEM (Figure 10e,f). These fibers exhibited in the IR spectrum (Figure 11b) characteristic bands associated with H-bonded to OH stretching vibrations at 3281 cm^{-1}, CH_2 asymmetric stretching at 2921 cm^{-1}, CH_2 asymmetric stretching at 2892 cm^{-1}, CH_2 symmetric stretching vibration at 2851 cm^{-1}, C=O stretching of oxidized cellulose and carbonyl and carboxyl stretching in rosin at 1719 cm^{-1}, absorbed water in cellulose at 1629 cm^{-1}, CH_2 bending and asymmetric C-O stretching in cellulose at 1425 cm^{-1}, CH bending of cellulose at 1363 cm^{-1}, OH in-plane bending in cellulose at 1333 cm^{-1}, CH wagging of cellulose at 1312 cm^{-1}, OH out-of-plane bending in cellulose at 1246 cm^{-1}, OH in-plane bending in cellulose at 1202 cm^{-1}, asymmetric in-plane ring stretching in β-glycoside (C_1-O-C_4) in cellulose at 1101 cm^{-1}, C-O stretching in second alcoholic groups in cellulose at 1049 cm^{-1}, C-O stretching in first alcoholic groups in cellulose at 1026 cm^{-1}, C-O and C-H stretching at 1000 cm^{-1} and asymmetric out-of-phase ring stretching in β-glycoside (C_1-O-C_4) bond in cellulose at 896 cm^{-1} [33].

Similarly, a casein-formaldehyde resin (galalith) was the main component identified in the items ETN058, ETN060, and ETN073 (Figure 12a, Figures S2 and S3). The IR spectra included bands ascribed to N-H stretching at 3261 cm^{-1}, amide B (NH- stretching) at

3065 cm^{-1}, aliphatic C-H asymmetric stretching at 2918 and 2956 cm^{-1}, aliphatic C-H symmetric stretching at 2874 and 2849 cm^{-1}, C=O ester carbonyl band at 1731 cm^{-1}, amide I band (C=O stretching amide) at 1626 cm^{-1}, amide II band (N-H bending) at 1545 cm^{-1}, C-H bending at 1446 cm^{-1}, C-N stretching at 1397 cm^{-1}, interaction between the NH bending and CN stretching of the C-N-H group at 1238 cm^{-1}, and O-C-C stretching at 1074 cm^{-1} [34]. Bands at 1412, 874, and 713 cm^{-1} are assigned to calcite, the shoulder at 1565 and the band at 1515 cm^{-1} are assigned to calcium carboxylates, bands at 1152, 1074, and 1034 cm^{-1} are assigned to quartz and clayey minerals, and the broad and intense band at 540 cm^{-1} is attributed to stretching Fe-O bands in iron-red oxide (see Table 2 for assignments).

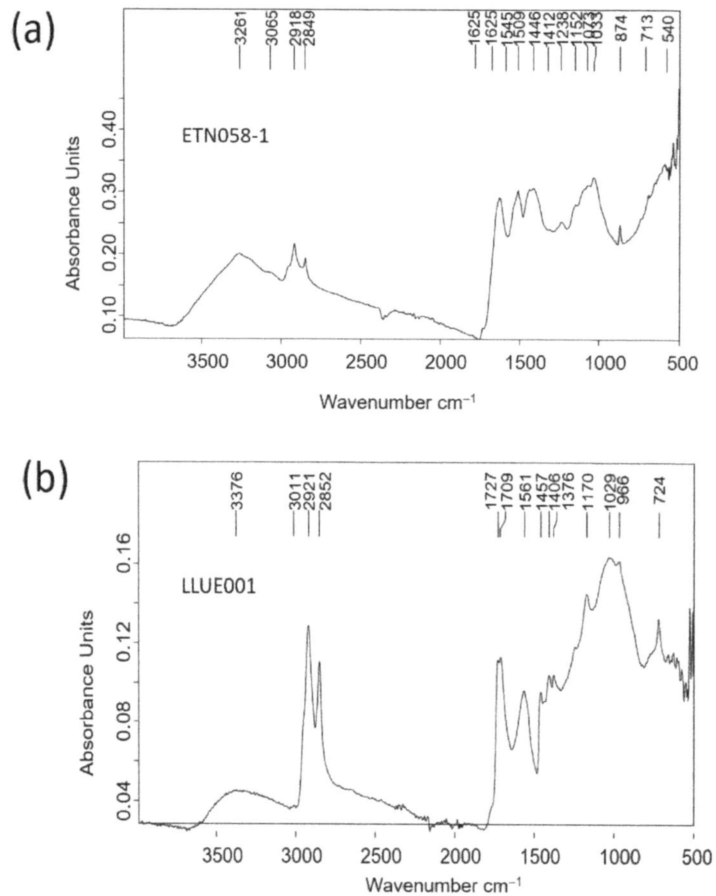

Figure 12. IR absorption spectra acquired in samples excised of the personal items: (**a**) nit comb ETN058-1; (**b**) organic remains inside a glass bottle with the inscription "Naranjil Purgante ideal" LLUE001.

Organic remains inside a glass bottle with the inscription "Naranjil Purgante ideal" LLUE001 exhibited characteristic bands of an oil associated with H-bonded OH stretching vibrations at 3376 cm^{-1}, alkenyl C=C stretching vibrations at 3011 cm^{-1}, CH_2 asymmetric stretching at 2921 cm^{-1}, CH_2 symmetric stretching at 2852 cm^{-1}, C=O asymmetric stretching of ester groups at 1727 cm^{-1}, CH_2 symmetric deformation at 1457 cm^{-1}, carboxylic group deformation at 1406 cm^{-1}, CH_3 symmetric deformation at 1376 cm^{-1}, C-(C=O)-O stretching of ester groups at 1170 cm^{-1}, O-C-C stretching of ester groups at 1029 cm^{-1}, -C=CH- out-of-plane bending at 966 cm^{-1}, and -CH_2- asymmetric bending (C-H rocking) at 724 cm^{-1} [35]. C=O asymmetric stretching (ν_α) of fatty monoacid groups is found at 1709 cm^{-1}, ν_α of fatty diacids appear at 1700^{sh} cm^{-1}, and ν_α of calcium carboxylate of fatty mono and diacids are found at 1561, 1549^{sh}, and 1509^{sh} cm^{-1} [36] (Figure 12b).

2.3. Adipoceres

Most of the personal items were impregnated with a white–greyish waxy substance that was associated with a well-known phenomenon in forensic sciences related to a series of specific chemical processes of the decomposition of the fatty matter of human and animal cadavers in burials (see Figures 5a,b,d, 6a–d and 7c) [16–25]. The adipoceres were unambiguously identified by FTIR spectroscopy in all the personal belongings. Three types of adipoceres were identified in the set of samples analyzed that differ in the composition of the compounds that form this complex substance. In most of the personal belongings, the adipoceres identified were characterized by the significant presence of the calcium salts of fatty acids among the compounds that form the mixture (see, as an example, Figure 13a and Figure S4 corresponding to cigarette paper F111-141 and rope handcuffs ETN045). Calcium salts were prevalent due to the easy adsorption of the fatty acids to calcium carbonate in the burial environments and the low solubilities in water [37]. Bands at 3563, 3353, and 1620 cm^{-1} were ascribed to the OH stretching and bending of crystalline water in hydrated calcium salts. The features from the hydrocarbon chains are observed at 2955 and 2915 cm^{-1} (antisymmetric stretching of CH_3 and CH_2 groups) and 2874 and 2847^{sh} cm^{-1} (symmetric stretching of CH_3 and CH_2 groups). The value of 2915 cm^{-1} for the ν_a CH_2 groups points to saturated fatty monoacid salts with chain lengths up to C14:0 [36]. Bands of carboxylate groups are found at 1573 and 1538 cm^{-1} (antisymmetric stretching) and 1465, 1432, and 1418 cm^{-1} (symmetric stretching and δ CH_2 scissoring doublet) characteristic of uni- and bidentate calcium fatty monoacids salts (Figure 13b) [38]. Sub-bands from calcium fatty monoacid salts ascribed to twisting vibration (τ CH_2) are found at 1378, 1350, 1336, 1318, and 1299 cm^{-1}. A pattern of bands at 1277, 1257, 1233, 1211, 1205^{sh}, and 1189 cm^{-1} is ascribed to wagging vibrations (ω CH_2). These multiplet series are typical of calcium fatty monoacid salts. Stretching vibration of (C-OH) group occurs at 1112 cm^{-1}. Unassigned band at 962 cm^{-1} and band at 719 cm^{-1} assigned to rocking vibration (ρ) of CH_2 groups are also observed [36].

Although the spectral profile of the IR spectrum acquired in the nit comb sample ETN058 exhibited the main absorption bands that were previously described, specific differences were found (Figure 14a). In particular, the band at 2664 cm^{-1} is assigned to hydroxyl–carbonyl interactions due to fatty monoacid dimers, and the band at 1698 cm^{-1} is attributed to the C=O stretching vibrations of carboxylic groups. Like in the calcium salts, the δ CH_2 scissoring vibration splits into two components at 1463 and 1411 cm^{-1} and the umbrella vibration at 1378 cm^{-1}. The low intensity (Figure 14b) of these bands indicates that the fatty acids in the adipocere have long chains like stearic (octadecanoic) acid (C18:0) or up to that. The twisting vibration is split throughout 1378–1310 cm^{-1}, whereas the wagging vibration is found in the range 1310–1186 [39]. According to Filipoulou et al. [36], the double split in the twisting region indicates length chains of C14:0-C18:0. In the wagging region the number of bands that form the multiplet increases with the hydrocarbon chain length following the rule $\omega = \frac{n}{2} - 1$, where n is the number of carbons in the hydrocarbon chain. Thus, the seven bands identified with two shoulders at 1259 and 1239 cm^{-1} in this region suggest that the main component of this adipocere is palmitic (hexadecenoic) acid (C16:0), accompanied by stearic acid (C18:0) and arachidic (eicosanoic acid) (C20:0). This result is confirmed by the calculated distancing between consecutive bands in the range 18–20 cm^{-1}.

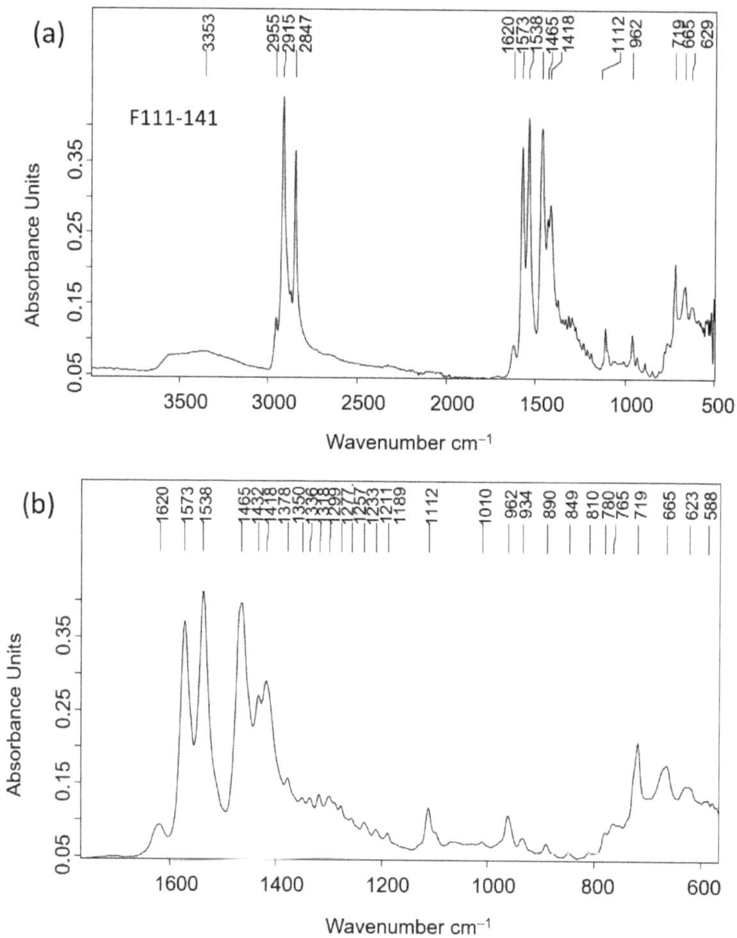

Figure 13. IR absorption spectra acquired in samples excised of the personal items: (**a**) cigarette paper in a matchbox F111-141; (**b**) detail of the 600–1800 cm^{-1} region.

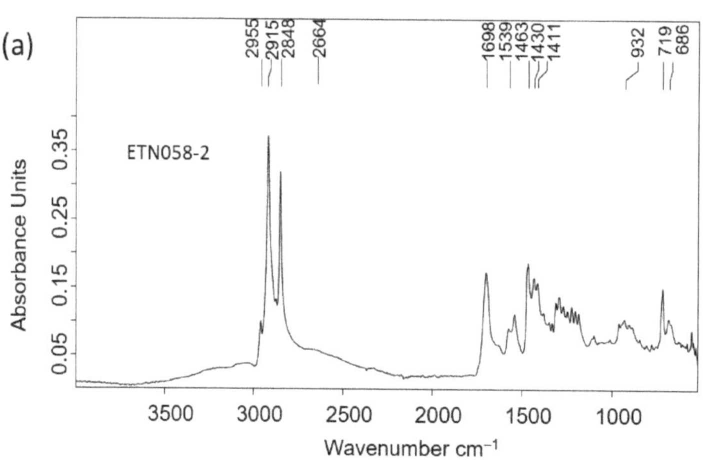

Figure 14. *Cont.*

(b)

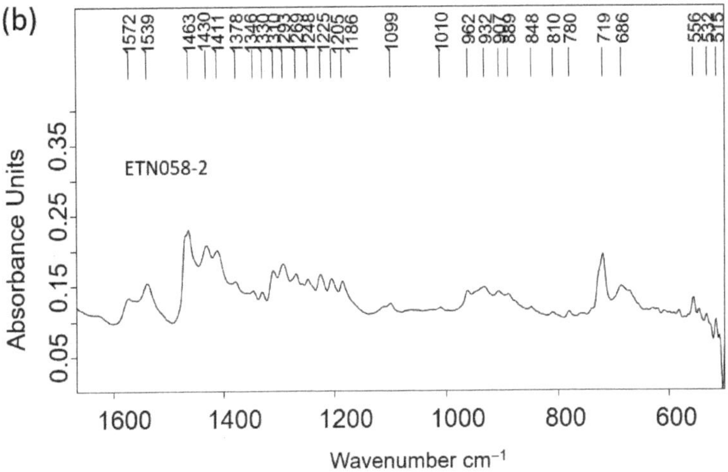

Figure 14. IR absorption spectra acquired in samples excised of the personal items: (**a**) nit comb ETN058; (**b**) detail of the 600–1800 cm^{-1}.

The IR spectrum ETN073-2 acquired in a sample of adipoceres excised from the metal spectacles frame is shown in Figure S5. The contact of adipoceres with the parts of the brass surface of the spectacles frame exposed has resulted in the formation of zinc salts of long-chain fatty monoacids, which are recognized by the occurrence of asymmetric stretching vibrations of the Zn-carboxylate group at 1591 (tetranuclear complex) and 1536 cm^{-1}, symmetric stretching at 1397 cm^{-1}, and doublet at 74^{sh} and 719 cm^{-1} ascribed to the rocking vibration of CH_2 groups, together with bands at 1464^{sh} and 1454 cm^{-1} assigned to the scissoring vibrations of CH_2 groups ascribed to zinc salts of C8:0 and C9:0 short-chain fatty diacids [36].

2.4. Corrosion Products

The electron images complemented with X-ray microanalysis and IR spectra have allowed the identification of some mineral corrosion products formed in certain metallic items by their characteristic micromorphology. Figure 15a,b show the microphotograph and the backscattered electron image acquired in the cross-section of a sample excised from the case of the cigarette lighter ETN009. The case is made with steel covered by a thin brass layer. Both parts show clear signs of alteration. In the outermost layer, the remains of what appears to be the original brass can be seen, forming a thin, highly fragmented sheet dispersed in a mixture of corrosion products and infiltrating exogenous materials that form a layer at ca. 30 µm. Those infiltrations could be favored by the formation of copper corrosion products that produce swelling and increase the porosity of the layer [40,41]. Beneath the brass, the upper part of the steel foil can be seen, with evidence of complete corrosion, which would have caused it to flake off from the inner unaltered steel core (not shown in the image) as result of the advance of the corrosion process. The arrow in Figure 15b indicates a thin (12 µm) band that presents ripple-like dislocations. This particular feature is tentatively associated with a differential elongation of the steel grains in the surface of the foil that could have occurred during the manufacture of the steel plate used to make the lighter case by the cold-rolling process [42]. Figure 15c shows fine-plate crystals of lepidocrocite (g-FeO(OH)) formed in the pores and fissures connected with the external soil. This crystalline material forms flower-like structures [43]. Figure 15d shows the semicrystalline goethite (a-FeO(OH)) formed in the case. The particles exhibit a cotton-ball characteristic morphology [44,45]. Alteration products of the metals that form the brass have also been identified by their characteristic morphology. The rounded platelet-shape of the particles found in the case surface are identified as hydrocerussite ($Pb_3(CO_3)_2(OH)_2$) (Figure 15e). Figure 15f shows submicron-sized crystals with a flake-like morphology. EDX analysis of these particles reported molar ratios of O/Cu = 4.2 and S/Cu = 1.1. Analogous crystal micromorphologies have been described for brochantite [46] and posnjankite [40] grown in copper patinas formed by exposition to the atmosphere and are very similar to covellite (CuS) [47].

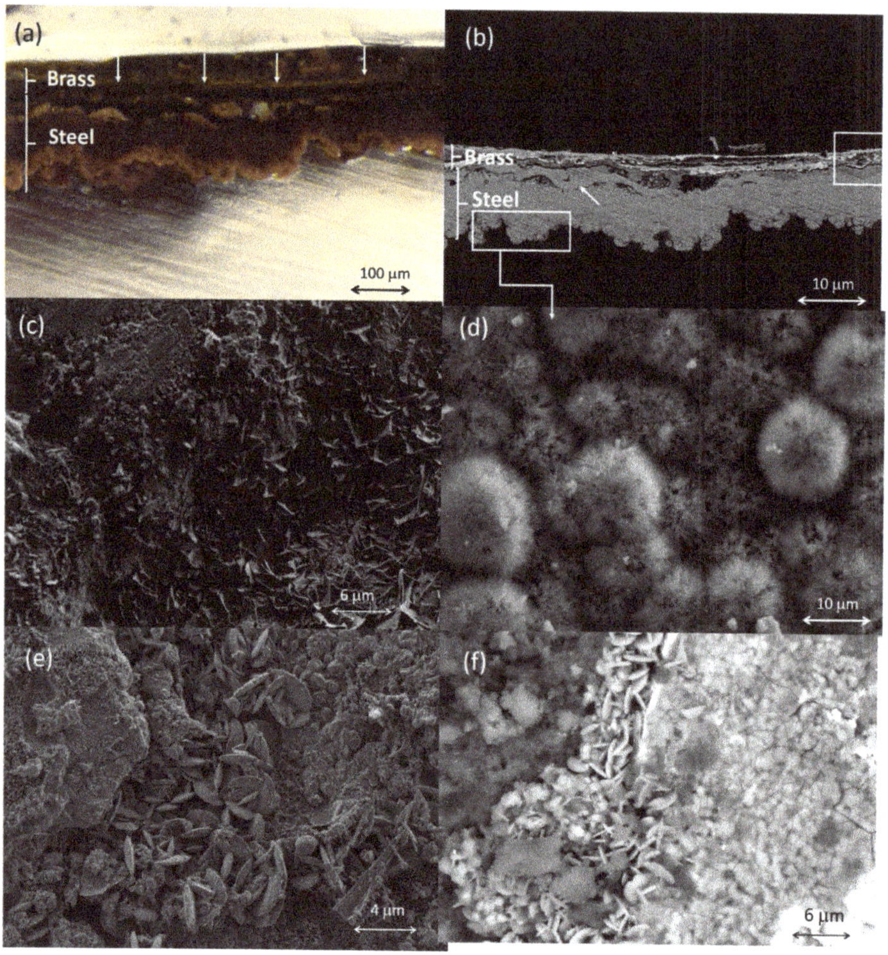

Figure 15. (**a**) Microphotograph of a sample excised from the case of the cigarette lighter ETN009 prepared as cross-section, where the arrows point out the upper surface of the cross-section in the sample; (**b**) backscattered electron image acquired in the cross-section shown in (**a**); detailed analysis of the upper framed area is shown in (**d**); (**c**) secondary electron image of crystals of lepidocrocite in the outer layers; (**d**) backscattered electron image of semicrystalline goethite formed in the inner layer; (**e**) secondary electron image of hydrocerussite crystals; (**f**) backscattered electron image of platy crystals of copper corrosion products.

Figure 16 shows in detail the complex stratified structure of the corrosion layers of the sample ETN009. The different layers and areas compositionally differentiated have been labeled from A to H. The overall elemental composition of the layers and areas is provided in Table S2. The two-dimensional distribution of the content of the different elements is graphically illustrated in the X-ray maps shown in Figure 17. Oxygen is present in all the layers in concentrations over 22 wt%, indicating that the corrosion process has progressed considerably and no or little unaltered brass or steel still remains. The A to C regions correspond to the brass quaternary alloy Cu-Zn-Pb-Sn, whereas areas D-H are located in the steel layer. Chlorine (~1%) is present in the outermost layer A. Sulfur is present along the cross-section with prevalence in B and C regions, with concentrations of 7.3 and 14.5%, respectively. That indicates that copper sulfate or sulfide species were formed. However, the Cu/S molar ratios calculated by the ZAF method do not allow any conclusion and could suggest that non-stoichiometric compounds are present in these layers. The significant presence of Fe in the outermost layers A to C indicates that the diffusion of Fe ions throughout the brass layer has taken place. Other foreign elements such as Al, K, and Si in the A, B, and D layers confirm that these strata are highly altered regions with easy access to the attacking solution and the exogenous materials. The lower D to H regions correspond to the steel structure of the lighter case. The interphase between the brass cover and the steel case (region D) has the lowest Fe content (59.0%) and some Cu, Zn, and Pb. These elements are also found in areas E and F, suggesting that diffusion from the outermost brass layer has occurred at some extent. The Fe content increases from 59.0% to 77.7% when it passes from layer D to H, maintaining a Fe/O molar ratio of 0.98–1.00. This molar ratio value, around 1, suggests the presence of mixtures of oxides and oxyhydroxides that prevalently approximate the structure of magnetite and goethite [48]. Small amounts of sulfur are present throughout the entire section of the steel sheet, ranging between 0.89 and 1.41 wt% in D-F regions and 0.2–0.39 wt% in the G-H regions.

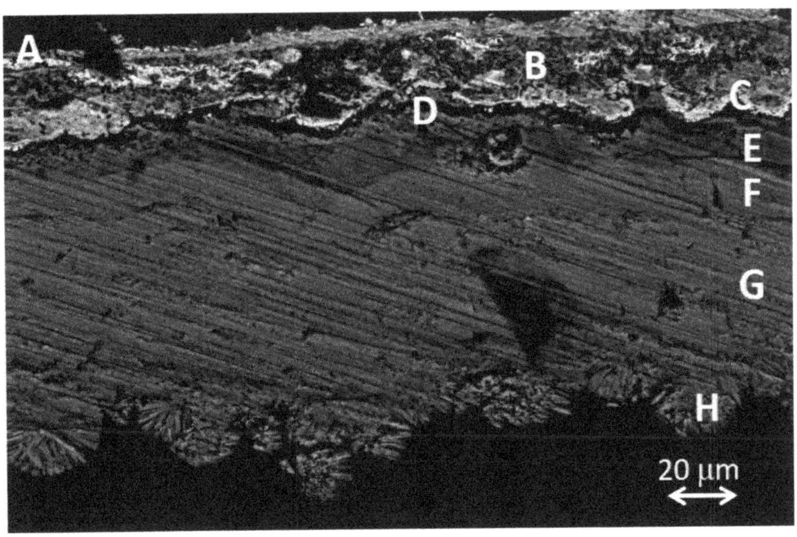

Figure 16. Backscattered electron image acquired in the cross-section of the cigarette lighter ETN009 (see area framed in Figure 15b). A, B and C correspond to the three layers and areas compositionally differentiated in the brass upper cover of the cigarette lighter; D, E, F, G and H correspond to the five layers and areas compositionally differentiated in the steel inner case of the cigarette lighter.

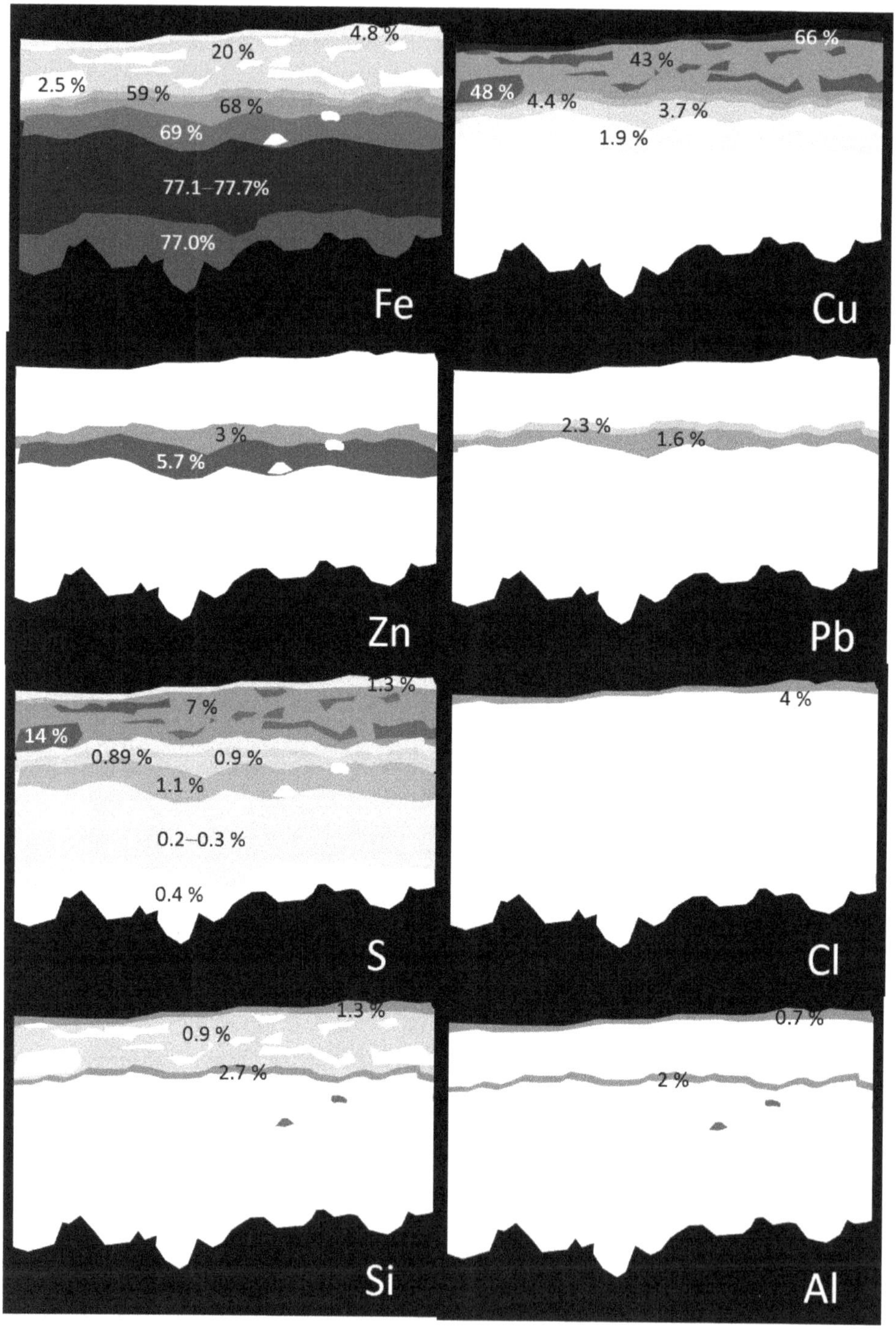

Figure 17. Elemental distribution maps acquired with FESEM–EDX on the backscattered electron image of Figure 16. Values in graphs indicate the weight % composition of the element in each specific area.

A similar cryptocrystalline morphology of hydrocerussite to that shown in Figure 15e was found in the corrosion layers of the pencil extender ETN010. The IR spectra acquired in samples of corrosion products formed on the surface of the metallic objects confirmed that these external corrosion layers were complex mixtures of corrosion salts, sometimes of inorganic–organic hybrid character. That is the case of the corrosion layer formed in the cigarette lighter ETN009 and the spoon ETN006. Figure 18a shows the IR spectrum acquired in the sample ETN009. The absorption bands of malachite $CuCO_3.Cu(OH)_2$ dominate the spectrum, namely, the $2\nu_2 + \nu_4$ CO_3 stretching vibration at 2499 cm^{-1}, $\nu_1 + \nu_4$ CO_3 stretching vibration at 1810 cm^{-1}, ν_3 CO_3 asymmetric stretching vibration with maximum at 1375 cm^{-1}, ν_2 asymmetric deformation at 860 cm^{-1} with shoulder at 840 cm^{-1}, and ν_4 symmetric deformation at 741 cm^{-1} [49]. The presence of siderite from the case and calcite from soil particles integrated into the sample cannot be ruled out as they would overlap in the previous bands, given the proximity between the absorption band maxima. OH stretching and deformation bands at 3200 and 1670 cm^{-1}, respectively, and broad band at 1051 cm^{-1} associated with asymmetric stretch ν_3 SO_4 are ascribed to sulfates [50]. Bands of stretching vibrations of methyl and methylene groups in organic compounds occur at 2925 and 2852 cm^{-1}. The IR spectrum acquired in a sample of the corrosion layer formed on the spoon ETN006 surface is shown in Figure 18b. Calcite is identified by the bands at 1420, 878, and 718 cm^{-1}; bands of quartz and clays occur at 1151, 1102, and 1027 cm^{-1} (see Table 2 for assignment); hydroxyl stretching bands occur at 3262 cm^{-1}; hydroxyl bending bands at 897 and 745 cm^{-1} and Fe-O stretching bands at 630 cm^{-1} are ascribed to goethite [50,51]. Shoulders at 1372 and 1324 cm^{-1} are assigned to calcium, zinc, and copper oxalates [51]. Figure 18c shows a secondary electron image of a sample excised from the remains of the glasses frame ETN073. The image shows a notable amount of cryptocrystalline particles deposited on the surface of the spectacles frame. The EDX analysis performed in point 1 provides a P/Ca molar ratio of 1.81 and a P/Zn molar ratio of 1.14. (Figure 18c). These P/Ca and P/Zn ratios suggest the presence of scholzite ($CaZn_2(PO_4)_2{\cdot}2H_2O$).

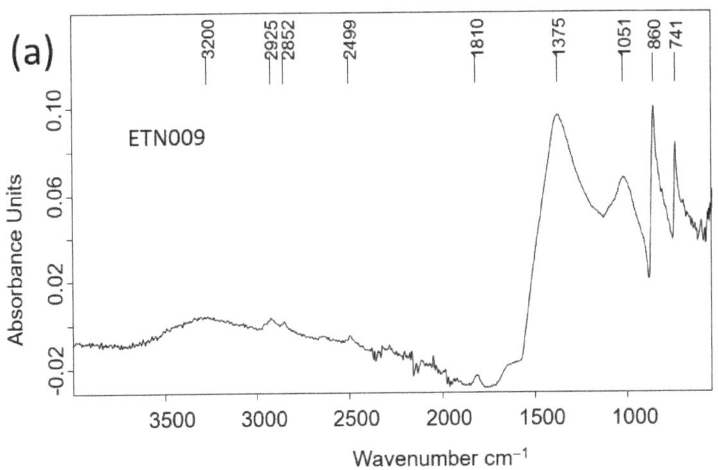

Figure 18. *Cont.*

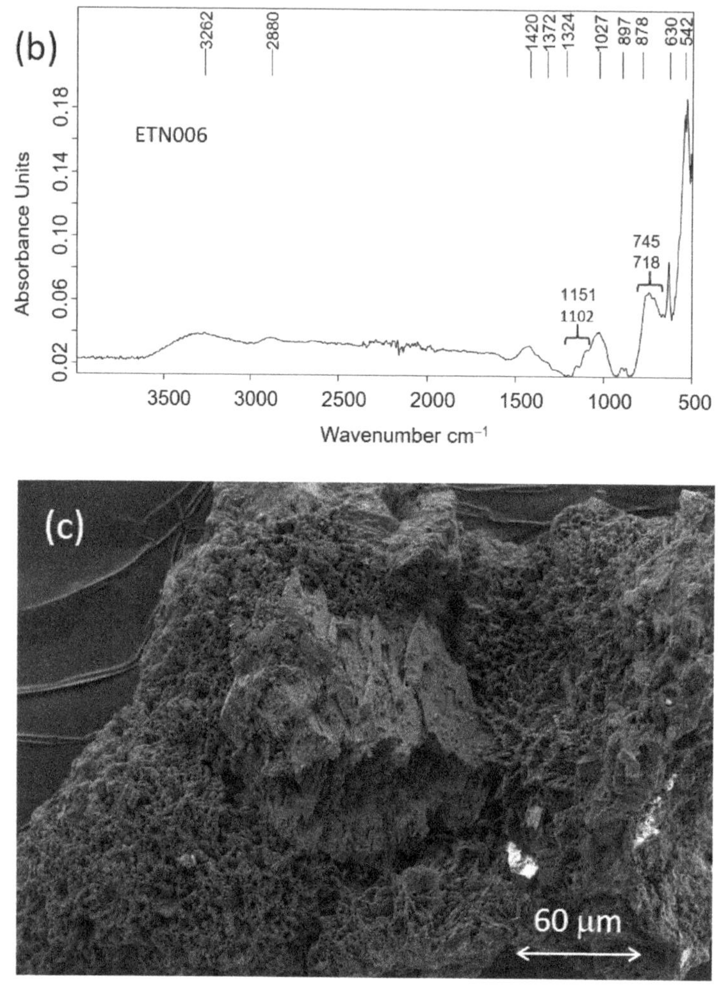

Figure 18. (**a**) IR absorption spectrum of cigarette lighter ETN009; (**b**) spoon ETN006; (**c**) secondary electron image acquired in a sample excised from the corrosion products of the metal spectacles frame ETN073.

2.5. Fungal Biodeterioration

In the image shown in Figure 19a, spots of a few tenths of a millimeter of diameter of microbiological nature can be observed in a sample excised from the organic materials covering the tag knife. The IR spectrum acquired in this sample (Figure 19b) exhibits a band profile characteristic of fungi colonies with stretching and deformation bands of OH groups at 3264, stretching at methyl and methylene groups in lipids at 2921 cm^{-1} and 2852 cm^{-1}, stretching vibrations at 1737 cm^{-1} ascribed at lipids, a band at 1630 cm^{-1} ascribed to amide I band in proteins, a band at 1559 cm^{-1} assigned to amide II band in proteins, deformation bands of C-H bounds at 1456^{sh}, 1403, and 1375 cm^{-1}, and an amide III band at 1321^{sh} cm^{-1} [52]. This last band and that other at 1630 cm^{-1} are also assigned to calcium oxalate's symmetric and asymmetric stretching vibrations, respectively. Intense bands at 1043 cm^{-1} and 976 cm^{-1} are ascribed to polysaccharides. The latter and unassigned weak bands in the range 700–900 cm^{-1} (microorganism fingerprint) are characteristic of fungi [53].

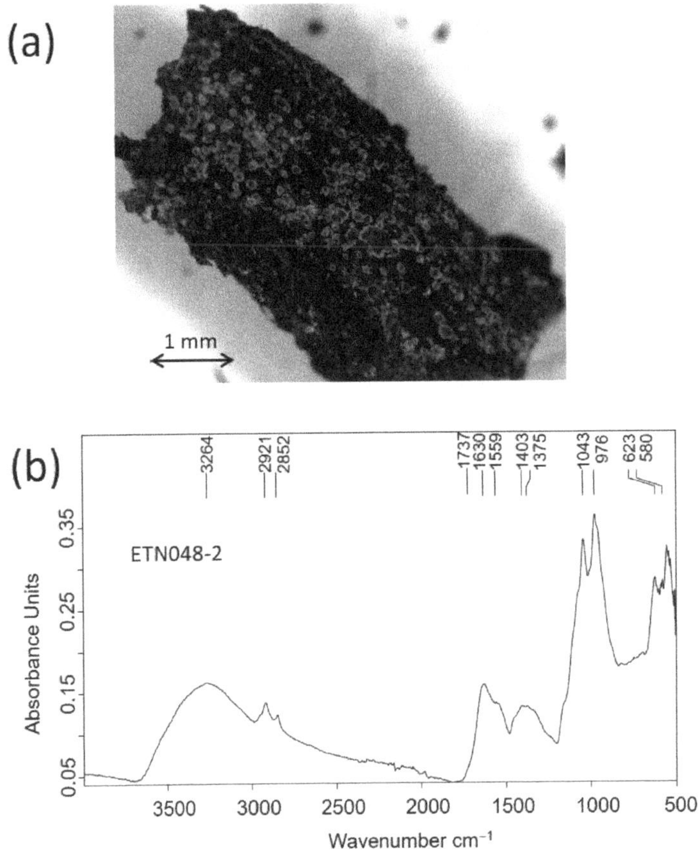

Figure 19. Knife ETN048: (**a**) microphotograph of the microorganisms formed in the surface of the sample of the knife; (**b**) IR absorption spectrum.

2.6. Grave Soil

In parallel to the personal items analysis, samples of the soil (SF2-01 to SF2-04) that filled the mass grave number 2 were analyzed using XRD (Figure 20), FTIR, and FESEM–EDX (Figure S6). The samples were taken in four points far from the corpse found in the saca placed at a depth of 4.75 m. The main compounds identified in the soil were calcite and quartz, accompanied by mica minerals, mainly of the phlogopite type, and kaolinite. The average elemental content of the soil, expressed as wt% and obtained from analysis in the area mode (see Materials and Methods section for details) using FESEM–EDX, was 41 $\pm$ 1 O %; 1.54 $\pm$ 0.03 Mg %; 6 $\pm$ 1 Al %; 19 $\pm$ 4 Si %; 0.3 $\pm$ 0.9 P %; 0.4 $\pm$ 0.4 S %; 1.7 $\pm$ 0.3 K %; 26 $\pm$ 5 Ca %; 0.4 $\pm$ 0.3 Ti %; 4 $\pm$ 1 Fe%. In addition, the spot analyses performed on individual grains enabled the identification of some of the most abundant minerals present in the soil from the experimental atomic percentages calculated using the ZAF method. Thus, potassium feldspars of experimental stoichiometry ($(K,Mg)_{1.00}(Al,Fe)_{0.98}Si_{3.26}O_8$), wollastonite ($Ca_{1.10}Si_{1.00}O_4$), zircon ($Zr_{0.99}Si_{1.00}O_4$), and gypsum ($Ca_{1.16}SO_4.2H_2O$) could be identified. The presence of calcium apatite/hydroxyapatite and chloride could be noted by the corresponding emission lines of P and Ca. The apatite found in samples taken from the grave soil may be due to the proximity to human remains, which would have been the sources of phosphorus and calcium [54]. Still, it may also have a geological origin.

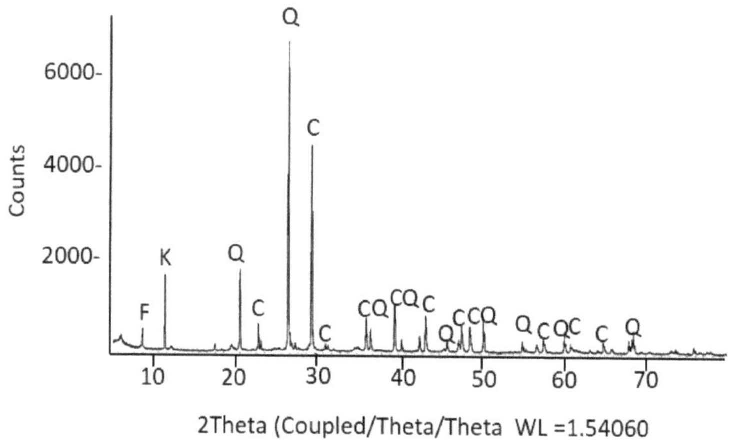

Figure 20. Diffractogram of a sample of the soil that filled the mass graves. C: calcite; Q: quartz; K: kaolinite; F: phlogopite.

3. Discussion

3.1. Connecting the Materials Identified and the Manufacturing Technique

The primary purpose of this study was to identify the materials that composed the personal items collection as, in most cases, it was not possible to identify them using organoleptic methods due to the poor state in which they were found.

The high level of oxygen detected in the sample MCP029 from the aluminum devotional medallion raises the question of whether this metal had a protective anodizing treatment. Anodizing is an oxidizing treatment of the metal surface that can be driven either electrochemically or chemically to form an aluminum oxide film much more stable and thicker than the aluminum oxide film formed naturally. The generated anodized oxide layer is thick, dense, and passive enough to avoid corrosion and abrasion of the metal surface. The Bengough–Stuart process, developed in 1923, was the first anodizing method consisting of a chromic acid–base treatment, and it was used to protect seaplane parts from corrosion [55]. That and other anodizing treatments became popular in the early 1960s [56,57]. The surface of anodized aluminum is characterized by two different morphologies depending on whether the formed oxide film is slightly soluble or completely insoluble in the electrolyte used. The former (i.e., sulfuric, phosphoric, chromic, and oxalic acids) produce porous films. In contrast, non-porous barrier films are obtained with the latter (i.e., neutral boric acid solution, ammonium borate, or tartrate aqueous solutions (pH 5–7), ammonium tetraborate in ethylene glycol, and organic acids such as citric, malic, or glycolic acids). The porous anodized surfaces have an outer nanoporous layer and an inner non-porous thinner layer that acts as a barrier in contact with the metal surface. Accelerated aging experiments simulating natural weathering have demonstrated that the corrosion of a non-treated aluminum surface is characterized by micro-cracks and pitting on the micrometer scale, whereas anodized surfaces exhibit a more diminished pattern in the nanometer scale [58]. Thus, this different behavior can provide clues about the manufacturing method used for making the devotional medallion MCP029. Figure 21 shows the secondary electron image of the sample MCP029 in which fractures measuring several tens of micrometers in length and depth are visible on the sample's surface, indicating that this object was not subjected to an anodizing treatment. That would be consistent with the period in which it is supposed to have been manufactured, that is, the beginning of the 20th century until the beginning of the Spanish Civil War.

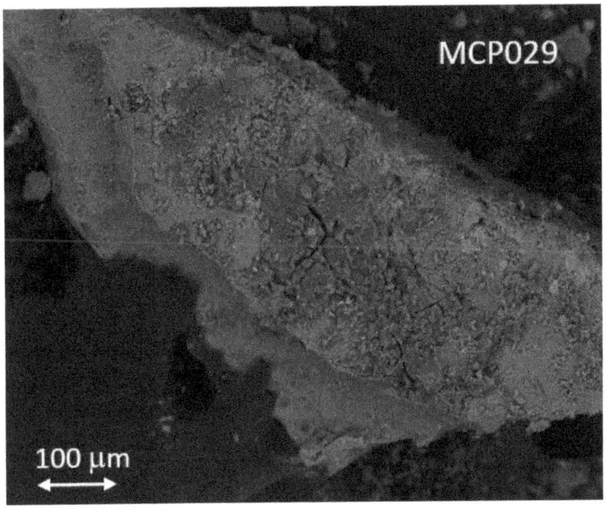

Figure 21. Secondary electron image of sample MCP029 corresponding to the aluminum devotional medallion.

Copper alloys have prevailed among the metallic items. A silver copper alloy was used to make the ring ETN033. It is well known that the solubility of copper in silver is about 8–10 wt% at 780 °C (eutectic temperature) [59]. Nevertheless, when silver is alloyed with copper, only one proportion is known to result in a homogeneous melt, which is 72% silver and 28% copper (alloy standard 1). Proportions other than that give rise to segregation problems in the object. If the proportion of silver is higher than alloy standard 1, the central part of the object becomes enriched in silver, and the edges harden more quickly. On the contrary, if the silver proportion falls below 72%, the object's core becomes Ag-poor, and the edges become Ag-rich [60]. For this reason, the alloys used in jewelry are around these values. According to Hiscox and Hopkins [60], the silver alloys usually used in jewelry are (Figure 22a) 77 parts Ag, 6 parts Cu (standard 2); 40 parts Ag, 4 parts Cu (standard 3); 4 parts Ag, 1 part Cu (standard 4), and, more specifically, alloys for chains are prepared with 3 parts Ag and 2 parts Cu (standard 5). The diagram depicting the Cu content vs. the Ag content in Figure 22a shows that the ring's chemical composition is close to standards 2 and 3, with the highest silver contents. However, it is also observed that the ring has a lower silver content than these standard alloys. Suppose the composition of silver is compared separately, in that case, standard 4 is the closest to the experimental composition obtained for silver in the ring, but, in contrast, the copper content is notably lower than standard 4. This loss can be associated with leaching and reprecipitation processes in the burial soil. These processes are common in archaeological objects found in burials, and they have been widely studied and described. The formation of cuprite (Cu_2O), tenorite (CuO), atacamite [$CuCl_2{\cdot}3Cu(OH)_2$] chlorargyrite ($AgCl$), and stromeyerite [$CuAgS$] are among the corrosion products most frequently found [59,61–65]. Most of these researches have demonstrated that, in general, copper is more prone to undergo leaching. According to Ingo [65], the intimate contact between copper and silver grains in the Ag-Cu alloys, which have different electrochemical potentials, promotes couples strongly conducive to corrosion. In those microscopical systems, the copper phases become the anodic area, and preferential dissolution of copper occurs. These processes affecting selectively copper in Ag-Cu alloys could justify the discrepancies found in the experimental composition of the ring and the composition of the standard alloys, in particular, standard 4.

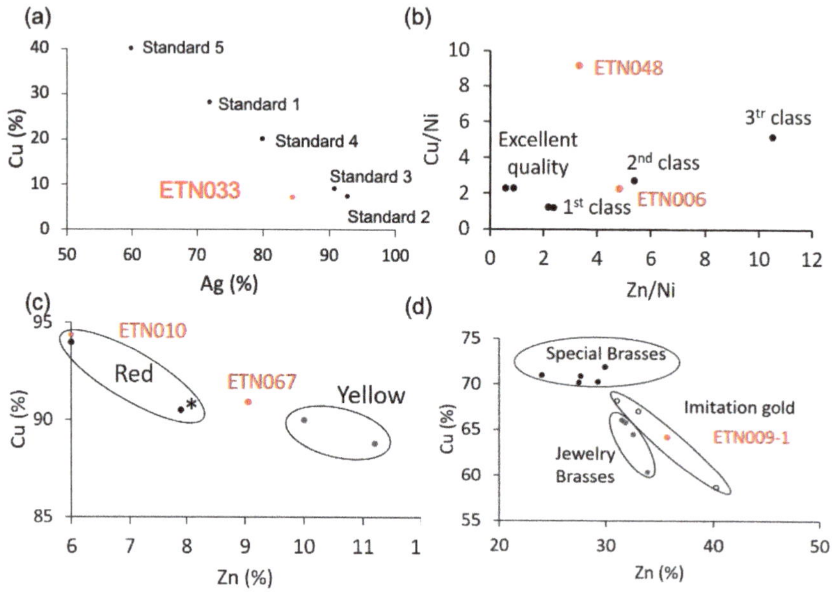

Figure 22. (**a**) Copper content versus silver content of standard Ag-Cu alloys and sample ETN033 excised of a silver ring; (**b**) Cu/Ni vs. Zn/Ni of standard alloys and samples ETN006 and ETN048 excised of a spoon and a knife; (**c**) Cu wt% vs. Zn wt% of standard jewelry brass alloys, and samples ETN010 and ETN067 excised from the pencil extender and the medal (* the alloy includes 1.6 wt% Pb); and (**d**) Cu wt% vs. Zn wt% of standard jewelry brass alloys, standard special brass alloys, imitation gold alloys, and sample ETN009-1 excised from the cigarette lighter. Black dots that correspond to experimental compositions corresponding to a specific reference alloy are framed with black continuous lines; red dots correspond to the experimental values found in the samples.

The samples ETN006 and ETN048 from the spoon and the knife were identified as copper ternary alloys composed of copper, zinc, and nickel. The proportion found among the three elements points towards German silver. This alloy is well known for being comparable to silver in terms of its visual appearance, improving its hardness and unalterability. Properties such as ductility, fusibility, or color vary according to the proportion in which the three metals of the alloy are combined. According to Hiscox and Hopkins [60], the usual composition of German silver consists of 50–66 parts of copper, 19–31 parts of zinc, and 13–18 parts of nickel despite the best composition for manufacturing spoons, forks, and knives consisting of 50 parts copper, 25 parts zinc, and 25 parts nickel. Other proportions that improve or reduce the alloy properties have also been used that incorporate small amounts of lead, iron, manganese, or tin. Figure 22b shows the diagram Cu/Ni vs. Zn/Ni in which the most common compositions of German silver are depicted according to the quality of the cutlery. The position of the sample ETN006 ascribes this alloy to the group of the second class. The sample ETN048 excised from the knife does not match any German silver alloy. This could be due to the sample's poor quality and thick corrosion layer enriched in zinc and copper alteration products, which distort the result. This may lead to an overestimation of the copper content to the detriment of the actual zinc content, as shown by the position occupied by this sample in the diagram.

A ternary alloy Cu-Zn-Pb was identified in sample ETN010 excised from the pencil extender and sample ETN067 from the medal. A silver coating has also been identified in the pencil extender. The composition of both samples is depicted in Figure 22c, which shows the diagram of Cu % content vs. Zn % content. The diagram also shows the composition of common Cu-Zn and Cu-Zn-Pb alloys used in jewelry with characteristic red and yellow-reddish colors [60]. The chemical composition of sample ETN010 accurately matches that of one of the two red alloys, whereas the ETN067 sample exhibits an intermediate composition between red and yellow alloys. The lead content found in the samples would be explained by providing greater malleability during manufacturing. The pencil extender, also called pencil lengthener, is a small instrument used to facilitate the use of pencils when the small length hinders their use. The extenders are made out of metal or wood. Engineers, drafters, and artists currently used extenders to maximize the use of their favorite pencils. Due to its shape, the binary Cu-Zn composition was likely adapted by including a small amount of lead in the alloy, prioritizing malleability over other properties such as color to adapt it to the industrial process of manufacturing the body of the objects by molding or casting.

On the other hand, barite and a cobalt pigment were also identified in the medal, which is associated with the rest of the original pictorial decoration of the medal. Thus, it is possible that barite was used to decorate the white parts of the medal, and cobalt pigment was used to decorate the blue parts. Nevertheless, barite is a well-known pigment added as an extender by the paint manufacturers to provide opacity [66]. Identifying a blue cobalt pigment in the same sample as barite does not exclude that the latter was merely used as an extender of the cobalt blue pigment used for decorating in blue the medal.

The elemental composition found in the sample ETN009-1 excised from the metallic case of the cigarette lighter corresponds to a quaternary brass alloy Cu-Zn-Sn-Pb. In contrast, other internal parts of the lighter were made of steel (ETN009-2) (see Table 1). Brasses have been widely used in jewelry (costume jewelry) to make jewels at a much lower cost. Ordinary brass is more rigid than copper and better able to withstand stress. Annealed brass can be stamped and percussion processed. Concerning composition, copper and zinc are alloyed in brass at all proportions, with their properties varying depending on the temperature at which the alloy is produced. The alloys with the most significant resistance contain 20 to 30% Zn. If the concentration is raised to 60%, the resistance decreases, preventing mechanical uses. There are very ductile and malleable brasses, which, together with the different varieties of color, make them very suitable for artistic applications. The malleability varies with the composition of the two main elements, Cu and Zn, the temperature, and the addition of small quantities of certain metals such as lead and tin. Some alloys can be rolled at any temperature, and others can be used only when hot. Alloys with a 15 to 30% zinc content are suitable for drawing into wire. On the other hand, a series of alloys of copper and zinc have been developed that have been generically called gold imitations. These brasses have been widely used in jewelry (costume jewelry) to make jewelry at a much lower cost. These include chrysocolla, Mannheim gold, and mosaic gold. Figure 22d shows a diagram of Cu wt% vs. Zn wt% in which the copper and zinc content of a series of common brasses of use in jewelry and watchmaking (Stoberg, Hamilton, and watchmaking brass), special brasses (special French brass, French brass for molding and casting, English brass, Heegermuhl, and Ausburg brass), and gold imitations (chrysocolla, Manhein gold, and musive gold) in use in the first half of the 20th century are depicted [60]. The brass of the cigarette lighter ETN009-1 case has a high zinc content in the concentration range in which gold imitation alloys are also found. However, these alloys are usually ternary, including tin as a minor element in some cases and lead in others. Therefore, it is more plausible that the ETN009-1 alloy corresponds to a common quaternary brass such as Hamilton brass, which has a similar copper percentage (Hamilton brass, 64.5% Cu, 32.5% Zn, 0.3% Sn, and 2.7% Pb).

Silver was also detected in spot analyses from samples of the pencil extender (ETN010). Silver must have been applied as a coating on the brass body of the pencil extender to embellish it. At the beginning of the 20th century, the silver plating of brass objects was carried out using three different methods: dry, wet, and electroplating. The dry method consisted of rubbing the metal surface of the object with a paste that included silver salts (i.e., three parts silver chloride, three parts sodium chloride, two parts chalk, and six parts potash; or 1 part silver nitrate and three parts potassium cyanide; or three parts of silver chloride, eight parts of sodium tartrate, and 1 part of sodium chloride). The brass object was carefully cleaned with diluted nitric acid and ammonia and dried with sawdust. Finally, the silver layer was protected with lacquer. The resulting silver layer is not very durable, so this method only applies to objects that will not suffer much wear, such as barometer and thermometer scales. The wet method could be carried out following a cold or hot treatment. Cold treatment was carried out by immersing the object in a bath containing a liquid that, according to Boettegen's formula, included two parts of silver hyposulfite, 1 part of ammonium chloride, and 20 parts of water. Another bath formula specific to brass consisted of 20 parts silver nitrate, 120 parts potassium cyanide, 30 parts washed gypsum, and 1000 parts water. A hot silver plating of brass was carried out at 80–90 °C [60]. The objects were immersed in enameled containers. The bath consisted of 10 parts of silver nitrate dissolved in 500 parts of water, which were mixed with 35 parts of potassium cyanide dissolved in 500 parts of water [60]. Until the end of the 1950s, the industrialization of electroplating would not begin in Spain since electrolytic coatings were only used in a very artisanal way in small manual installations, except in some state-owned companies of the National Institute of Industry (INI) such as SEAT automotive factory. The electroplating was carried out in electrolytic baths of silver and potassium cyanide. The anode was a silver bar of size proportional to that of the brass object that acted as a cathode [67,68]. The analysis of the ETN010 pencil extender has not provided any experimental evidence to determine which of these three methods was used to manufacture this object. However, the use for which it was intended required the application of a layer of wear-resistant plate, so it is very likely that the electroplating method carried out in artisan workshops was used. Still, the cold or hot wet plating method has also not been ruled out.

In the ring ETN042 and the button MCP001, calcium carbonate has been identified as the main component. Bivalve (Mollusca: Bivalvia) shells are composed of 95–99.9% calcium ($CaCO_3$) and calcium–magnesium carbonates ($CaMg(CO_3)_2$) present in the form of calcite, aragonite, and dolomite in an organic matrix of conchiolin proteins with inorganic trace elements that vary according to the species and the surrounding environment [69]. Thus, the geogenic formation process in the first two and the biogenic in the last explains the identification of two different polymorphs of calcium carbonate, calcite, and aragonite, respectively. The IR absorption spectra obtained from the three samples can provide additional information beyond recognizing the constituent mineral species. This additional information relates to these materials' degree of atomic order. It has been demonstrated that the atomic disorder in a calcium carbonate structure is closely related to the origin; that is, the process of calcium carbonate formation determines the final structure of the mineral: geogenic (limestone, chalk, and marble rocks), anthropogenic (mortar and plaster products of pyrotechnologies) or biogenic (eggshell, mollusk shells, coral, foraminifera). In mortars and plasters, lime is first produced by heating geogenic calcium carbonate to high temperatures. When the lime is applied in the construction, the lime is allowed to hydrate and absorb CO_2 from the atmosphere. In the biogenic formation of calcium carbonate structures such as nacre, the bio(macro)molecules interplay with the precursor amorphous calcium carbonate (ACC) and drive the process toward the final crystal structures [70].

An indicator of the degree of the atomic disorder can be obtained by studying CO_3 deformation vibrations ν_2 and ν_4 that exhibit the IR spectra acquired in calcium carbonate of different origins as the ν_4 band is more influenced by the atomic disorder than the ν_2 band. This can be quantified by measuring the intensity ratio I_2/I_4 of both ν_2 and

ν_4 bands [71]. The higher the value of I_2/I_4, the greater the degree of disorder in the crystal structure. All this has given rise to an archaeometrical research line focused on distinguishing among calcites formed under diverse conditions based on the measuring of the I_2/I_4 ratio [72–75]. According to the literature [71–73,75], well-ordered calcite materials exhibit sharp and intense ν_2 and ν_4 bands. The I_2/I_4 ratio for well-ordered geogenic sparry calcite is around 2.0–2.2, and geogenic calcite ranges from 2.3 to 3.4. In anthropogenic calcite, the ν_4 band maintains its shape without widening but diminishes the intensity to the point that in the amorphous calcite, the ν_4 band does not appear [70]. The I_2/I_4 ratios found in anthropogenic mortars and plasters from lime formed at elevated temperatures (kiln at 700 °C or over) range widely between 3 and 7. Chu et al. [72] point out several causes: the original limestone may have been partially burned during the lime manufacture, and the resultant mortar is a mixture of original geogenic crystals and neoformed anthropogenic calcite grains. Limestone aggregates may have also been added to improve their binding properties. In addition, neoformed calcite in the mortar can dissolve and reprecipitate over time in the form of more ordered crystals. In biogenic materials, the ν_4 band widens symmetrically around the band maximum more than the ν_2 band. The I_2/I_4 ratios of biogenic calcite ranges from 2 to 3.6. In the biogenic formation of calcium carbonate materials, the short-range orders play a key role in crystalline polymorph selection. In mollusks, colloidal nanoseeds of immature ACC with random crystallographic orientations fuse by oriented attachment to form the mature tablet of aragonite [69].

The experimental I_2/I_4 values were 2.27 and 2.43 for the ETN053 ring and MCP001 biogenic nacre, respectively. In the two cases, the values found are in the low part of the experimental ranges found in the literature, which indicates that the calcite and aragonite are structurally well ordered at the atomic scale. This result suggests that the burial has not altered the nacre structurally. Therefore, the alterations caused must be associated with the interaction of the external environment. The organic matter coming from the decomposition of the deceased corpses has formed an ionomeric phase promoted by the soil moisture that impregnates the ring and has released the calcium ions of the surface and, finally, has formed carboxylate complexes [29]. In the button, the corrosion processes occur through interlaminar decohesion produced by the sorption of water present as soil moisture, which produces a significant loss in the mechanical resistance of the material.

Another interesting group of objects is formed by the two nit combs ETN060 and ETN058 and the spectacles frame ETN073. The IR spectra acquired in these samples indicate that a casein-formaldehyde (CF) resin was used to make these materials. CF is not consistently recognized in organoleptic examinations of heritage goods as it has a similar appearance to other types of semi-synthetic resins such as cellulose nitrate or acetate, hence, the importance of developing precise identification methods. A thorough examination of the IR spectra acquired on the samples reveals that the CF resins identified on all three objects were dyed. This result was confirmed by FESEM–EDX analysis (Figure S7 and Table S3). The dying with an earth pigment provided the tortoiseshell appearance to these personal items [66].

Until 1936, the range of plastic materials with practical applications was limited. However, using some of them as imitation materials of other more expensive materials from natural origin had already become widespread. An example is CF resin, which was used to make perfect imitations of semi-precious materials such as mother of pearl, tortoiseshell, ivory, horn, and ebony. Although the CF resin was synthesized in 1897 by the German printer Adolph Spitteler and his associate W. Krische, the milk plastic process was not patented until 1911 by the British company Syrolit Ltd. This plastic material became popular because, once manufactured, it could be ironed and cleaned wet or dry, making it able to manufacture buttons, knitting needles, fountain pens, and hair combs. The development of other petroleum-based polymers, such as polyvinyl chloride, polyethylene, polyester, acrylic, and melamine, in the first half of the 20th century, was progressively limiting the applications of this material [76]. This plastic material could be dyed with colorants during the processing of the object [77]. The coloring of CF plastic is carried out before the extrusion, although this process can be performed after manufacturing using many recipes. To dye the object after extrusion, the material, with the final shape already given, had to be wrapped in muslin and immersed in a bath of dye heated to 90 °C for at least 15 min [78]. The description of both procedures provides clues as to what procedure could have been used to stain the personal objects analyzed. Staining using a 90 °C hot bath requires water-soluble dyes, which would rule out the earth pigments and calcium carbonate extenders identified in the samples analyzed. In addition, these stains are superficial. This also seems to rule out this dyeing system for the objects analyzed.

Three natural products were identified in the personal item sharpener ETN012, rope handcuffs ETN045, and the contents of the little bottle LLUE001. Manufacture by the tin industry of manual pencil sharpeners began in the second half of the 19th century. At the same time, mass pencil production took place. Mechanical sharpeners were the subject of numerous patents registered worldwide between 1850 and 1900. At the beginning of the 20th century, various sharpeners were available according to needs, and their designs were perfected over time. The first portable pencil sharpeners were small, metal, and conical in shape adapted to the graphite pencil. Among the materials used in the first half of the 20th century for making sharpeners, metal, wood, and semi-artificial resins are included. Although there are no reports of pencil sharpeners made with CF or other semi-synthetic resin, this polymer was likely used to manufacture this object given the large number of applications it had at this time, some of which, such as combs or buttons, required, like the pencil sharpener, a specific mechanical and abrasion resistance. The analysis has resolved this item by demonstrating that, in this case, wood was chosen.

Esparto grass experienced two moments of splendor in the 20th century, coinciding with the two great world wars. The first, between 1914 and 1920, lasted until the 1930s. In both cases, the shortage and concomitant rise in the price of other fibers such as jute and hemp caused esparto grass to gain value. The rope handcuffs ETN045 found have the traditional characteristics of a simple three-strand braid. Each strand is made of five or six esparto fibers.

More interesting is the information provided by the item LLUE001. The name "Naranjil", engraved on the glass bottle, is the trademark of a castor oil flavored with orange marketed by the company V. Llorens Sastre based in Valencia. Naranjil was a castor oil used as a purgative. It was registered in 1925 by Llorens and was on the market for most of the 20th century (Figure S8). Large pharmaceutical companies had not yet been widely settled in Spain then. It was common to find pharmaceutical products on the market for medicinal purposes based on natural products processed by doctors and pharmacists who patented and marketed them. Thus, a Naranjil competing trademark was the castor oil from the Camós Laboratory [79]. This product was produced in a traditional pharmacy in Algemesí (Valencia, Spain) that was run for decades by Mr. Fernando Camós Vidal, who died in 1986 [80].

3.2. Alteration Processes

The particular burial conditions have determined the main chemical and biochemical mechanisms that have altered the state of the materials comprising the personal items. Metallic objects have suffered electrochemical corrosion processes typical of burial environments in which the soil is cyclically saturated with humidity that subsequently decreases, giving way to a dry phase. In the humid phases, a thin layer of water is fixed to the surface of the metal, carrying salts and dissolved oxygen and acting as an electrolyte. The metallic ions released in this process reprecipitate on the object's surface as salts with the available anionic species, carbonates, chlorides, sulfates, etc. Composition of the soil, pH, microstructure, and presence of cracks are factors that contribute to make complex the corrosion mechanisms.

Figure 23a,c show a scheme of the electrochemical processes that have driven the steel and brass corrosion processes in the cigarette lighter ETN009. According to the literature [81], in iron and steel objects, the cathodic process is mainly undergone by oxygen that reacts with H^+ ions to form water or with water to form hydroxyl ions, and the reduction of hydrogen is a minor process. On the other hand, the anodic process of the oxidation of iron to Fe^{2+} initially takes place in the surface of the unaltered metal in a separate point to the cathodic reaction. Once the process has started, the advance of the redox process depends on the environmental conditions. The pH of the medium, along with the O_2 content, is an important factor that influences the behavior of the corrosion process. At a pH lower than six and the O_2 level is high, the oxidation of Fe^{2+} ions to Fe^{3+} is slow, and most Fe^{2+} remains in solution. At low levels of O_2 and pH values higher than six, the precipitated $Fe(OH)_2$ is rapidly hydrolyzed to $Fe(OH)_3$. Over time, the latter converts to goethite and can pass to magnetite [82]. With respect to the environment, the soil is a complex medium that contributes to influence the metal corrosion with different factors of geochemical type such as soil mineralogy, pH, electrical resistivity, and hydrological type such as water flow through ground layers. Thus, depending on the chemical and mineralogical composition of soil, the Fe^{2+} ions can precipitate and form $Fe(OH)_2$ or can combine with exogenous ions from the soil such as Cl^-, $SO_4{}^{2-}$, or $CO_3{}^{2-}$ yielding a corrosion layer with heterogeneous composition dominated by the more stable compounds [82]. Indeed, many analytical studies on the corrosion of archaeological iron artefacts agree that the main phases are goethite, magnetite, and maghemite (g-Fe_2O_3) and, less frequently, siderite and chloride ($Fe_{2(}OH)_3Cl$ and akageneite (b-FeO(OH)). Neff et al. [83] establish that these materials form a compact and dense layer called the dense product layer (DPL), which forms from the unaltered metal and progressively gets thicker. Over the DPL is found the transition medium (TM) that contains both corrosion products and soil materials.

The multi-layered structure found in the ETN009 sample can be described according to the Neff's model (Figure 23b). In the central part of the original steel foil of the case (G region), the Fe content is at ca 77%, maintaining a Fe/O molar ratio at ca 1. That indicates that the steel foil now is completely evolved into a steel DPL consisting of a mixture of oxides, hydrated oxides, oxyhydroxides, and carbonates that prevalently approximate the structure of magnetite and goethite in its outermost part. The small amounts of sulfur between 0.20 and 1.41 wt% suggests that the DPL also includes small amounts of iron sulfides and/or sulfates. Cotton-ball semicrystalline goethite prevails in the inner region L in contact with a great internal fissure [48].

A highly porous TM region (represented by areas D, E, and F) is recognized over the steel DPL, characterized by the presence of Cu, Zn, and Pb diffused from the upper brass layer and the progressive reduction of the Fe content until 59 wt%. That confirms the existence of decuprification, dezincification, and deleadification. The identification of S, Cl, Al, Si, K, Cu, and Zn together with Fe and O suggests that the steel TM has been formed by the cementing of the corrosion products and soil particles.

The outermost brass layer constitutes another singularity and adds complexity to this case study because two dissimilar metallic materials, brass and steel, are in electrical contact in the same electrochemical environment, and galvanic corrosion can occur. Here, Fe has a more negative electrochemical potential and becomes the anode, corroding faster than when steel is not in brass contact [41]. However, the results obtained in the brass indicate that this layer has also undergone intense corrosion processes that have produced various products. The O/Cu and S/Cu molar ratio values found in the remains of the original brass layer do not allow the type of compounds to be identified since they do not fit any particular composition. Despite this, it can be concluded that the material that forms this layer is probably a complex mixture with the prevalence of copper carbonates and oxides and some copper sulfate/sulfide.

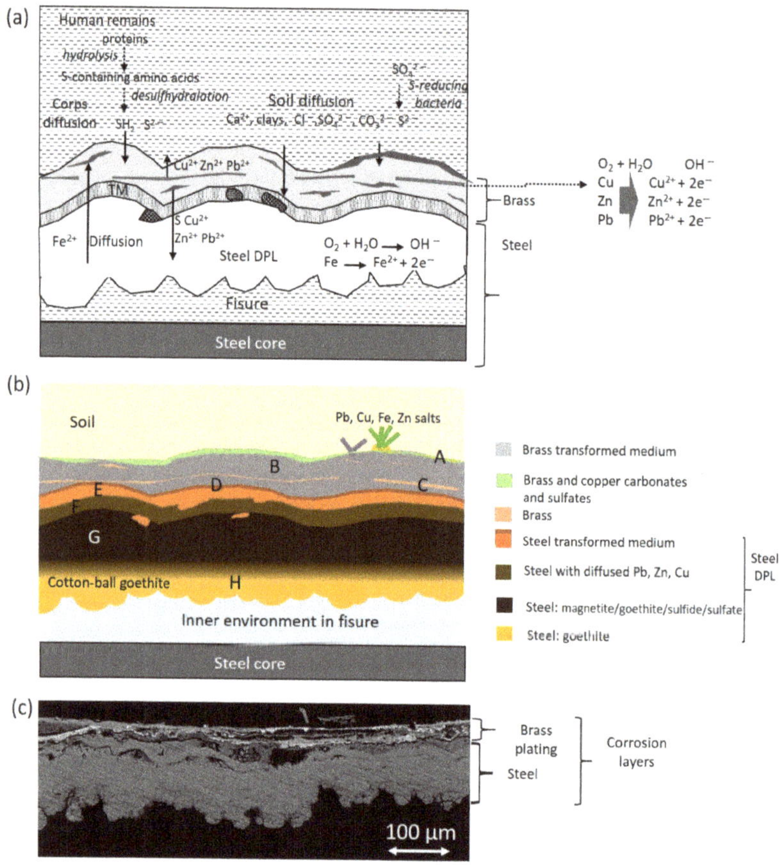

Figure 23. (**a**) Scheme of the processes that drive the corrosion and proposed redox and diffusion processes that have taken place in the case in the lighter ETN009; (**b**) regions with characteristic chemical composition identified by EDX and distribution of the brass and steel DPLs and TMs; (**c**) backscattered electron image of the studied sample that corresponds to the corrosion layers excised from the ETN009 case.

The fact that copper sulfide/sulfate compounds were formed as alteration products in the mass graves is an interesting finding due to the rarity of these compounds in burial environments, and this is probably linked to the proximity of the personal items to human remains. That raises two questions: first, the source of this element, and second, the mechanism by which Cu-S compounds are formed. Regarding the source of S, a possible origin would be the dissolution of gypsum in the soil, which has been confirmed by analyzing several samples taken from different points in the mass graves. These sulfates could combine directly with the copper ions extracted from the brass layer by ion-exchange processes to form a copper sulfate such as brochantite. A second pathway forms sulfides from the soil sulfates through dissimilatory reducing microorganisms, i.e., sulfate-reducing bacteria [84–86]. This second mechanism is frequent in iron objects found in waterlogged archaeological sites and results in the formation of pyrite (FeS_2) [84,87]. The iron reduction is coupled with organic matter or S^{2-} oxidation. Similarly, the human remains in the mass grave could provide a microenvironment for the artifacts close to them. The easy oxidation undergone by the organic matter of the soft parts mediated by microorganisms could promote the reduction of SO_4^{2-} to S^{2-}. Then, the S^{2-} could combine with copper ions, resulting in the unstable covellite (CuS) that further formed chalcanthite (CuS_2) and subsequently was oxidized to the most stable copper sulfate brochantite ($Cu_4SO_4(OH)_6$) in the presence of soil moisture [88]. This mechanism has been proposed by Muro et al. [89], who identify brochantite in copper objects found in a burial tumulus containing human remains. An alternative source of S^{2-}, tentatively proposed here, could be the H_2S release due to the decomposition of the proteins from the human remains close to the artifacts. Some proteins include amino acids such as cysteine and cystine that contain sulfur atoms. During the decomposition of the protein, these amino acids undergo a desulfhydration process that reduce them to form ammonia, thiols, pyruvic acid, hydrogen sulfide, and sulfides. Hydrogen sulfide release is favored in anaerobic environments such as grave soils [90]. If the environmental conditions in the grave turn aerobic over time, the transformation of sulfide in sulfate occurs [91]. On the other hand, it is well known from many studies of the corrosion mechanisms of copper, bronze, and other copper alloys exposed to the polluted air that exposition to hydrogen sulfide forms copper sulfides and specifically forms brochantite with high levels of humidity up to 70% RH [88].

The morphology exhibited by the brass shown in Figures 16 and 23c suggests that the original brass layer would have become disjointed and fragmented due to local differences between the development of the corrosion process. The in situ formation of corrosion products combined with the diffusion processes of exogenous and endogenous species would have contributed to the micro-locally differentiated corrosion processes and rates and the dislocation of the brass layer. The original brass layer now is distorted and discontinuous (areas C) and completely converted to corrosion products characterized by high levels of Cu and O and the presence of S at a lesser extent (averaged molar ratios O/Cu = 2.9 and S/Cu = 0.6). A TM (areas A and B) associated with the brass wraps the remains of the original brass layer and is characterized by a significant Cu content, the presence of Zn and Pb diffused from the brass layer, Fe diffused from the case, and soil tracer elements such as Cl, S, Al, and Si coming from the soil.

The cryptocrystalline particles deposited in the sample ETN073 from the spectacles frame (Figure 18c) identified as scholzite would have been formed from the Zn^{2+} ions diffused from the metal glasses frame as result of dezincification in parallel to the calcium hydroxyapatite dissolution from the bones of the human remains disposed close to the spectacles. Similar processes have been reported in a burial tumulus where the deceased remains resulted in the formation of libethenite ($Cu_2(PO_4)(OH)$) and sampleite ($NaCaCu_5(PO_4)4Cl.5H_2O$) [92].

The presence of calcium and copper oxalates, observed in some objects such as the spoon ETN006 and the knife ETN053, is associated with the action of microorganisms that have colonized these objects. It is possible that the presence of the remains of organic matter due to the alimentary use of these objects, which during burial could have been an ideal nutrient for certain fungi, has favored the development of this alteration mechanism, not observed in the rest of the objects.

However, without a doubt, the mechanism of alteration that has stood out in this set of objects has been the presence of adipoceres. It is worth highlighting the identification of two main types of adipoceres with a differentiated chemical composition: the nit comb ETN058 is a representative example of acidic adipoceres characterized by the predominance of fatty acids. The cigarette paper in a matchbox F111-141 and the rope handcuffs ETN045 are examples of the second type of adipoceres characterized by the majority presence of calcium salts of fatty monoacids. The fatty acids in contact with the surface of brass belongings have formed, at lesser extent, more specific adipoceres where zinc carboxylic complexes are prevalent. These last adipoceres have contributed to the corrosion of brass objects. The IR spectrum ETN073-2 acquired in a sample of adipoceres from the brass spectacles frame provides evidence of such a corrosion mechanism. Also noteworthy is the predominance of the monoacid acid and monoacid salt type compared to the diacid type, which is only found associated with Zn salts formed from brass. The lower presence of zinc salts compared to calcium salts is due to the more ubiquitous nature of calcium ions in a calcite-rich soil. Another aspect worth noting is the absence of copper or iron salts, which might be expected given the copper- and iron-rich composition of many of the personal items and which contrasts with the unexpected identification of zinc salts if it is considered that the latter is a minor component of the brass alloys. This fact is explained by the ease with which brasses with high Zn content of over 20% undergo dezincification when in contact with acidic or alkaline solutions. Among the factors that could promote dezincification in the brass personal belongings found in the mass graves can be the formation of earth deposits on the brass surface or at crevices, allowing differential aeration cells to form mildly aggressive aqueous soil mediums with a high chloride content or a high or low pH [93,94]. In the case of the zinc salts found in contact with the eyeglass frame, it is evident that this process involved a prior dezincification stage, which could have been facilitated by the presence of moisture in the soil and quicklime incorporated after depositing each saca, which would significantly raise the pH of the medium, favoring the corrosion of the brass by dezincification.

The extent of the adipoceres infiltrating the items and the possibility, as just described, of prolonging irreversible alteration processes in both organic and metallic materials required the adoption of specific intervention treatments. Sanitization was the first step in the conservation process. The larger superficial deposits of adipoceres that detached easily from the item were mechanically removed on a case-by-case basis. However, many objects were infiltrated of adipoceres to different extents, especially those made of organic materials. In such instances where mechanical removal was not feasible, a natural and fully sustainable product, melaleuca oil (tea tree essential oil), was used. This product, when prepared as a hydroalcoholic solution (70% ethanol and 30% water), can act as a biocide due to its fungicidal and bactericidal properties [95].

4. Materials and Methods

4.1. Optical Microscope

Stereoscopic microscope Leica GZ6 (X10-X50) and Leica DM2500 P (Leica Microsystems, Heidelberg, Germany) were used for preparing and examining cross-section samples, respectively. IC80HD Leica Digital FireWire Camera (DFC) with Leica Application Suite (LAS) software version 3.3.0 has been used to acquire and process digital images.

4.2. FESEM–EDX

A FESEM Zeiss (Orsay Physics, Fuveau, France; Kleindiek, Reutlingen, Germany; Oxford Instruments, Abingdon, UK) model Auriga compact equipment was used for the FESEM–EDX examination. The X-ray microanalysis was performed using an Oxford-X Max X-ray microanalysis system coupled to the FESEM. Operation conditions in the FESEM–EDX are as follows: a voltage of 20 kV, a current beam of 3.76–3.83 μA, and resolution of 127 eV at 5.9 keV. X-ray detector operates at a working distance of 6–7 mm. The electron beam was always disposed perpendicularly to the polished surface of the cross-sections.

The same samples examined under an optical microscope were mounted on aluminum disks and carbon-coated before acquiring images and X-ray spectra. EDX analysis was performed with the Aztec software version 3.3sp1. Quantitative compositions were made from X-ray spectra acquired in point-mode acquisition. The wt% values of the different elements were calculated by applying ZAF method of correction of interelemental effects on intensity values. Three X-ray spectra from the selected area, of a few hundred of µm square, were acquired for quantitative analyses. The stoichiometry of the mineral was deduced from the molar ratio among the identified elements, which is provided by the quantitative processing of the detected X-ray counts. Other details concerning operation conditions are provided in Table S4 [96–98].

4.3. FTIR Spectroscopy

The mineral and molecular composition of the personal items was characterized by Fourier Transform Infrared Spectroscopy (FTIR) in attenuated total reflectance (ATR) with a Vertex70 spectrometer (Bruker Optik GmbH 2012, Ettlingen, Germany) with MKII Golden Gate with an ATR accessory (Specac®, Orpington, UK). This instrument is equipped with an FR-DTGS (fast recovery deuterated triglycine sulfate) temperature-stabilized coated detector and an MKII Golden Gate Attenuated Total Reflectance (ATR) accessory (Specac Ltd., Orpington, UK). The IR spectra were acquired by collecting 32 scans at a resolution of 4 cm^{-1}. The spectral range scanned was 500–4000 cm^{-1}. The spectra were processed using the OPUS software version 7.0 (Bruker Optik GmbH 2012, Ettlingen, Germany). The samples were directly analyzed or, if necessary, powdered in a small agate mortar, and a small amount of the powdered mixture was deposited on the diamond window.

4.4. XRD

Diffractometer Bruker model D8 Advanced A25 equipped with a Lynxeye fast detector and an X-ray tube that emits Cu Ka radiation. The operating conditions for the X-ray emission tube were 40 kV and 40 mA, and the acquisition conditions for the patterns were 5–80° 2q with an exposition time of 0.8 s.

4.5. Preparation of Samples

FESEM–EDX analyses were performed from the samples directly mounted on aluminum disks and C coated. The cross-section of the sample ETN009 was prepared by embedding in polyester resin (Glasspol S.L, Valencia, Spain) and polished with SiC foils with grid sizes of 220, 500, 2400, and 4000 (Struers, Germany) in a Struers LaboPol-20 instrument. FTIR spectroscopic analyses were performed on samples powdered in an agate mortar.

4.6. Ethical Considerations and Procedures

This study is based on exhumed objects from the research works carried out in the Municipal Cemetery of Paterna by the ArqueoAntro Scientific Association, a team dedicated to Management Archaeology, Historical Memory, Missing Children Cases, Training Courses, and Forensic Expertise. This association is hereby established under Spanish Organic Law 1/2002, of 22 March, and complementary regulations, with legal personality and full capacity to act, and with no profit motive. All exhumations were developed following the forensics protocols and legal authorizations and with the consent and participation of the family associations of victims [11]. Additionally, the project from which this study derives involved an interdisciplinary curatorial team comprising experts in exhumation processes, heritage management, public memory policies, and museum management [9]. Most importantly, the victims' families were part of this project, which allowed the technical team to work, analyze, and publish our research on the exhumed objects. Regarding ethical responsibilities involved in handling, analyzing, and publishing data from mass graves and given the sensitive nature of the materials [15], the studio was implemented with maximum respect for the victims and their living relatives. Furthermore, the technical team established a close relationship with the families and relatives, who followed on-site year after year the exhumation works so that all the operations and scientific and technical procedures had their approval [23].

5. Conclusions

This research opens a new scenario in which multi-technique approaches based on the combination of fast and quasi-non-invasive techniques such as optical microscopy, FESEM–EDX, XRD, and FTIR spectroscopy are a suitable alternative to characterize materials and alteration processes in instances in which the analysis of a large number of archaeological objects with a diversity of materials of both organic and inorganic nature and alterations is demanded. Thus, inorganic materials such as nacre, limestone, pigments, and different brasses and silver plating have been identified. The organic materials identified are wood, esparto, cotton yarns, castor oil, and casein formaldehyde. The study has also provided data on how these objects were manufactured. In addition, typical metal corrosion processes have been identified with other unusual ones, such as the formation of brochantite and scholzite due to the proximity to the deceased corpses in decomposition. The generation of adipoceres during the decaying of the corpses has resulted in the formation of metal–carboxylate complexes and, mediated by microorganisms, metal oxalates, prevalently with calcium ions. Therefore, the analyses performed have provided the restorers with information on how to apply the most appropriate intervention treatments.

The study also provided clues of the forensic type, which are of interest to historians in finding out what life was like for the owners when they were imprisoned. The carefully carved stone rings reveal how the prisoners spent their long hours of imprisonment. Some of them entertained themselves with manual work that was not devoid of artistic value. The pencil extender and sharpener found provide similar information, suggesting that they wrote down thoughts and ideas, simply wrote letters to inform their families about their situation, or wrote farewell letters in the knowledge of their fateful end. Other items have more mournful implications. Unlike the previous items, it is unlikely that a prisoner would have been allowed to have a rope made of esparto grass, confirming as has been documented that it was used as handcuffs to immobilize the prisoner during transfer and execution. The small "Naranjil" bottle could have been used for medicinal purposes. However, it could also have been used to insert a piece of paper with the name of the prisoner to allow them to be identified in the mass grave once the execution had taken place. That made it easier for the family to make a subsequent claim. That seems to have been a common practice among the prisoners at that time, given the practices of the Francoist army [99,100], and this research has contributed to confirming such a hypothesis.

Supplementary Materials: The following supporting information can be downloaded at https://www.mdpi.com/article/10.3390/molecules30132783/s1. Figure S1: IR absorption spectrum acquired in sample ETN048 corresponding to a binding substance used for fixing the cotton yarns to the knife tag; Figure S2: IR absorption spectrum acquired in sample ETN060 corresponding to a nit comb; Figure S3: IR absorption spectrum acquired in sample ETN073 corresponding to the spectacles frame; Figure S4: IR absorption spectrum acquired in sample ETN045-2 excised from a rope; Figure S5: IR absorption spectrum acquired in sample ETN073-2 excised from spectacles frame; Figure S6: X-ray spectrum obtained in the sample SF2-01 from the soil of the mass grave number 2; Figure S7: X-ray spectrum obtained in the sample ETN060 excised from a nit comb. The emission line profile corresponds to an earth pigment; Figure S8: Label of the bottles LLUE001; Table S1: Description of the analyzed belongings; Table S2: Elemental composition (wt%) found in the different layers of the cross-section of sample ETN009; Table S3: Chemical composition corresponding to ETN060 sample excised from the nit comb. st: wt% sigma; Table S4: Experimental working conditions of FESEM–EDX.

Author Contributions: Conceptualization, M.T.D.-C.; methodology, M.T.D.-C.; writing—original draft preparation, M.T.D.-C.; supervision, M.T.D.-C. and T.P.O.; investigation, M.T.D.-C., T.P.O., R.C.R. and J.M.M.; data curation, M.T.D.-C. and T.P.O.; writing—review and editing, M.T.D.-C. and T.P.O. All authors have read and agreed to the published version of the manuscript.

Funding: This research was funded by MCIN/AEI/10.13039/501100011033 and Agencia Estatal de Investigación (AEI) grant PID2020-113022GB-I00.

Informed Consent Statement: Not applicable.

Data Availability Statement: The data presented in this study are available on request from the corresponding author.

Acknowledgments: The authors wish to state that they have carried out this analytical study with total dedication and commitment to serve as a tribute and reparation for the victims and alleviate the pain of the families. The following are also acknowledged: Eloy Ariza, Andrea Moreno, Tono Vizcaíno, and Miguel Mezquida, curators of the temporary exhibition "Arqueología de la memoria. Las fosas de Paterna"; the personnel of the Museo de Prehistoria de Valencia and L'ETNO in València; the partnership platform of the victims of Francoism of the Mass Graves of Paterna; the scientific partnership ArqueoAntro; Manuel Planes Insausti and Jose Luis Moya López technical supervisors of the Electron Microscopy Service of the Universitat Politècnica de València.

Conflicts of Interest: The authors declare no conflicts of interest.

Abbreviations

The following abbreviations are used in this manuscript:

l'ETNO	Museums of Ethnography
GC–MS	Gas chromatography–mass spectrometry
OM	Optical microscopy
SEM	Scanning electron microscopy
FESEM–EDX	Field emission X-ray microanalysis scanning electron microscopy
FTIR	Fourier transform infrared spectroscopy
XRD	X-ray diffraction
C8:0	Octanodioic acid
C9:0	Nonanodioic acid
C14:0	Tetradecanoic acid
C16:0	Hexadecanoic acid
C18:0	Octadecanoic acid
C20:0	Eicosanoic acid
ACC	Amorphous calcium carbonate
ZAF	Z—stopping power, backscattering factor and X-ray production power (Z), absorption (A) and fluorescence (F)
CF	Casein formaldehyde
DPL	Dense product layer
TM	Transition medium
ATR	Attenuated total reflectance
FR-DTGS	Fast recovery deuterated triglycine sulfate

References

1. Gabarda, V. El cementeri de Paterna com a exemple de monumentalització popular, Postguerres/Aftermaths of war. In *Postguerres/Aftermaths of War*; Abelló, T., Cattini, G.C., Gavin, V., Ibarz, J., Santacana, C., Solé, O., Vives, A., Eds.; Universitat de Barcelona: Barcelona, Spain, 2020; pp. 169–187.

2. Ley 52/2007, de 26 de Diciembre, por la que se Reconocen y Amplían Derechos y se Establecen Medidas en Favor de Quienes Padecieron Persecución o Violencia Durante la Guerra civil y la dictadura [Law 52/2007, 26th December, Which Recognizes and Expands Rights and Establishes Measures in Favor of Those Who Suffered Persecution or Violence During the Civil War and the Dictatorship]. Available online: https://www.boe.es/eli/es/l/2007/12/26/52/con (accessed on 24 February 2025).
3. Orden PRE/2568/2011, de 26 de Septiembre, por la que se Publica el Acuerdo del Consejo de Ministros de 23 de Septiembre de 2011, por el que se Ordena la Publicación en el Boletín Oficial del Estado del Protocolo de Actuación en Exhumaciones de Víctimas de la Guerra Civil y la Dictadura [Order of September 26, Which Publishes the Agreement of the Council of Ministers of September 23, 2011, Which Orders the Publication in the Official State Gazette of the Protocol for Action in Exhumations of Victims of the Civil War and the Dictatorship]. Available online: https://www.boe.es/eli/es/o/2011/09/26/pre2568 (accessed on 24 February 2025).
4. Ley 14/2017, de 10 de Noviembre, de Memoria Democrática y Para la Convivencia de la Comunitat Valenciana. [Law 14/2017, of November 10, on Democratic Memory and for the Coexistence of the Valencian Community]. Available online: https://www.boe.es/eli/es-vc/l/2017/11/10/14/con (accessed on 24 February 2025).
5. Decreto 1/2023, de 13 de enero, del Consell, por el cual se Desarrolla la Ley 14/2017, de 10 de Noviembre, de la Generalitat, de Memoria Democrática y para la Convivencia de la Comunitat Valenciana. [Decree 1/2023, of 13 January, of the Consell, by which Law 14/2017, of November 10, of the Generalitat, on Democratic Memory and for the Coexistence of the Valencian Community, Is Developed]. Available online: https://dogv.gva.es/es/eli/es-vc/d/2023/01/13/1/ (accessed on 24 February 2025).
6. Díaz-Ramoneda, E.; Vila, A.; Sancho, S.; Calpe, A.; Iglesias, J.; Mezquida, M. Les fosses de Paterna, testimonis de la maquinària repressiva del règim franquista al País Valencià. *Rev. D'arqueologia Ponent* **2021**, *31*, 239–258. [CrossRef]
7. Múgica, J.; Pasíes, T.; Canal, R. Las fosas de Paterna: Un proyecto de conservación sostenible y puesta en valor en el Museu de Prehistòria de València. *PH Boletín Inst. Andal. Patrim. Histórico* **2023**, *31*, 320–321. [CrossRef]
8. Múgica, J. Las fosas de Paterna: Un proyecto de Conservación y Puesta en Valor en el Museu de Prehistòria de València. Master's Thesis, Universitat Politècnica de València (UPV), Valencia, Spain, 25 July 2023. Available online: https://riunet.upv.es/entities/publication/4462e493-882e-45bd-8013-a8625ca3569c (accessed on 9 April 2025).
9. Moreno, A.; Vizcaíno, A.; Mezquida, M.; García, M.J.; Gadea, I. (Eds.) *Mass Graves of Francoism. Archaeology, Anthropology and Memory*; Diputació de València-Museu de Prehistòria de València: Valencia, Spain, 2023. Available online: https://mupreva.org/web_mupreva_dedalo/publicaciones/1631/va (accessed on 9 April 2025).
10. Moreno, A.; Vizcaíno, A.; Mezquida, M.; Ayán, X.M. *Las Fosas del Franquismo. Arqueología, Antropología y Memoria*; Diputació de València-Museu de Prehistòria de València: Valencia, Spain, 2023. Available online: http://mupreva.org/pub/1600/va (accessed on 9 April 2025).
11. Moreno, A.; Mezquida, M.; Schwab, M.E. Exhumaciones de fosas comunes en el País Valenciano: 10 años de intervenciones científicas. *Ebre 38. Rev. Int. Guerra Civil.* **2021**, *11*, 125–152. [CrossRef]
12. Gabarda, V. *El Cost Humà de la Repressió al País Valencià (1936–1956)*; Universitat de València: València, Spain, 2020.
13. Arqueologia de la Memòria. Arqueologia de la Memòria. Les Fosses de Paterna. Available online: https://mupreva.org/exposiciones/79/va (accessed on 9 April 2025).
14. Museo de Prehistoria de Valencia. Available online: https://mupreva.org/home/?q=va (accessed on 9 April 2025).
15. Moreno, A.; Ariza, E.; Mezquida, M. Tratamiento, significado y gestión de la cultura material exhumada en contextos de represión y conflicto: Las fosas del franquismo. In *Arqueología y Memoria Histórica: Retos Actuales de la Investigación*; Rodríguez, O., Ed.; University of Seville: Seville, Spain, 2025; pp. 57–91.
16. Moses, R.J. Experimental adipocere formation: Implications for adipocere formation on buried bone. *J. Forensic Sci.* **2012**, *57*, 589–595. [CrossRef]
17. Fiedler, S.; Graw, M. Decomposition of buried corpses, with special reference to the formation of adipocere. *Naturwissenschaften* **2003**, *90*, 291–300. [CrossRef] [PubMed]
18. Takatori, T.; Ishiguro, N.; Tarao, H.; Matsumiya, H. Microbial production of hydroxy and oxo fatty acids by several microorganism as a model of adipocere formation. *Forensic Sci. Int.* **1986**, *32*, 5–11. [CrossRef] [PubMed]
19. Forbes, S.L.; Stuart, B.H.; Dent, B.B. The effect of the burial environment on adipocere formation. *Forensic Sci. Int.* **2005**, *154*, 24–34. [CrossRef]
20. Forbes, S.L.; Stuart, B.H.; Dent, B.B. The identification of adipocere in grave soils. *Forensic Sci. Int.* **2002**, *127*, 225–230. [CrossRef]
21. Takatori, T. Investigations on the mechanism of adipocere formation and its relation to other biochemical reactions. *Forensic Sci. Int.* **1996**, *80*, 49–61. [CrossRef]
22. Lerchi, A.; Krap, T.; Eppenberger, P.; Pedergnana, A. Implementation of adipocere fingerprinting in archaeology by applying a forensic approach. *TrAC Trends Anal. Chem.* **2022**, *157*, 116801. [CrossRef]
23. Magni, P.A.; Lawn, J.; Guareschi, E.E. A practical review of adipocere: Key findings, case studies and operational considerations from crime scene to autopsy. *J. Forensic Leg. Med.* **2021**, *78*, 102109. [CrossRef] [PubMed]

24. Kumar, T.M.; Monteiro, F.N.; Bhagavath, P.; Bakkannavar, S.M. Early adipocere formation: A case report and review of literature. *J. Forensic Leg. Med.* **2009**, *16*, 475–477. [CrossRef]
25. Ubelaker, D.H.; Zarenko, K.M. Adipocere: What is known after over two centuries of research. *Forensic Sci. Int.* **2011**, *208*, 167–172. [CrossRef]
26. Derrick, M.R.; Stulik, D.C.; Landry, J.M. *Infrared Spectroscopy in Conservation Science*; Getty Publications: Marina del Rey, CA, USA, 1999.
27. Chalmers, H.G.M.; Edwards, M.H. (Eds.) *Infrared and Raman Spectroscopy in Forensic Science*; John Wiley & Sons: Hoboken, NY, USA, 2012.
28. Gunasekaran, S.; Anbalagan, G.; Pandi, S. Raman and infrared spectra of carbonates of calcite structure. *J. Raman Spectrosc.* **2006**, *37*, 892–899. [CrossRef]
29. Izzo, F.C.; Kratter, M.; Nevin, A.; Zendri, E. A critical review on the analysis of metal soaps in oil paintings. *ChemistryOpen* **2021**, *10*, 904–921. [CrossRef]
30. Madejová, J.; Gates, W.P.; Petit, S. IR spectra of clay minerals. In *Infrared and Raman Spectroscopies of Clay Minerals (Developments in Clay Science Series)*; Gates, W., Kloprogge, J.T., Madejova, J., Bergaya, F., Eds.; Elsevier: Oxford, UK, 2017; Volume 8, pp. 107–149.
31. Pizzo, B.; Pecoraro, E.; Alves, A.; Macchioni, N.; Rodrigues, J.C. Quantitative evaluation by attenuated total reflectance infrared (ATR-FTIR) spectroscopy of the chemical composition of decayed wood preserved in waterlogged conditions. *Talanta* **2015**, *131*, 14–20. [CrossRef] [PubMed]
32. Schwanninger, M.J.C.R.; Rodrigues, J.C.; Pereira, H.; Hinterstoisser, B. Effects of short-time vibratory ball milling on the shape of FT-IR spectra of wood and cellulose. *Vibrat Spectrosc.* **2004**, *36*, 23–40. [CrossRef]
33. Kostadinovska, M.; Spirovska, Z.J.; Taylor, T. A procedure for identifying cellulose fibers in paper artifacts. In Proceedings of the 4th International Virtual Conference on Advanced Scientific Results SCIECONF-2016, Ostrava, Czech Republic, 6–10 June 2016; Technical University of Ostrava: Ostrava, Czech Republic, 2016; pp. 156–162.
34. Cucci, C.; Bartolozzi, G.; Marchiafava, V.; Picollo, M.; Richardson, E. Study of semi-synthetic plastic objects of historic interest using non-invasive total reflectance FT-IR. *Microchem. J.* **2016**, *124*, 889–897. [CrossRef]
35. Van der Weerd, J.; Van Loon, A.; Boon, J.J. FTIR studies of the effects of pigments on the aging of oil. *Stud. Conserv.* **2005**, *50*, 3–22. [CrossRef]
36. Filopoulou, A.; Vlachou, S.; Boyatzis, S.C. Fatty acids and their metal salts: A review of their infrared spectra in light of their presence in cultural heritage. *Molecules* **2021**, *26*, 6005. [CrossRef]
37. Frye, G.C.; Thomas, M.M. Adsorption of organic compounds on carbonate minerals. 2. Extraction of carboxylic acids from recent and ancient carbonates. *Chem. Geol.* **1993**, *109*, 215–226. [CrossRef]
38. Lu, Y.; Miller, J.D. Carboxyl stretching vibrations of spontaneously adsorbed and LB-transferred calcium carboxylates as determined by FTIR internal reflection spectroscopy. *J. Colloid. Interface Sci.* **2002**, *256*, 41–52. [CrossRef]
39. Shurvell, H.F. Spectra—Structure Correlations in the Mid- and Far-Infrared. In *Handbook of Vibrational Spectroscopy*; Chalmers, J.M., Ed.; John Wiley & Sons, Ltd.: Chichester, UK, 2006.
40. Watanabe, M.; Toyoda, E.; Handa, T.; Ichino, T.; Kuwaki, N.; Higashi, Y.; Tanaka, T. Evolution of patinas on copper exposed in a suburban area. *Cor Sci.* **2007**, *49*, 766–780. [CrossRef]
41. Ashkenazi, D.; Inberg, A.; Langgut, D.; Hendler, N.; Cvikel, D. Brass–iron couple and brass–iron–wood ternary system of metal objects from the Akko 1 shipwreck (Israel). *Cor Sci.* **2016**, *110*, 228–241. [CrossRef]
42. Aghamohammadi, H.; Jamaati, R. Effect of cold single-roll drive rolling on the microstructural evolution and mechanical properties of ferritic stainless steel. *J. Mat. Res. Technol.* **2024**, *29*, 2679–2688. [CrossRef]
43. Thalib, S.; Ikhsan, M.; Fonna, S.; Huzni, S.; Ridha, S. Identification of corrosion product on medium carbon steel under the exposure of Banda Aceh's atmosphere. In *IOP Conference Series: Materials Science and Engineering*; IOP Publishing: Bristol, UK, 2003; Volume 352, p. 012004.
44. Antunes, R.A.; Costa, I.; Faria, D.L.A. Characterization of corrosion products formed on steels in the first months of atmospheric exposure. *Mat. Res.* **2003**, *6*, 403–408. [CrossRef]
45. Moreno, J.D.; Bonilla, M.; Adam, J.M.; Borrachero, M.V.; Soriano, L. Determining corrosion levels in the reinforcement rebars of buildings in coastal areas. A case study in the Mediterranean coastline. *Constr. Build. Mater.* **2015**, *100*, 11–21. [CrossRef]
46. Fitzgerald, K.P.; Nairn, J.; Atrens, A. The chemistry of copper patination. *Cor Sci.* **1998**, *40*, 2029–2050. [CrossRef]
47. Mindat. Available online: https://www.mindat.org/gm/1144 (accessed on 18 April 2025).
48. Grevey, A.L.; Vignal, V.; Krawiec, H.; Ozga, P.; Peche-Quilichini, K.; Rivalan, A.; Mazière, F. Microstructure and long-term corrosion of archaeological iron alloy artefacts. *Herit. Sci.* **2020**, *8*, 57. [CrossRef]
49. Gettens, R.J.; West Fitzhugh, E. Malachite and Green Verditer. In *Artists' Pigments: A Handbook of Their History and Characteristics*; Roy, A., Ed.; Archetype: London, UK, 1993; Volume 2, pp. 183–202.
50. Rruff DataBase. Available online: https://rruff.info/about/about_general.php (accessed on 5 March 2025).

51. Doménech-Carbó, M.T.; Álvarez-Romero, C.; Doménech-Carbó, A.; Osete-Cortina, L.; Martínez-Bazán, M.L. Microchemical surface analysis of historic copper-based coins by the combined use of FIB-FESEM-EDX, OM, FTIR spectroscopy and solid-state electrochemical techniques. *Microchem. J.* **2019**, *148*, 573–581. [CrossRef]
52. Naumann, A.; Navarro-González, M.; Peddireddi, S.; Kües, U.; Polle, A. Fourier transform infrared microscopy and imaging Detection of fungi in wood. *Fungal Genet. Biol.* **2005**, *42*, 829–835. [CrossRef] [PubMed]
53. Fischer, G.; Braun, S.; Thissen, R.; Dott, W. FT-IR spectroscopy as a tool for rapid identification and intra-species characterization of airborne filamentous fungi. *J. Microbiol. Meth.* **2006**, *64*, 63–77. [CrossRef] [PubMed]
54. Scott, D.A.; Maish, J.P. A Lydian Bed of Iron, Bronze and Copper-Technical Examination of a Metallurgical Masterpiece. *Stud. Conserv.* **2010**, *55*, 3–19. [CrossRef]
55. Edwards, J. *Coating and Surface Treatment Systems for Metals*; Finishing Publications Ltd. and ASM International: Geauga County, OH, USA, 1997; pp. 34–38.
56. Sheasby, P.G.; Pinner, R. *The Surface Treatment and Finishing of Aluminum and Its Alloys*, 6th ed.; ASM International & Finishing Publications: Geauga County, OH, USA, 2001; Volume 2.
57. Schuman, T.P. Protective Coatings for Aluminum Alloys. In *Handbook of Environmental Degradation of Materials*, 3rd ed.; Kutz, M., Ed.; Elsevier Inc.: Amsterdam, The Netherlands, 2018.
58. Liang, J.; Yue, W.; Gu, Y.; Liu, J.; Wang, C.; Ma, H. Improving Corrosion Resistance and Corrosive Wear Resistance of Aluminum Alloy Drill Pipe by Surface Nanocrystallization and Micro-arc Oxidation. *J. Mater. Eng. Perform.* **2018**, *27*, 4462–4472. [CrossRef]
59. Ingo, G.M.; Riccucci, C.; Faraldi, F.; Pascucci, M.; Messina, E.; Fierro, G.; Di Carlo, G. Roman sophisticated surface modification methods to manufacture silver counterfeited coins. *Appl. Surf. Sci.* **2017**, *421*, 109–119. [CrossRef]
60. Hiscox, G.D.; Hopkins, A.A. *Recetario Industrial [Industrial Recipe Book]*; Club de Golf Mexico: Mexico City, Mexico, 1997.
61. Doménech-Carbó, M.T.; Di Turo, F.; Montoya, N.; Catalli, F.; Doménech-Carbó, A.; De Vito, C. FIB-FESEM and EMPA results on Antoninianus silver coins for manufacturing and corrosion processes. *Sci. Rep.* **2018**, *8*, 10676. [CrossRef]
62. Keturakis, C.J.; Notis, B.; Blenheim, A.; Miller, A.C.; Pafchek, R.; Notis, M.R.; Wachs, I.E. Analysis of corrosion layers in ancient Roman silver coins with high resolution surface spectroscopic techniques. *App Surf. Sci.* **2016**, *376*, 241–251. [CrossRef]
63. Di Fazio, M.; Di Turo, F.; Medeghini, L.; Fabrizi, L.; Catalli, F.; De Vito, C. New insights on medieval Provisini silver coins by a combination of non-destructive and micro-invasive techniques. *Microchem. J.* **2019**, *144*, 309–318. [CrossRef]
64. Fabrizi, L.; Di Turo, F.; Medeghini, L.; Di Fazio, M.; Catalli, F.; De Vito, C. The application of non-destructive techniques for the study of corrosion patinas of ten Roman silver coins: The case of the medieval Grosso Romanino. *Microchem. J.* **2019**, *145*, 419–427. [CrossRef]
65. Ingo, G.M.; Balbi, S.; de Caro, T.; Fragalà, I.; Angelini, E.; Bultrini, G. Combined use of SEM-EDS, OM and XRD for the characterization of corrosion products grown on silver roman coins. *Appl. Phys. A* **2006**, *83*, 493–497. [CrossRef]
66. Mayer, R. *Materiales y Técnicas del Arte [Materials and techniques of Art]*; Ediciones Akal: Madrid, Spain, 2005; Volume 28.
67. La Galvanotecnia y su Industria en España Entre los Años 1950–2005. [Electroplating and Its Industry in Spain Between 1950–2005] Institut Metalurgic de Cataluña. Available online: https://www.centrem.cat/ecomu/upfiles/aiasdestacat/fitxers/A1256.pdf (accessed on 9 March 2025).
68. Julve, E. Historia de la galvanotecnia y técnicas afines. *An. Química RSEQ* **2009**, *105*, 227.
69. Nudelman, F. Nacre biomineralisation: A review on the mechanisms of crystal nucleation. *Semin. Cell Dev. Biol.* **2015**, *46*, 2–10. [CrossRef] [PubMed]
70. Khouzani, M.F.; Chevrier, D.M.; Güttlein, P.; Hauser, K.; Zhang, P.; Hedin, N.; Gebauer, D. Disordered amorphous calcium carbonate from direct precipitation. *Cryst. Eng. Comm.* **2015**, *17*, 4842–4849. [CrossRef]
71. Gueta, R.; Natan, A.; Addadi, L.; Weiner, S.; Refson, K.; Kronik, L. Local atomic order and infrared spectra of biogenic calcite. *Angew. Chem. Int. Ed.* **2007**, *46*, 291–294. [CrossRef]
72. Chu, V.; Regev, L.; Weiner, S.; Boaretto, E. Differentiating between anthropogenic calcite in plaster, ash and natural calcite using infrared spectroscopy: Implications in archaeology. *J. Archaeol. Sci.* **2008**, *35*, 905–91186. [CrossRef]
73. Beniash, E.; Aizenberg, J.; Addadi, L.; Weiner, S. Amorphous calcium carbonate transforms into calcite during sea-urchin larval spicule growth. *Proc. R. Soc. Lond. Ser. B* **1997**, *264*, 461–465. [CrossRef]
74. Regev, L.; Poduska, K.M.; Addadi, L.; Weiner, S.; Boaretto, E. Distinguishing between calcites formed by different mechanisms using infrared spectrometry: Archaeological applications. *J. Archaeol. Sci.* **2010**, *37*, 3022–3029. [CrossRef]
75. Doménech-Carbó, M.T.; Osete-Cortina, L. Another beauty of analytical chemistry: Chemical analysis of inorganic pigments of art and archaeological objects. *ChemTexts* **2016**, *2*, 1–50. [CrossRef]
76. Kaner, J.; Ioras, F.; Ratnasingam, J. Performance and stability of historic casein formaldehyde. *e-Plastory J. Plast. Hist.* **2017**, *2*. Available online: http://e-plastory.com (accessed on 10 March 2025).
77. Erinoid Ltd. *Erinoid Book of Dye Recipes*; Lightpill Mills: Stroud, UK, 1951.
78. Erinoid Ltd. *Colour in Casein Booklet*; Lightpill Mills: Stroud, UK, 1951.

79. Comercio Histórico Valencia. Available online: https://comercioshistoricosdevalencia.blogspot.com/2020/06/productos-de-epoca-el-aceite-de-ricino.html (accessed on 10 March 2025).
80. Algemesí en la Memoria. Available online: https://algemesienlamemoria.blogspot.com/ (accessed on 10 March 2025).
81. Turgoose, S. The corrosion of archaeological iron during burial and treatment. *Stud. Conserv.* **1985**, *30*, 13–18. [CrossRef]
82. Selwyn, L.S.; Sirois, P.I.; Argyropoulos, V. The corrosion of excavated archaeological iron with details on weeping and akaganeite. *Stud. Conserv* **1999**, *44*, 217–232. [CrossRef]
83. Neff, D.; Dillmann, P.; Bellot-Gurlet, L.; Béranger, G. Corrosion of iron archaeological artefacts in soil: Characterisation of the corrosion system. *Corros. Sci.* **2005**, *47*, 515–535. [CrossRef]
84. Chang, J.; Li, Y.; Lu, H. The Morphological Characteristics of Authigenic Pyrite Formed in Marine Sediments. *J. Mar. Sci. Eng.* **2022**, *10*, 1533. [CrossRef]
85. Canfield, D.E.; Raiswell, R.; Bottrell, S. The Reactivity of Sedimentary iron minerals towards sufide. *Am. J. Sci.* **1992**, *292*, 659–683. [CrossRef]
86. Burdige, D.J.; Komada, T. Iron redox cycling, sediment resuspension and role of sediments in low oxygen environments as sources of iron to water column. *Mar. Chem.* **2020**, *223*, 103793. [CrossRef]
87. Doménech-Carbó, M.T.; Guasch-Ferré, N.; Álvarez-Romero, C.; Castillo-Belinchón, R.; Pérez-Mateo, S.; Buendía-Ortuño, M. Study of the Geological Context of the 7th–6th Century BC Phoenician Era Shipwreck "Mazarrón 2" (Murcia, Spain). *Minerals* **2024**, *14*, 778. [CrossRef]
88. Hjelm-Hansen, N. Cleaning and stabilization of sulphide-corroded bronzes. *Stud. Conserv.* **1984**, *29*, 17–20. [CrossRef]
89. Muros, V.; Scott, D.A. The occurrence of brochantite on archaeological bronzes: A case study from Lofkënd, Albania. *Stud. Conserv.* **2018**, *63*, 113–125. [CrossRef]
90. Ioan, B.G.; Manea, C.; Hanganu, B.; Statescu, L.; Solovastru, L.G.; Manoilescu, I. The chemistry decomposition in human corpses. *Rev. Chim.* **2017**, *68*, 1352–1356. [CrossRef]
91. Forbes, S.L. Decomposition Chemistry in a Burial Environment. In *Soil Analysis in Forensic Taphonomy*; Tibbett, M., Carter, D., Eds.; CRC Press, Taylor and Francis Group: New York, NY, USA, 2008; p. 203.
92. Muros, V.; Parry, V.; Whyte, A. Conservation Laboratory Research Projects. *Orient. Inst. News Notes* **2002**, *175*, 1–8.
93. Khoshnaw, F.; Gubner, R. (Eds.) *Corrosion Atlas Case Studies. Part II: Corrosion Topics*; Elsevier: Oxford, UK, 2019; pp. xliii–lxviii.
94. Boyatzis, S.C.; Fragkos-Livanios, L.; Giannoulaki, M.; Filopoulou, A. Infrared spectroscopy reveals the reactivity of fatty acids on copper surfaces and its implications for cultural heritage objects. *Herit. Sci.* **2023**, *11*, 196. [CrossRef]
95. Martínez, R.; Rocabayera, R. Els olis essencials. Conservació preventiva en el control microbiològic. Unicum. *Rev. L'escola Super. Conserv. I Restauració Béns Cult. Catalunya* **2011**, *10*, 135–141. Available online: https://unicum.escrbcc.cat/index.php/unicum/article/view/121/240 (accessed on 27 May 2025).
96. Reed, S.J.B. *Electron Microbe Analysis and Scanning Electron Microscopy in Geology*; Cambridge University Press: Cambridge, UK, 1996.
97. Veritá, M.; Basso, R.; Wypysk, M.T.; Koestler, R.J. X-ray microanalysis of ancient glassy materials: A comparative study of wavelength dispersive and energy dispersive techniques. *Archaeometry* **1994**, *36*, 241–251. [CrossRef]
98. Kuisma-Kursula, P. Accuracy. Precision and Detection Limits of SEM–WDS. SEM–EDS and PIXE in the Multi-Elemental Analysis of Medieval Glass. *X-Ray Spectrom.* **2000**, *29*, 111–118. [CrossRef]
99. Moreno, A.; Mezquida, M.; Ariza, E. No solo cuerpos: La cultura material exhumada de las fosas del franquismo en Paterna. *Sagvntvm. Papeles Lab. Arqueol. Valencia.* **2021**, *53*, 213–235. [CrossRef]
100. Moreno Martín, A.; Vizcaíno, T. *Arqueología de la Memoria. Las Fosas de Paterna. Guía de la Exposición*; Diputació de València, Museu de Prehistòria de València: Valencia, Spain, 2023. Available online: http://mupreva.org/pub/1632/va (accessed on 16 April 2025).

molecules

MDPI

Article

Modern Analytical Chemistry Meets Heritage Books: Analysis of Volatile Organic Compounds (VOCs) from Two Books Preserved at the Biblioteca Capitolare of Busto Arsizio

Chiara Chiodini [1], Pierangela Rovellini [2], Matteo Chiodini [1], Luca Giacomelli [3,*], Daniela Baglio [2] and the 5B IISS Torno Working Group [†]

[1] IISS Torno, 20022 Milan, Italy
[2] Innovhub SSI, 20133 Milan, Italy
[3] Polistudium SRL, 20121 Milan, Italy
* Correspondence: luca.giacomelli@polistudium.it
† Listed in Appendix A.

Abstract: The development of sensitive, non-invasive methods is essential for the preservation and study of heritage books, allowing insights into their historical production processes and conservation needs. Volatile organic compound (VOC) analysis provides a valuable, non-destructive approach to assess paper composition and degradation in historical volumes. In this study, we analyzed VOC emissions from two books preserved at the Biblioteca Capitolare of Busto Arsizio, Italy: a 16th-century Latin grammar book and a 19th-century mathematics handbook for measurement conversions. Using headspace solid-phase microextraction (HS-SPME) and gas chromatography–mass spectrometry (GC-MS), VOCs were sampled after 24 h of storage at room temperature. The results revealed distinct degradation markers: Straight-chain aldehydes, indicative of lipid oxidation, were more prevalent in the 16th-century book, reflecting the higher quality and durability of its rag-based paper. In contrast, elevated furfural levels in the 19th-century book suggest accelerated cellulose hydrolysis typical of wood pulp paper. Additionally, the presence of menthol and anethole in both volumes points to the use of bacteriostatic agents for preservation. These findings not only highlight differences in material composition but also underscore the importance of tailored conservation approaches for historical documents from different eras.

Keywords: heritage preservation; volatile organic compounds; headspace SPME-GC/MS; paper degradation markers; non-invasive analysis

1. Introduction

The preservation of historical paper artifacts, including books and manuscripts, is vital to safeguarding cultural heritage [1–5]. These items embody a tangible connection to past knowledge, artistic practices, and social values, making their conservation essential for maintaining a link to history. However, due to varying materials and production techniques, books from different historical periods face diverse challenges of degradation [1,6]. Monitoring and understanding these degradation processes is therefore critical to developing conservation strategies that extend the longevity and accessibility of these valuable artifacts.

Paper degradation typically occurs through two primary mechanisms: acid-catalyzed hydrolysis and oxidative degradation [7–9]. Books made from wood pulp, especially those produced after the 19th century, are susceptible to rapid degradation. The acidic compounds in wood pulp catalyze the breakdown of cellulose fibers, producing byproducts like furfural, a marker of cellulose and hemicellulose hydrolysis. This degradation is associated with a significant decline in paper quality. In contrast, historical papers produced from plant-based rags, as in books from the 16th century, contain fewer acidic components and are thus more resistant to hydrolysis. Instead, these older papers tend to degrade via the oxidation of residual lipids, releasing aldehydes like octanal and nonanal as byproducts, which serve as markers of oxidative degradation.

In this study, we analyzed volatile organic compounds (VOCs) emitted from two books preserved at the Biblioteca Capitolare of Busto Arsizio in Italy (Figure 1) [10]. The first volume, a Latin grammar book by Gian Alberto Bossi, was published in the late 16th century and widely used to teach noble families, with its content still relevant in some classrooms today. The second volume, a mathematics handbook published in 1860, was designed to standardize measurement conversions in post-unification Italy. The books represent distinct production materials and processes, offering insights into different degradation pathways and preservation needs.

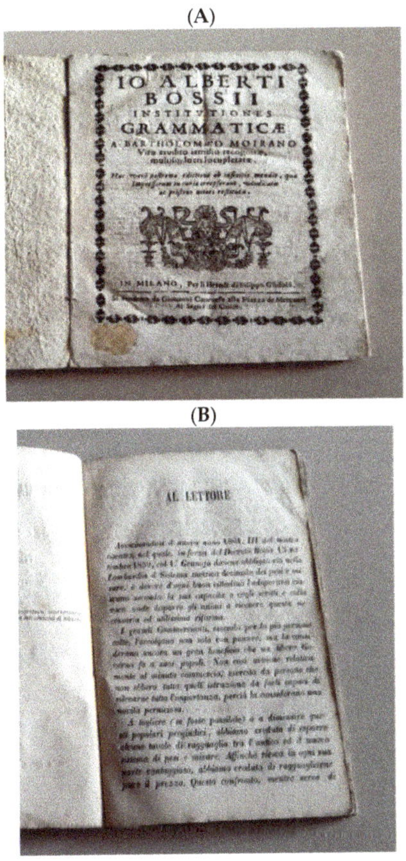

Figure 1. The two heritage books examined in this study. Panel (**A**): XVI century book. Written by Gian Alberto Bossi, this Latin grammar became a bestseller and was widely used by teachers instructing students from noble families. Preserved in Busto Arsizio, it remains relevant today, with examples still used in modern classrooms. Panel (**B**): XIX century book. This 1860 book on mathematical conversions, acquired at an antique market, was widely used in early post-unification Italy to standardize local measurements according to the new national system.

Headspace solid-phase microextraction (HS-SPME) coupled with gas chromatography–mass spectrometry (GC-MS) enables non-invasive VOC analysis, preserving the structural integrity of valuable documents [9]. This approach allows for the identification of compounds linked to specific degradation mechanisms, such as aldehydes from lipid oxidation and furfural from cellulose hydrolysis. The technique provides conservators and researchers with a means to monitor the chemical health of historical documents, minimizing risk to the artifacts.

This study aims to identify and compare degradation markers in two heritage books from different centuries to assess their respective preservation states. Using non-invasive HS-SPME/GC-MS analysis, we analyzed the VOCs emitted from a 16th-century Latin grammar book and a 19th-century mathematics handbook preserved at the Biblioteca Capitolare of Busto Arsizio, Italy [9,11]. By examining specific VOC profiles, such as aldehydes and furfural, this research seeks to provide a clearer understanding of the degradation pathways affecting each type of paper, offering insights relevant to conservation science. This collaborative work between IIS G. Torno, Biblioteca Capitolare of Busto Arsizio, and InnovHub SSI highlights the importance of partnerships in heritage science, facilitating the preservation of historical artifacts for future generations.

2. Results

The analysis of VOCs emitted from the two historical books revealed distinct profiles that reflect differences in their material composition and degradation pathways. Each book's VOC profile highlights its preservation state, providing insights into how historical manufacturing techniques influence long-term chemical stability. The main compounds that allow differentiation of the degradation processes are listed in Table 1.

Table 1. Retention times (minutes), similarity index (SI) obtained using NIST-20 Mass Spectral Library, and theoretical linear retention index (LRI) (NIST-20 library) calculated by injecting a mixture of linear alkanes in the range C10-C30 under the same instrumental conditions and relative peak areas (%) of volatile organic compounds (VOCs) detected in the 16th-century Latin grammar book and the 19th-century mathematics handbook. The table highlights the presence of key compounds associated with oxidative degradation (e.g., aldehydes and alkanes) in the 16th-century book and acid-catalyzed hydrolysis markers (e.g., furfural) in the 19th-century book, reflecting distinct degradation pathways due to differences in paper composition and manufacturing techniques. Confidence in compound identification was ensured using a combination of spectral similarity index (SI) and comparison of theoretical and experimentally calculated linear retention indices (LRIs). VOC analysis was conducted in duplicate for each book, and Table 1 reports the average relative peak areas (%). To evaluate meaningful differences between the degradation processes of the two books, total aldehyde, total hydrocarbon, and furfural values were compared using a two-tailed t-test. All three parameters showed statistically significant differences at the 95% confidence level ($p < 0.05$) and are marked with an asterisk (*).

	Retention Time (min)	SI (NIST-20)	LRI (Calculated)	LRI (Theoretical)	Area %	
Compound					*XVI Century Book*	*XIX Century Book*
Octanal	17.15	771	1304	1289 (±9)	0.42	0.46
Tetradecane	18.98	909	1401	1400	12.48	17.81
Nonanal	19.36	879	1424	1391 (±8)	2.36	2.25
Pentadecane	20.91	786	1502	1500	0.97	1.05

Table 1. *Cont.*

Compound	Retention Time (min)	SI (NIST-20)	LRI (Calculated)	LRI (Theoretical)	Area %	
					XVI Century Book	*XIX Century Book*
2-ethylhexanol	21.01	844	1506	1491 (±5)	1.61	1.57
Furfural	21.09	830	1511	1461 (±11)	0.11	0.31 *
Decanal	21.42	879	1538	1498 (±8)	4.76	3.90
Hexadecane	22.79	913	1601	1600	10.45	10.06
Menthol	24.05	911	1673	1637 (±6)	4.67	6.30
Octadecane	26.19	825	1802	1800	1.66	1.38
Butoxyethoxy ethanol	26.73	894	1827	1796 (±13)	4.35	0.71
Anethole	27.49	698	1845	1817 (±5)	1.05	1.43
Diphenyl ether	30.61	809	2129	2017 (±15)	1.01	1.42
Diisopropylnaphthalene	32.32	719	2230	2228 (±14)	4.65	5.06
Diethyl phthalate	35.58	745	2394	2365 (±7)	0.77	1.05
Benzophenone	38.27	678	2508	2450 (±16)	0.59	0.36
Diisobutyl phthalate	38.49	916	2525	2536 (±10)	5.76	1.45
Total aldehyde					7.54 *	6.61
Total hydrocarbons					25.56	30.30 *

2.1. *VOC Profile of the 16th-Century Book*

The 16th-century Latin grammar book exhibited a VOC profile rich in alkanes and aldehydes, which are indicative of lipid oxidation—a slower degradation pathway associated with high-quality rag-based paper (Figure 2) [8,9]. The profile was dominated by tetradecane (12.48% of total VOCs) and hexadecane (10.45%), long-chain alkanes likely originating from residual organic components in the plant-based rag fibers. These alkanes suggest the presence of stable, organic compounds within the paper, contributing to its durability. Additionally, moderate levels of aldehydes, such as nonanal (2.36%) and decanal (4.76%), support the presence of oxidative degradation as the primary pathway, while octanal was detected at a lower concentration of 0.42%.

Furfural, a typical marker of acid-catalyzed hydrolysis, was found in minimal amounts (0.11%), underscoring the rag paper's chemical resistance to acid degradation [12]. Certain compounds, such as butoxyethoxy ethanol (4.35%), menthol (4.67%), and anethole (1.05%), may suggest historical conservation efforts, although no documentation is available for these specific books. Both menthol and anethole possess bacteriostatic properties, which could hypothetically support such a role, but their presence may also stem from ambient exposure or unrelated sources [13,14]. This VOC profile underscores the durability of the 16th-century book, with a slow, oxidative degradation pathway that aligns with the chemically stable nature of rag-based paper.

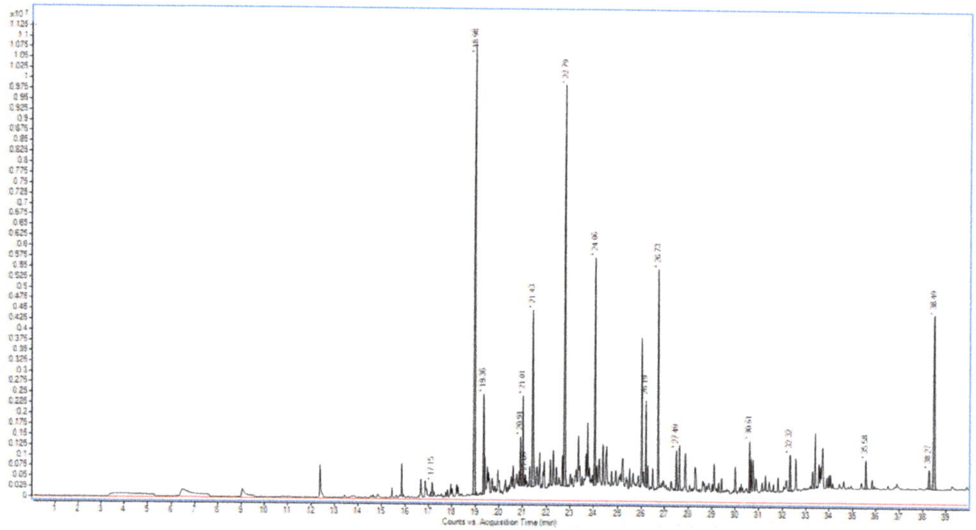

Figure 2. GC spectrum of the VOC profile for the 16th-century Latin grammar book (black line) and blank exposure analysis (red line). Y-axis label: Total ion current.

2.2. *VOC Profile of the 19th-Century Book*

The 19th-century mathematics handbook displayed a VOC profile that reflects the composition and degradation characteristics of wood pulp-based paper (Figure 3) [8,9]. The profile was rich in long-chain alkanes, with tetradecane comprising 17.81% of total VOCs and hexadecane comprising 10.06%. These alkanes are common in wood pulp paper and likely represent residual organic compounds within its structure. In addition to alkanes, furfural was detected at a concentration of 0.31%, indicating acid-catalyzed hydrolysis, which is prevalent in wood pulp due to its acidic nature.

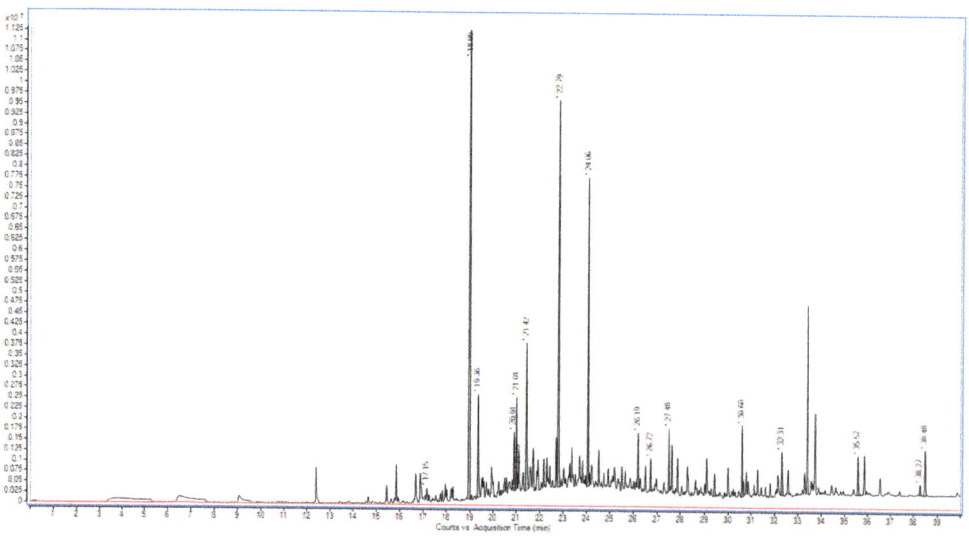

Figure 3. GC spectrum of the VOC profile for the 19th-century mathematics handbook (black line) and blank exposure analysis (red line). Y-axis label: Total ion current.

The profile also included notable levels of menthol (6.30%) and synthetic compounds such as diphenyl ether (1.42%) and diethyl phthalate (1.05%). The higher menthol content suggests possible preservation treatments to counteract microbial damage, while synthetic additives like diethyl phthalate may have been introduced to extend the lifespan of the paper. The presence of furfural and synthetic compounds suggests a more chemically vulnerable paper type, susceptible to rapid degradation under acidic conditions, which may have prompted additional preservation efforts [12].

2.3. Comparative Analysis of VOC Profiles and Implications for Conservation

A comparative analysis of VOCs based on peak area integration highlights the distinct degradation mechanisms of each book (Table 1). The 16th-century book exhibited high levels of long-chain alkanes (e.g., tetradecane at 12.48% and hexadecane at 10.45%) and moderate levels of aldehydes (e.g., decanal at 4.76% and nonanal at 2.36%), indicating that oxidative degradation is the primary pathway, consistent with the stability of rag-based paper. The minimal concentration of furfural (0.11%) further supports the chemical resilience of this material against acid hydrolysis, demonstrating the durability of rag-based paper in historical artifacts.

In contrast, the 19th-century book showed a higher concentration of furfural (0.31%), reflecting its vulnerability to acid-catalyzed hydrolysis due to the inherent acidity of wood pulp paper [12]. This increased susceptibility to hydrolytic degradation in the 19th-century book underscores the rapid deterioration that acidic materials face, especially in environments where preservation conditions may fluctuate. The relatively lower levels of aldehydes in this book compared to the 16th-century book suggest that oxidation is a secondary degradation pathway, less significant in the chemically unstable environment created by acidic wood pulp.

These findings are relevant for heritage science as they reveal how historical manufacturing methods and material choices directly impact the longevity of cultural artifacts. The chemically stable nature of rag-based paper used in the 16th-century book suggests that such materials are inherently better suited for long-term preservation, with oxidative degradation occurring at a relatively slow pace. This durability makes rag-based artifacts more resilient to environmental fluctuations and less dependent on rigorous conservation efforts.

Conversely, the findings for the 19th-century book emphasize the inherent vulnerability of wood pulp paper to acidic degradation. The high furfural content highlights a pressing need for targeted conservation strategies, especially for collections dominated by 19th-century and early 20th-century materials [12]. Preservation efforts for such books may require more stringent environmental controls to limit acid hydrolysis, such as maintaining stable humidity levels, reducing temperature fluctuations, and possibly applying deacidification treatments to neutralize the acidic compounds within the paper fibers.

Additionally, the presence of menthol and synthetic compounds in both books points to historical or contemporary preservation interventions, reflecting an ongoing awareness of the need to mitigate microbial damage and extend paper longevity [13–15]. These findings underline the role of active preservation treatments and monitoring, especially for vulnerable wood pulp-based collections, and suggest that conservation strategies can benefit from regular VOC profiling to assess the effectiveness of these treatments over time.

3. Discussion

This comparative VOC profile demonstrates how historical paper composition and manufacturing methods significantly influence the preservation state of cultural artifacts. The findings advocate for conservation approaches tailored to the specific degradation mechanisms of each paper type, ensuring that valuable historical documents are preserved for future generations. This study highlights the potential of VOC analysis through HS-SPME-GC/MS as a non-invasive tool for assessing the preservation state and degradation mechanisms of historical books. By examining VOC emissions from two books—a 16th-century Latin grammar and a 19th-century mathematics handbook—we identified distinct chemical profiles that reflect the impact of material composition and historical production methods on the longevity of paper-based artifacts.

The 16th-century book, constructed from high-quality rag-based paper, exhibited a VOC profile dominated by aldehydes and long-chain alkanes, indicative of oxidative degradation—a slower process that aligns with its durable material. In contrast, the 19th-century book, made from wood pulp, showed elevated furfural levels, a marker of acid-catalyzed hydrolysis, pointing to rapid degradation characteristic of acidic wood pulp paper. These differences underscore the need for tailored conservation strategies to mitigate the degradation of historically significant books, particularly those vulnerable to acid-catalyzed decay.

While the VOC profiles primarily reflect degradation of the paper substrate—namely, aldehydes from lipid oxidation and furfural from cellulose hydrolysis—minor contributions from inks, adhesives, or sizing agents cannot be fully excluded. No illustrations or visible ink corrosion were observed, and the detected VOCs did not include typical ink-related markers such as aromatic amines or phthalates [8,12,16,17]. Prior studies have reported compounds like isopropyl esters and dimethylformamide as possible indicators of inks or industrial additives in historical papers [16,17]. Thus, while ink degradation likely had minimal impact in the present analysis, future investigations using non-invasive techniques (e.g., FTIR, XRF, hyperspectral imaging) may help clarify the role of non-cellulosic components in VOC emissions [16,17].

In addition to degradation markers, compounds such as menthol, diphenyl ether, and diethyl phthalate were detected. These may reflect past preservation treatments or synthetic additives but could also result from environmental exposure or sample handling. Similar compounds have been reported in historical papers as possible contaminants or additives [8,12,13]. In the absence of direct historical or material evidence, we interpret these findings with caution.

Of note, this project was originally developed as part of a high school educational program, which shaped its exploratory and didactic focus. While scientific rigor was maintained in data acquisition and interpretation, some experimental details and long-term data storage practices typical of more advanced analytical workflows were beyond the intended scope. Nonetheless, we believe this study offers a valuable proof-of-concept and contributes meaningful insights into non-invasive diagnostic approaches for heritage materials, particularly in highlighting how accessible analytical techniques like HS-SPME-GC/MS can support paper-based conservation efforts.

Indeed, this project was made possible through a collaborative effort between public education and private industry, specifically involving teachers and students from IIS G. Torno (Castano Primo, Italy) and the heritage science researchers at InnovHub SSI (Milan, Italy). This partnership not only facilitated access to unique historical materials and advanced analytical techniques but also provided educational and hands-on scientific experiences for students. The collaboration exemplifies how interdisciplinary efforts between academic institutions and industry can enrich both scientific research and cultural preservation efforts.

By fostering such partnerships, we can continue to develop and refine non-invasive methods to support the long-term preservation of cultural heritage, ensuring that these artifacts remain accessible to future generations.

4. Materials and Methods

4.1. Sample Collection and Storage

Two historical books—a 16th-century Latin grammar by Gian Alberto Bossi and a 19th-century mathematics handbook, sourced from the Biblioteca Capitolare of Busto Arsizio in Italy—were selected for VOC analysis, according to [9,11]. Each book was stored in a 12 L glass desiccator, equipped with a silicone septum to accommodate the insertion of an SPME fiber. To minimize potential contamination, storage of the books was carried out in a new glass desiccator.

4.2. Headspace Solid-Phase Microextraction

The VOCs emitted were captured using HS-SPME. A 50/30 μm divinylbenzene/carboxen/polydimethylsiloxane (DVB/CAR/PDMS) fiber (Supelco, Bellefonte, PA, USA) was selected for its broad adsorption range, covering both polar and non-polar compounds. A new SPME fiber was used and preconditioned by heating it to 270 °C for 1 h to remove any potential contaminants, ensuring sample integrity. The same SPME fiber was then used to perform all the analyses (blanks and samples). The SPME fiber was preconditioned by heating it to 270 °C for 1 h to remove any potential contaminants, ensuring sample integrity.

To enhance headspace VOC concentration, the books were positioned upright with their central pages opened and slightly fanned out to maximize surface exposure within the chamber. The selected pages exhibited the highest textual density; neither volume included illustrations or distinct ink formulations. The selected pages exhibited the highest textual density; neither volume included illustrations or distinct ink formulations. The HS-SPME extraction was carried out at 20 °C for 24 h. This duration was selected based on the procedure described in [10], where a 24-h extraction was demonstrated to yield reproducible and representative signals across a wide range of VOCs.

Prior to sampling, each desiccator was evacuated to remove any potential environmental contaminants and then sealed and left undisturbed for 24 h at an ambient temperature of 20 °C for a blank exposure.

4.3. Gas Chromatography–Mass Spectrometry Analysis

After VOC collection, the fiber was immediately inserted into an Agilent 7890B gas chromatograph coupled with a 5977B mass spectrometer (Agilent Technologies, Santa Clara, CA, USA) for compound desorption and analysis. The chromatographic separation was conducted using a polar polyethylene glycol (PEG) capillary column (50 m in length, 0.2 mm internal diameter, 0.4 μm film thickness), ideal for separating complex organic mixtures. The oven temperature was initially set at 40 °C and then gradually increased at a rate of 10 °C per minute until reaching 250 °C, where it was held for 5 min. Helium served as the carrier gas at a constant flow rate of 1.0 mL/min.

The mass spectrometer operated in electron ionization (EI) mode at 70 eV, scanning mass-to-charge ratios (m/z) from 40 to 450 amu to capture a comprehensive VOC profile. Identification was based on comparing observed spectra to the NIST Mass Spectral Library, version 2020 (National Institute of Standards and Technology, Gaithersburg, MD, USA), using a minimum spectral similarity index (SI) threshold of 600. Then, the compound identification is considered a correct identification if the maximum similarity score of a query mass spectrum fulfills this threshold. In order to improve the identification process, the linear retention indices (LRIs) were calculated. This latter parameter was calculated by injecting a mixture of linear alkanes, from C10 to C30, under the same instrumental conditions as the samples. A difference between the calculated and tabulated RLIs of about 50 units is considered acceptable, using a polar gas chromatographic column. The results obtained are shown in Table 1. Using SI and LRI simultaneously improves the accuracy of the library search, focusing particularly on aldehydes, furans, alkanes, and alcohols as primary markers of lipid oxidation and cellulose hydrolysis, indicative of the primary degradation pathways in historical paper.

4.4. Semiquantitative Analysis of VOCs

To compare VOC concentrations between the two books, semiquantitative analysis was performed. Each compound's chromatographic peak area was integrated and expressed as a percentage of the total ion chromatogram. The results are shown in Table 1 as the average of two independent analyses on each book. The control of the duplicate analyses was evaluated on the basis of the relative areas %, which did not differ from each other by more than 10%. This relative quantification enabled the comparison of VOC abundances, offering insights into specific degradation processes. Furfural, a byproduct of cellulose hydrolysis, served as an indicator of acid-catalyzed degradation, predominantly observed in the 19th-century volume. In contrast, the 16th-century book showed higher levels of straight-chain aldehydes, such as octanal, nonanal, and decanal, which are typical products of lipid oxidation, suggesting a greater oxidative stability. These peak area percentages provided a comparative basis for assessing each book's preservation state.

4.5. Quality Control and Blank Runs

Quality control was ensured by weekly monitoring of the empty desiccator using 24-h blank SPME fiber exposures to detect potential background contamination. Each fiber was reconditioned before sampling, and blank runs were conducted after each analysis to verify the absence of carry-over. These measures ensured the accuracy and reproducibility of the VOC profiles obtained.

Author Contributions: Conceptualization, C.C. and M.C.; methodology, P.R. and D.B.; formal analysis, All; investigation, All; writing—original draft preparation, L.G.; writing—review and editing, All; supervision, L.G. All authors have read and agreed to the published version of the manuscript.

Funding: This research received no external funding.

Institutional Review Board Statement: Not applicable.

Informed Consent Statement: Not applicable.

Data Availability Statement: The data supporting the findings of this study are available from the corresponding author upon reasonable request. Due to the educational scope of the project and the lack of ongoing access to the original historical books, supplementary files such as full chromatograms, match score tables, or raw GC-MS signal data were not archived for external sharing.

Conflicts of Interest: The authors declare no conflicts of interest.

Appendix A

The 5B IISS Torno Working Group.

Susanna Andrani, Simone Barzizza, Samuele Busti, Pietro Carturan, Luca Casucci, Mathias Chiodini, Noemi Freddi, Micaela Gallazzi, Andrea Maino, Sara Menescardi, Federica Milanti, Alessandro Pagano, Alessio Perlini, Mattia Rizzo, Davide Valente.

References

1. Yan, Y.; Tang, Y.; Yang, Y. Chemical Conservation of Paper-Based Cultural Heritage. *Molecules* **2024**, *30*, 122. [CrossRef] [PubMed]
2. Botti, S.; Bonfigli, F.; Mezi, L.; Flora, F. Monitoring of Cleaning Treatments for Paper Heritage with Raman Spectroscopy Mapping. *Molecules* **2023**, *29*, 15. [CrossRef] [PubMed]
3. Botti, S.; Bonfigli, F.; Nigro, V.; Rufoloni, A.; Vannozzi, A. Evaluating the Conservation State of Naturally Aged Paper with Raman and Luminescence Spectral Mapping: Toward a Non-Destructive Diagnostic Protocol. *Molecules* **2022**, *27*, 1712. [CrossRef] [PubMed]

4. Nemoykina, A.L.; Shabalina, A.V.; Svetlichnyi, V.A. Restoration and conservation of old low-quality book paper using aqueous colloids of magnesium oxyhydroxide obtained by pulsed laser ablation. *J. Cult. Herit.* **2019**, *39*, 42–48. [CrossRef]
5. Croitoru, C.; Roata, I.C. Ionic Liquids as Reconditioning Agents for Paper Artifacts. *Molecules* **2024**, *29*, 963. [CrossRef] [PubMed]
6. Sequeira, S.O.; Pasnak, E.; Viegas, C.; Gomes, B.; Dias, M.; Cervantes, R.; Pena, P.; Twarużek, M.; Kosicki, R.; Viegas, S.; et al. Microbial assessment in a rare Norwegian book collection: A One Health approach to cultural heritage. *Microorganisms* **2024**, *12*, 1215. [CrossRef] [PubMed]
7. Zhang, M.; Zhao, J.; Wang, S.; Dai, Z.; Qin, S.; Mei, S.; Zhang, W.; Guo, R. Carbon quantum dots for long-term protection against UV degradation and acidification in paper-based relics. *ACS Appl. Mater. Interfaces* **2024**, *16*, 5009–5018. [CrossRef] [PubMed]
8. Gaspar, E.M.; Santana, J.C.; Lopes, J.F.; Diniz, M.B. Volatile organic compounds in paper—An approach for identification of markers in aged books. *Anal. Bioanal. Chem.* **2010**, *397*, 369–380. [CrossRef] [PubMed]
9. Verticchio, E.; Frasca, F.; Cavalieri, P.; Teodonio, L.; Fugaro, D.; Siani, A.M. Conservation risks for paper collections induced by the microclimate in the repository of the Alessandrina Library in Rome (Italy). *Herit. Sci.* **2022**, *10*, 80. [CrossRef] [PubMed]
10. Clark, A.J.; Calvillo, J.L.; Roosa, M.S.; Green, D.B.; Ganske, J.A. Degradation product emission from historic and modern books by headspace SPME/GC-MS: Evaluation of lipid oxidation and cellulose hydrolysis. *Anal. Bioanal. Chem.* **2011**, *399*, 3589–3600. [CrossRef] [PubMed]
11. Stylianou, M.; Barlet, C.; Andreou, C.; Agapiou, A. Assessment of volatile emissions by aging books. *Environ. Sci. Pollut. Res.* **2024**, *31*, 17670–17677. [CrossRef] [PubMed]
12. Battistel, D.; Padovani, I.; Dallo, F.; Barbante, C.; Zendri, E.; Balliana, E. Evaluation of the volatile organic compound emissions in modern and naturally aged Japanese paper. *J. Cult. Herit.* **2018**, *33*, 18–29. [CrossRef]
13. Łojewski, T.; Sawoszczuk, T.; Łagan, J.M.; Zięba, K.; Barański, A.; Łojewska, J. Furfural as a marker of cellulose degradation: A quantitative approach. *Appl. Phys. A* **2010**, *100*, 873–884. [CrossRef]
14. Ludwiczuk, A.; Kieltyka-Dadasiewicz, A.; Sawicki, R.; Golus, D.; Ginalska, G. Essential oils of some *Mentha* species and cultivars, their chemistry and bacteriostatic activity. *Nat. Prod. Commun.* **2016**, *11*, 1015–1018. [CrossRef] [PubMed]
15. Galgano, M.; Capozza, P.; Pellegrini, F.; Cordisco, M.; Sposato, A.; Sblano, S.; Camero, M.; Lanave, G.; Fracchiolla, G.; Corrente, M.; et al. Antimicrobial activity of essential oils evaluated in vitro against *Escherichia coli* and *Staphylococcus aureus*. *Antibiotics* **2022**, *11*, 979. [CrossRef] [PubMed]
16. Lattuati-Derieux, A.; Bonnassies-Termes, S.; Lavédrine, B. Identification of volatile organic compounds emitted by a naturally aged book using solid-phase microextraction/gas chromatography/mass spectrometry. *J. Chromatogr. A* **2004**, *1026*, 9–18. [CrossRef] [PubMed]
17. Lattuati-Derieux, A.; Bonnassies-Termes, S.; Lavédrine, B. Characterisation of compounds emitted during natural and artificial ageing of a book: Use of headspace-solid-phase microextraction/gas chromatography/mass spectrometry. *J. Cult. Herit.* **2006**, *7*, 123–133. [CrossRef]

molecules

MDPI

Article

Poly(Acrylic Acid)/TiO_2 Nanocomposite Hydrogels for Paper Artwork Cleaning and Protection

Sabina Botti [1,*], Francesca Bonfigli [1], Rosaria D'Amato [1], Jasmine Rodesi [2,†] and Maria Gabriella Santonicola [2]

1 ENEA C.R. Frascati, Photonics Micro- and Nano-Structures Laboratory, Physical Technologies and Security Division, Nuclear Department, Via E. Fermi 45, 00044 Frascati, Italy; francesca.bonfigli@enea.it (F.B.); rosaria.damato@enea.it (R.D.)

2 Department of Chemical Engineering Materials Environment, Sapienza University of Rome, Via del Castro Laurenziano 7, 00161 Rome, Italy; rodesi.jasmine@gmail.com (J.R.); mariagabriella.santonicola@uniroma1.it (M.G.S.)

* Correspondence: sabina.botti@enea.it

† Current address: Institute for Medical and Human Genetics, Charité—Universitätsmedizin Berlin, 13353 Berlin, Germany.

Abstract: Paper-based artworks are prone to natural aging processes driven by chemical and biological processes. Numerous treatments have been developed to mitigate deterioration and prevent irreversible damage. In this study, we investigated the use of poly(acrylic acid)/TiO_2 composite hydrogels, combining their cleaning and protective functions in a minimally invasive treatment. Hydrogels allow for controlled water flow and photocatalytic TiO_2 nanoparticles enhance the hydrogel's efficacy by enabling the removal of oxidation products and inactivating biological contaminants. Furthermore, this innovative material can act as a protective coating against UV-induced aging, preserving both color and stability of the paper. Raman spectroscopy and confocal laser scanning microscopy imaging techniques were employed to evaluate the treatments, allowing for us to differentiate between hydrolytic and oxidative aging processes. Our findings demonstrate that papers coated with poly(acrylic acid)/TiO_2 composite hydrogels exhibit significant reductions in oxidative markers, an enhanced color stability, and an overall improved resistance to degradation compared to uncoated samples.

Keywords: Raman spectroscopy; cultural heritage; paper aging diagnostics; Raman spectral imaging; hydrogel paper cleaning

1. Introduction

Nowadays, physical and chemical technologies are applied in the study, conservation, and restoration of ancient artifacts. It is of paramount importance to acquire expertise in performing restoration treatments that slow down the aging effects of cultural artifacts preserving the artifact's unique characteristics. In this regard, the development of non-destructive techniques to assess an artifact's condition before and after treatment is a crucial step to achieve this goal. Spectroscopic techniques have played a pivotal role, particularly in the field of book heritage [1–9].

The natural degradation process of paper over time results in both chemical and physical alterations. The primary constituent of paper, cellulose, undergoes two main degradation mechanisms through hydrolytic and oxidative reactions [10]. Hydrolysis can occur in both acidic and basic environments causing depolymerization, whereas oxidation causes functional groups to bind to the cellulose, also disrupting its structure. The rate at which these processes occur is affected by several factors:

- Intrinsic characteristics of the paper, including the raw materials used during production and manufacturing methods.
- Nature of materials applied to the paper such as inks, pigments, binders, etc.
- Possible presence of pathogens and air pollutants in the storage environment, as well as light exposure.

To preserve valuable artwork, it is essential to slow down and prevent degradation phenomena by implementing controlled restoration and protection techniques. Among the various paper restoration operations, cleaning plays a fundamental role. The aim of cleaning is to remove all cellulose degradation products as well as contaminant particles from the paper surface—in other words, materials that can alter the paper appearance and cause further aging.

Cleaning treatments involving immersion in water are widely employed, as water effectively solubilizes most of the foreign particles on the surface, penetrating deeply in between cellulose fibers. Moreover, water promotes the formation of hydrogen bonds between cellulose fibrils, improving the bending resistance of paper sheets. However, washing treatments bring up some concerns: the uncontrolled water flow may meet more degradable constituents or trigger unwanted processes, potentially causing irreversible alterations to the appearance of the artifact and the historical information it conveys.

Lately, materials of a certain texture such as hydrogels have been used to control volume and speed of water flow. Hydrogels are not reactive and retain large volumes of water and/or other solvents, releasing them on the paper surface upon contact. These solvents dissolve the target substances, which are subsequently re-adsorbed by the hydrogels together with the solvents used.

The use of water alone in hydrogels has proven highly effective to clean water-soluble materials commonly found in book artifacts, such as polysaccharides (e.g., starch, gum arabic) and protein-based substances (e.g., gelatin, animal glues). For the removal of non-water-soluble substances, alternative solvents can be employed. Nevertheless, hydrogels make localized treatments on specific areas possible, significantly reducing exposure of operators to potentially harmful vapors. The use of hydrogels as supporting materials is well documented in the literature [11,12].

In this study, we employed a poly(acrylic acid) (PAAc) hydrogel, commercially available under the trade name Carbopol (CBP), loaded with TiO_2 nanoparticles for restoration purposes. Derived from polyacrylic acid, CBP can form gels with good viscosity already at low concentrations (1–1.5% w/v) [13,14], which guarantees reduced adhesive capacity and, hence, easy removal. The incorporation of TiO_2 nanoparticles enhances the hydrogel's cleaning efficacy due to their photocatalytic properties. TiO_2 nanoparticles enable the inactivation of bacteria, fungi, and algae, as well as the removal of organic contaminants [12,14–17]. Moreover, TiO_2 is considered safe and harmless to humans.

The synthesis of TiO_2 thin coatings typically employs high temperatures or mechanical conditions incompatible with paper substrates [18]. In a previous study, TiO_2–cellulose nanocomposite was deposited via spray techniques and was employed as paper consolidant to inhibit aging effects induced by Ultra Violet (UV) light, air pollutants, and microbial activity [19].

In this work, we incorporated laser synthesized TiO_2 nanoparticles with high chemical purity into a PAAc (CBP) matrix to create a composite hydrogel that could be applied on various paper samples [13,20] using a spatula. Our aim was to investigate the dual functionality of composite gel as cleaning agent and protective coating of paper artwork. Two experimental approaches were then followed:

In the first, the CBP/TiO_2 hydrogel was used as a cleaning agent. After application, the composite gel was activated by UV irradiation at 385 nm. When in contact with the porous surface of paper, the gel gradually releases water, facilitating cleaning. We conducted experiments with varying contact times to determine the optimal cleaning conditions that ensure removal of oxidation products and contaminants without compromising the structural integrity of the paper. After removal of the gel, residues can be found on the treated surfaces. Their presence was assessed via confocal Raman spectroscopy with surface scanning.

In a second experimental procedure, a thin layer of CBP/TiO_2 gel was applied to the paper without removal after UV activation. This approach aimed to evaluate its performance as a protective coating against light-induced aging.

The effects of the cleaning/protective procedures were analyzed by confocal Raman spectroscopy with surface scanning and confocal laser scanning microscopy (CLSM). In a previous study [7], we identified the following paper aging markers based on Raman measurements:

$$R_H = I_{1100cm^{-1}} / I_{1380cm^{-1}} \quad (1)$$

$$C_I = I_{2890cm^{-1}} / I_{1380cm^{-1}} \quad (2)$$

$$O_I = A_{1640\text{–}1850cm^{-1}} / A_{1500\text{–}1600cm^{-1}} \quad (3)$$

$$O_T = A_{1500\text{–}2800cm^{-1}} / A_{700\text{–}3000\ cm^{-1}} \quad (4)$$

where I and A indicate intensity and area of peaks in the selected Raman shift range, respectively.

These markers account separately for the hydrolysis and oxidation processes on cellulose, the main constituent of paper structure. Hydrolysis, which shortens cellulose chains, leads to a decrease of the R_H marker, proportional to the degree of polymerization, and of the crystallinity index C_I [7]. Oxidation processes, which involve the addition of functional groups to the cellulose backbone, lead to an increase of the O_T marker, followed by an increase of the O_I marker, which is directly proportional to the area of carbonyl bands ($A_{1640\text{–}1850}$), representing the final oxidation stages of cellulose and, therefore, giving insight on the advancement state of the whole oxidation process.

Their variation in percentage was used to evaluate the cleaning/protective procedure efficiency. Comparisons between treatments involving the CBP/TiO_2 composite gel activated with UV irradiation and the unmodified CBP gel showed a significant reduction in oxidation markers, whereas polymerization degree and crystallinity markers remained unaffected, showing that paper integrity was preserved.

Additionally, our measurements demonstrated that the CBP/TiO_2 coating effectively counter acted oxidation, slowed down hydrolysis maintaining color stability, and avoiding aging yellowing.

2. Results

2.1. *Characterization of Cleaning Process with Raman Micro-Spectroscopy and Confocal Laser Microscopy*

In our experiments, we analyzed different paper samples. Figure 1 shows white light optical images of a 19th-century paper sample before (Figure 1a) and after (Figure 1b) the cleaning treatment. The green square marks the 200 μm × 200 μm area scanned via Raman spectral imaging mode. Figure 1c shows the typical Raman spectra detected in a point of the square.

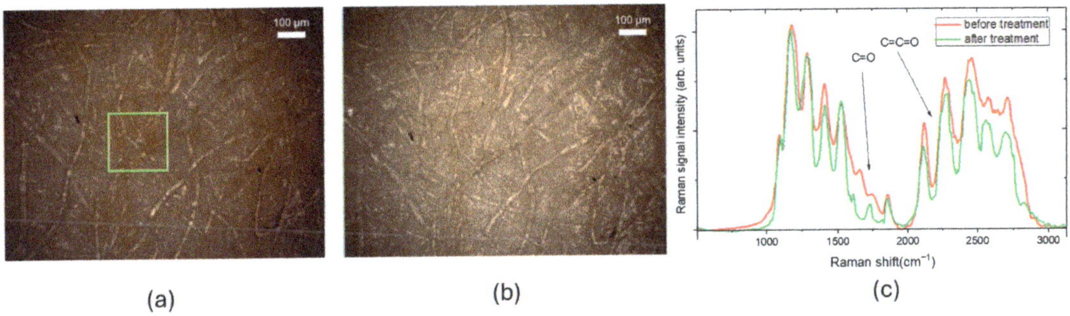

Figure 1. Optical images in bright field of XIX paper. The green rectangle indicates the area of 200 µm × 200 µm scanned with a step size of 5 µm in Raman spectral imaging mode using green excitation (λ = 532 nm) and 10X objective (**a**) before the cleaning treatment; (**b**) after the cleaning treatment; (**c**) examples Raman spectra acquired before and after the cleaning treatment.

The main peaks observed in the Raman spectra are attributed to vibrational modes of carbon-based species. These include C–C stretching and C–O–H and C–C–H vibrations in the 1000–1350 cm^{-1} range, as well as lower intensity in-plane bending vibrations of H–C–H and O–C–H at 1425 cm^{-1} [7,9,21,22]. The peaks with wavenumber greater than 1500 cm^{-1} are due to formation of oxidized molecular groups (excluding CH-CH_2 stretching peaks at 2850–3000 cm^{-1}). In particular, the bands at 1640 and 1740 cm^{-1} are, respectively, ascribed to C=O stretching in carbonyl and carboxyl groups. The band at 1550 cm^{-1} is ascribed to symmetric C=C stretching in C=C–O structures, while the peak at 1850 cm^{-1} represents an overlap between C=O stretching and allene C=C=C stretching. Vibrations in the 2000–2100 cm^{-1} region are attributed to C=C=O stretching of ketene groups. Finally, the peaks around 2550 cm^{-1} are associated with overtones and combinations of carboxylic group frequencies.

As shown in Figure 1c, following the cleaning treatment, the intensity of C=C-O and C=O group peaks decreases indicating that the gel application removes the oxidation products from the paper surface.

The cleaning treatments were carried out by applying the hydrogel onto the paper samples, with contact times of up to 1 h. Some additional tests were performed with a contact time of 2 h, but the changes in the markers were not significantly different from those observed after 1 h and transferring and excessive amount of water onto the paper surface was also a risk. During the cleaning treatment involving the CBP/TiO_2 nanoparticles, the gel was irradiated with UV light for the duration of the treatment.

For each spectrum in the Raman map, both before and after gel application, the values of the markers defined in (1)–(4) were calculated, obtaining a distribution of marker values for each map.

As an example, Figure 2 shows the O_T marker maps, in false colors from lower values in green to higher values in yellow-red, co-localized with the optical image of the XIX century paper sample: a large dark stain is visible in Figure 2a, which becomes lighter after cleaning, as shown in Figure 2b. Contact time was 1 h. By monitoring the peaks associated with the formation of oxidized molecular groups using the O_T marker, we can clarify the activity of CBP. Following the cleaning treatment, the yellow-red areas of the paper, corresponding to higher O_T values, decreased.

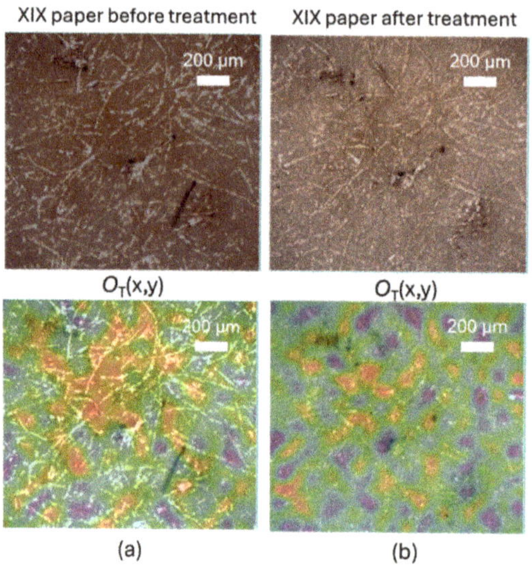

Figure 2. Optical images of XIX paper (**upper** panels) and O_T marker value colocalized maps. (**a**) Before and (**b**) after 1 h CBP cleaning treatment. The color code is blue, green, red, yellow for increasing value of O_T marker.

For each marker map, we calculated the average marker values before and after the CBP cleaning treatments, with and without the addition of TiO_2 nanoparticles. Figure 3 illustrates the observed changes in the marker values for the two paper samples used in our study: a XIX century paper sample as well as a 2021 laser printer paper that had been exposed to ambient laboratory light for three years.

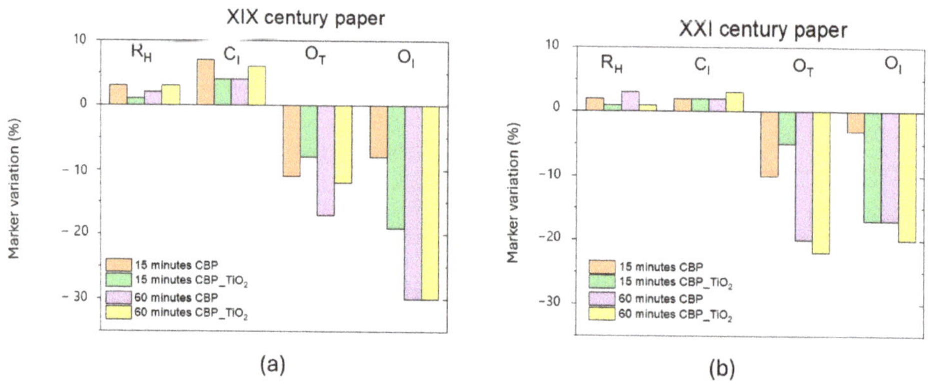

Figure 3. (**a**) Ageing marker variation in percentage for the XIX century paper, (**b**) the same for the XXI century paper. This laser printer 2021 paper was exposed to ambient light for three years.

Although the most pronounced effects on marker values were observed in the XIX century paper, both treatments resulted in an increase in the C_I and R_H markers. This finding can be attributed to the formation of hydrogen bonds between cellulose fibers facilitated by the hydrogel's water content. It was shown that after washing and drying processes the formation of additional hydrogen bonds can occur, enhancing some mechanical properties of the paper, such as bending resistance.

Oxidation markers decreased significantly (−20% O_T and −30% O_I after 1 h of treatment with CBP with and without TiO_2 addition). The reduction in O_I, which is proportional to the carbonyl content, suggests that oxidation products containing C=O groups were primarily removed. This is particularly noteworthy because such carbonyl-containing groups contribute to the formation of chromophore molecules responsible for the yellowish color typical of aged paper.

Although the application of the CBP/TiO_2 hydrogel leads to a noticeable decoloration of the paper (see Figures 1 and 2), there is no great improvement in the decrease of oxidation markers with the use of the composite hydrogel instead of the CBP. In this respect, the advantage to use the composite gel is that the presence of TiO_2 nanoparticles provides a biocidal effect.

Figure 4 shows the CLSM images obtained in fluorescence mode, showing both red and green fluorescence channels, as well as their merge, from a selected region of the 19th-century paper sample before (Figure 4a) and after (Figure 4b) the CBP/TiO_2 cleaning treatment. The graph on the right displays the intensity profiles of the red fluorescence signal, measured along the yellow arrows.

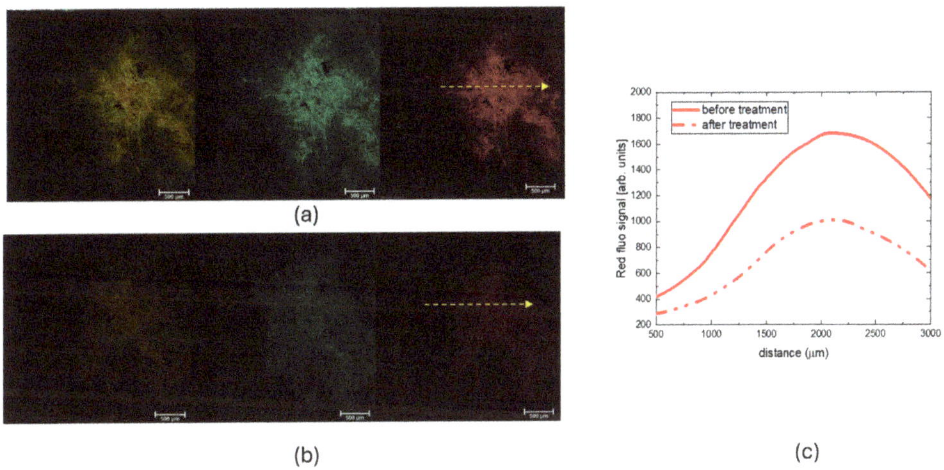

Figure 4. CLSM images (objective 4×) of a selected spot on Brehm paper. (**a**) Before cleaning treatment; (**b**) after cleaning treatment with CBP/TiO_2. The figures show red and the green fluorescence channels and their overlay in yellow color (**c**) Graph of the red fluorescence signal intensity profiles detected along the yellow arrows of the spot before and after CBP/TiO_2 treatment.

It is known that one of the effects of paper aging is the formation of products like simple sugars, cellulose oligomers, and phenolic compounds originating from the degradation of cellulose, hemicellulose, and lignin [23,24]. These species have a strong luminescence with a broad band; therefore, the observed decrease in fluorescence intensity, as evidenced by the CLSM measurements, indicates that, besides oxidized groups, the CBP/TiO_2 treatment eliminates from the paper surface also the cellulose degradation products.

Figure 5a reports the comparison between the Raman spectra of CBP/TiO_2 coated XIX century paper (red curve), uncoated paper (green curve), and CBP/TiO_2 gel (brown curve). As evidenced by the appearance of a new peak around 2800 cm^{-1}, ascribed to CH_2 vibration in CBP, the Raman spectrum measured from the coated paper sample can be expressed as a linear combination (classical least square fitting, CLS) of the CBP/TiO_2 gel spectrum (PC1 component) and the uncoated paper (PC2 component). The coefficients (scores) of the linear combination are reported in a x,y plane in Figure 5b as red diamonds.

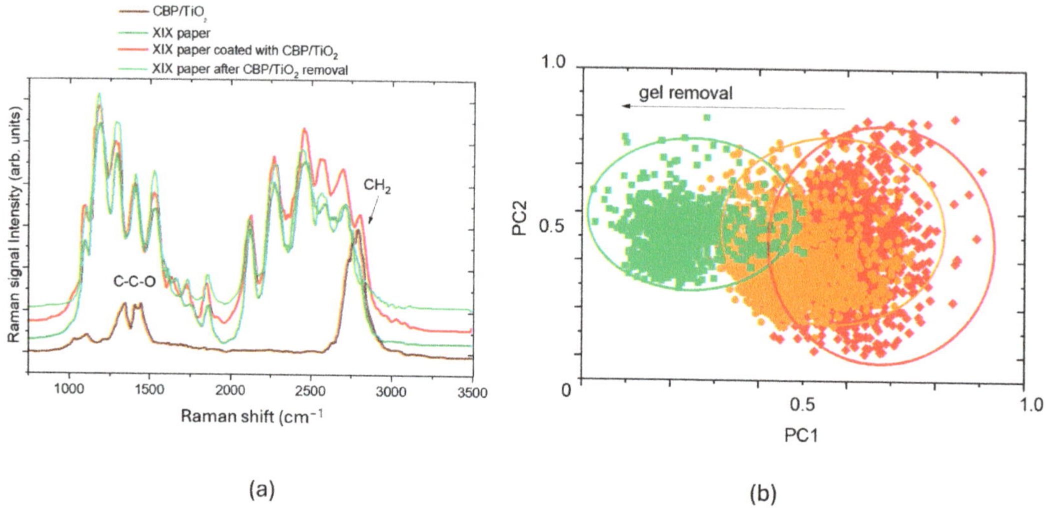

Figure 5. (**a**) Raman spectra of CBP/TiO_2 coated XIX century paper (red curve), XIX century paper sample after gel removal (light green curve), XIX century paper (green curve), and CBP/TiO_2 gel (brown curve); (**b**) score plot related to map of CBP/TiO_2 coated XIX century paper (red diamonds); CBP/TiO_2 coated XIX century paper after a first run of gel removal (orange dots) and a second run of gel removal (green squares). Ellipses are only a guide for the eyes.

When we progressively removed the gel by spatula, we observed a decrease in CH_2 peak intensity and a concomitant decrease of PC1 component, allowing for us to monitor and evaluate the effectiveness of gel removal and determine whether it was sufficient or not.

The gel left on the paper could prolong the protective activity against aging agents. In this regard, additional tests were performed by applying a thin layer of the CBP/TiO_2 gel onto the paper, aiming to evaluate its performance as protective layer, as discussed in the following paragraph.

2.2. *Characterization of Protective Effect of CBP/TiO_2 Coating*

To evaluate the suitability of the CBP/TiO_2 coating as a protective layer for paper, we used two samples of modern (2024) laser printer paper. One sample was coated with a thin layer of CBP/TiO_2 gel applied with a spatula, while the other remained uncoated. Both samples were aged by ultraviolet A irradiation at 385 nm (UVA) for 31 h. The optical images, shown in Figure 6, were subsequently analyzed using ImageJ software to evaluate RGB value distributions of the images.

The RGB histogram of the coated and uncoated samples, shown in Figure 7a, indicates that the CBP/TiO_2 coating is a transparent layer that does not significantly alter the color of the paper.

As shown in Figure 7b, prolonged exposure to UVA irradiation leads to the yellowing of uncoated paper, as demonstrated by a shift towards higher RGB values, in particular of the red component. In contrast, RGB values of the coated paper after UVA irradiation (see Figure 7c,d) remain similar, showing that the CBP/TiO_2 coating maintains the paper color. This property can be exploited for the protection of the art works against light.

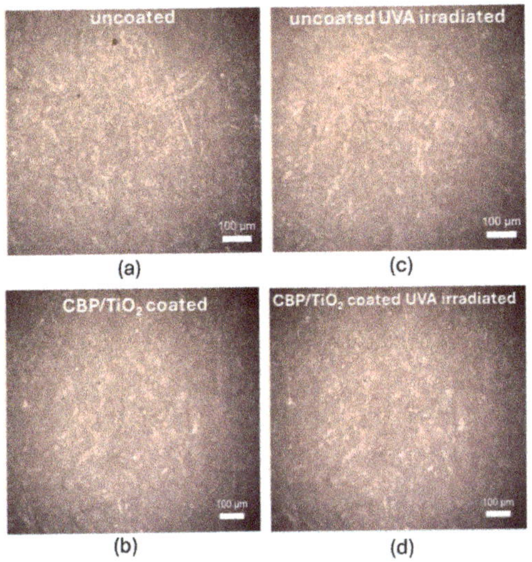

Figure 6. Optical images of 2024 modern paper: (**a**) uncoated and non-irradiated; (**b**) coated with a thin layer of CBP/TiO_2; (**c**) uncoated and UVA irradiated; (**d**) coated with a thin layer of CBP/TiO_2 and UVA irradiated.

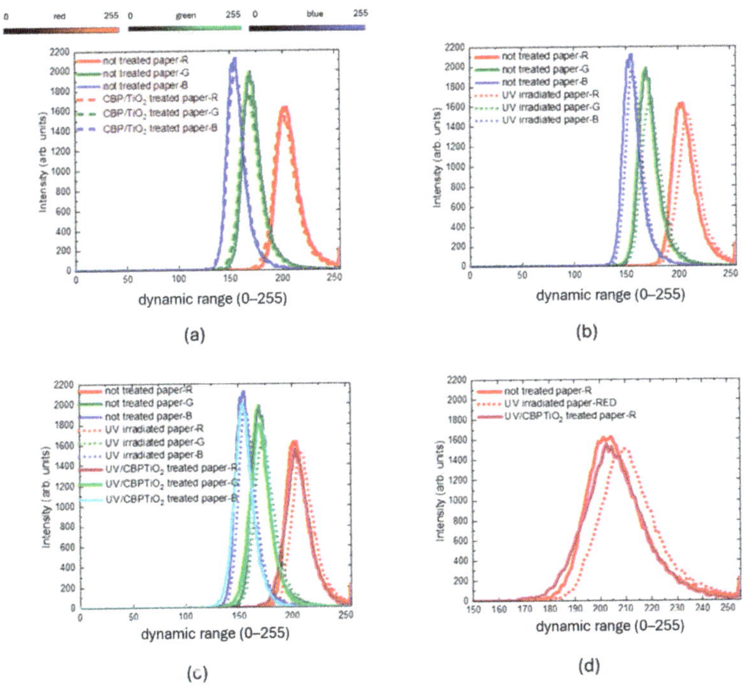

Figure 7. Red–green–blue value distributions of 2024 modern paper comparing (**a**) uncoated (solid lines) and CBP/TiO_2 coated paper (dashed line); (**b**) uncoated paper (solid lines) and uncoated paper UVA-irradiated (dotted lines); (**c**) uncoated paper (solid lines), uncoated UVA-irradiated paper (dotted lines), CBP/TiO_2 coated UVA-irradiated paper (brown, light green, and cyan solid lines). (**d**) Magnification of red value distribution of uncoated paper (solid line), uncoated UVA-irradiated paper (dotted line), uncoated UVA irradiated paper (brown solid line).

As shown in Figure 8a, the Raman spectra acquired from the 2024 paper before and after 31 h of UVA irradiation exhibit a significant difference, particularly in the C–O–C and CH_2 peaks, which are notably reduced after UVA exposure. The C-O-C peak related to the stretching of bond between monomers of cellulose, is proportional to the cellulose chain length (R_H marker) whereas the CH_2 peak is related to the order of fibrils (crystallinity degree, C_I marker). The decrease in these peaks, and hence of corresponding markers, indicates that the uncoated paper has undergone a light-induced degradation of cellulose via hydrolysis. The shortening of the cellulose chains is accompanied by an increase in disorder, measured by the decrease in the C_I marker, shown in Figure 8b.

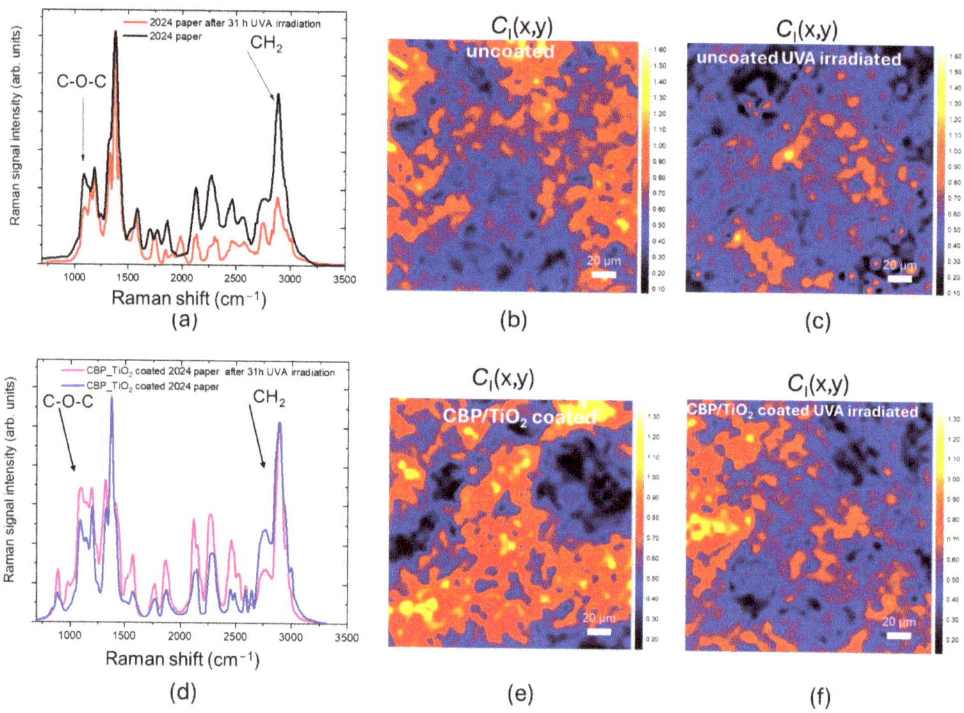

Figure 8. (**a**) Raman spectra acquired from 2024 paper before and after 31 h of UVA irradiation. (**b**,**c**) C_I marker value maps of uncoated 2024 paper non-irradiated and UVA-irradiated, respectively. (**d**) Raman spectra acquired from 2024 paper CBP/TiO_2 coated before and after 31 h of UVA irradiation. (**e**,**f**) C_I marker value maps of CBP/TiO_2 coated 2024 paper, non-irradiated and UVA-irradiated, respectively.

Differently, for the 2024 coated paper, the decrease in the C–O–C and CH_2 peaks is less pronounced (Figure 8d–f).

The protective effect of the coating against cellulose oxidation was also assessed. As shown in Figure 9a,b, after UVA exposure, the O_T marker increases, due to the incorporation of oxidized functional groups into the cellulose backbone.

As for the coated paper, the O_T marker decreases (Figure 9c,d), suggesting that the UVA irradiation may prolong the cleaning action of the gel.

The marker variation before and after 31 h in UVA irradiation is shown in percentage in Figure 10 for both uncoated and coated 2024 paper samples. The CBP/TiO_2 coating not only inhibits the oxidation and hydrolysis of cellulose but also inhibits the decrease of C_I.

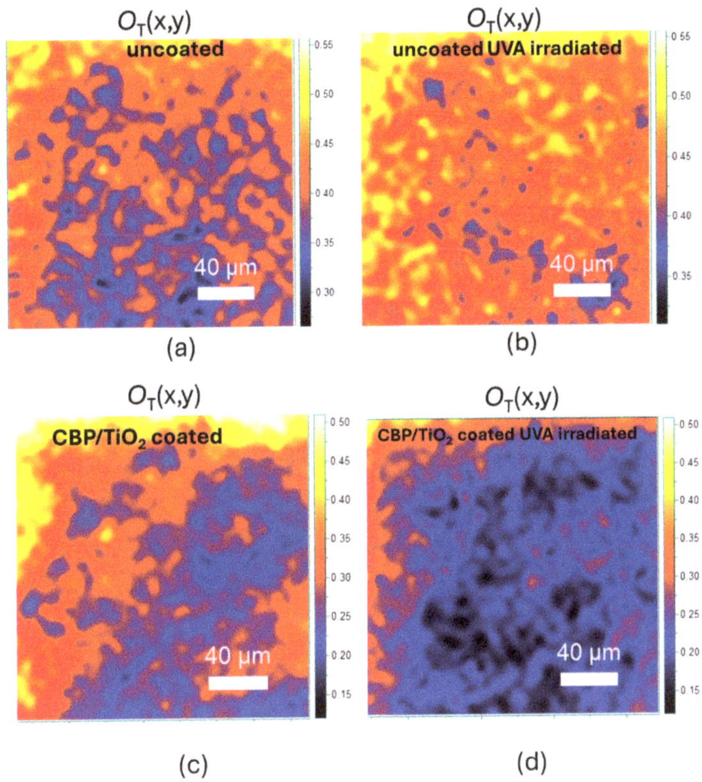

Figure 9. (**a**,**b**) O_T marker value maps of uncoated 2024 paper before and after 31 h of UVA irradiation, respectively. (**c**,**d**) O_T marker value maps of CBP/TiO_2 coated 2024 paper before and after 31 h of UVA irradiation, respectively.

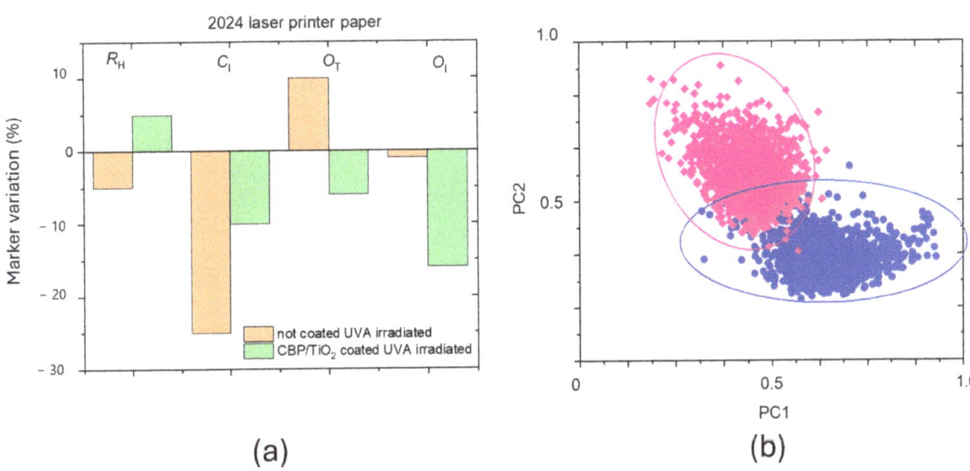

Figure 10. (**a**) Marker variation in percentage after 31 h of UVA irradiation for the uncoated and coated 2024 paper. (**b**) Score plots of coated paper before (blue points) and after irradiation (fuchsia points). The PC1 component is the CBP/TiO_2 gel Raman spectrum, while PC2 component is the paper one.

Figure 10b illustrates the score plots for coated samples before (blue dots) and after (fuchsia dots) UVA irradiation. The observed reduction in gel scores may be attributed to the gel depletion on the paper surface caused by UVA exposure. However, as observed in previous experiments [13], an alternative hypothesis to consider is the penetration of the gel layer beneath the outer fiber of the sample surface.

The same procedure was performed on the 2021 paper that had undergone the natural aging processes in the laboratory environment (see Figure 11a,b). In this case, as shown in Figure 11c for the red values of RGB distributions, the clearing effect of CBP/TiO_2 was preserved when exposed to UV radiation. The green and the blue distributions follow the same behavior of red values.

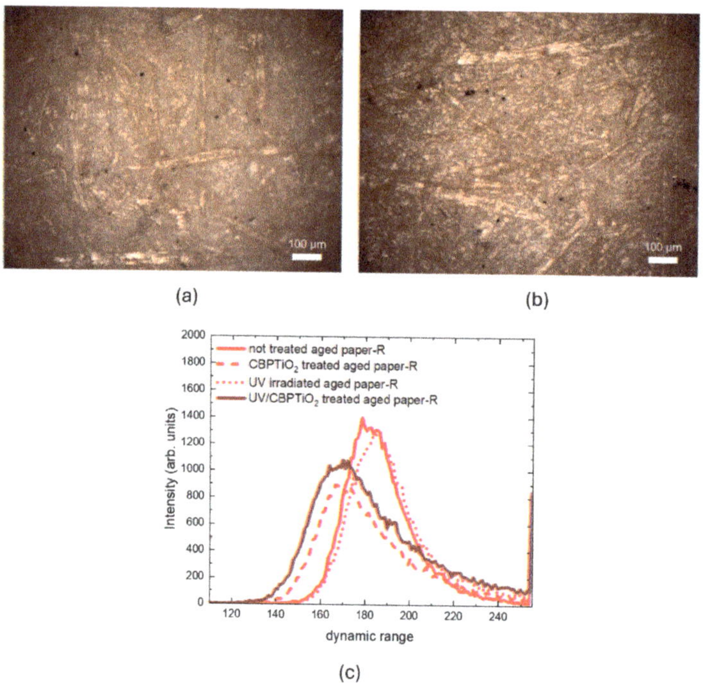

Figure 11. Optical images of 2021 paper naturally aged in laboratory: (**a**) uncoated; (**b**) coated with a thin layer of CBP/TiO_2. (**c**) Red value distributions of 2021 paper naturally aged in laboratory: uncoated paper (solid line), CBP/TiO_2 coated paper (dashed line), uncoated UVA-irradiated paper (dotted line), CBP/TiO_2 coated UVA-irradiated paper.

The O_T marker variation, shown in Figure 12a–d, highlights minimal degradation for samples coated with the CBP/TiO_2 composite. Specifically, the CBP/TiO_2 coating induces only a 3% increase in the O_T marker following UVA irradiation (Figure 12e). In comparison, the uncoated 2021 paper exhibits a 7% increase in the O_T marker, which is slightly lower than the value observed for modern paper. The difference can be explained by the fact that the 2021 paper sample had already been exposed to ambient light for three years, in which the oxidation process progressed rapidly during its initial stage and later slowed down, as is typical. It is worth noticing that in this case, the application of CBP/TiO_2 coating produces an increase in R_H and C_I markers; this can be due to the formation of hydrogen bonds between fibers as previously discussed.

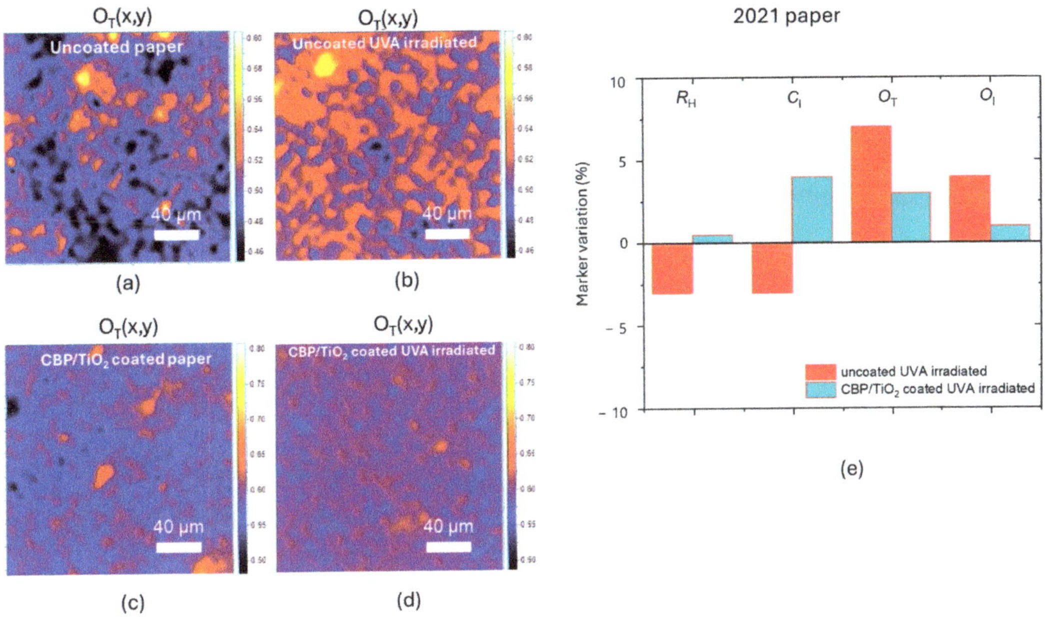

Figure 12. O_T marker value maps for (**a**) uncoated paper, (**b**) UVA-irradiated uncoated paper, (**c**) coated paper, (**d**) UVA-irradiated coated paper. (**e**) Marker variation in percentage after 31 h of UVA irradiation for the uncoated and CBP/TiO_2 coated 2021 paper exposed to ambient light of laboratory for three years.

Overall, the effect of CBP/TiO_2 coating on 2021 paper is to reduce the light induced oxidation and counteract the hydrolysis process.

To further assess the protective properties of the composite gel, an additional 2021 paper sample was prepared and coated with a thicker layer of CBP/TiO_2. Figure 13 shows CLSM images, in reflection mode, comparing the same region of uncoated (Figure 13a) and CBP/TiO_2 coated (Figure 13b) paper samples under identical acquisition conditions. A short pencil line, visible on the left side of the images, was used as spatial reference to facilitate microscope observation of the same area before and after treatment.

The slightly increased thickness of the coating layer does not alter the optical image resolution of the paper, as confirmed by the intensity profiles measured along the yellow arrow shown in Figure 13c. The samples were exposed to outdoor sunlight (22 W/m^2, 16 May 2022) for 30 min and 1 h. A slight but progressive decrease in the R_H index was observed, due to hydrolysis processes. The O_T marker showed a progressive decrease with increased irradiation time. On the other hand, the O_I marker exhibited an increase, suggesting that although the photoactivated TiO_2 decomposes organic oxidizing acids, the oxidation process continues, forming additional C=O bonds [7].

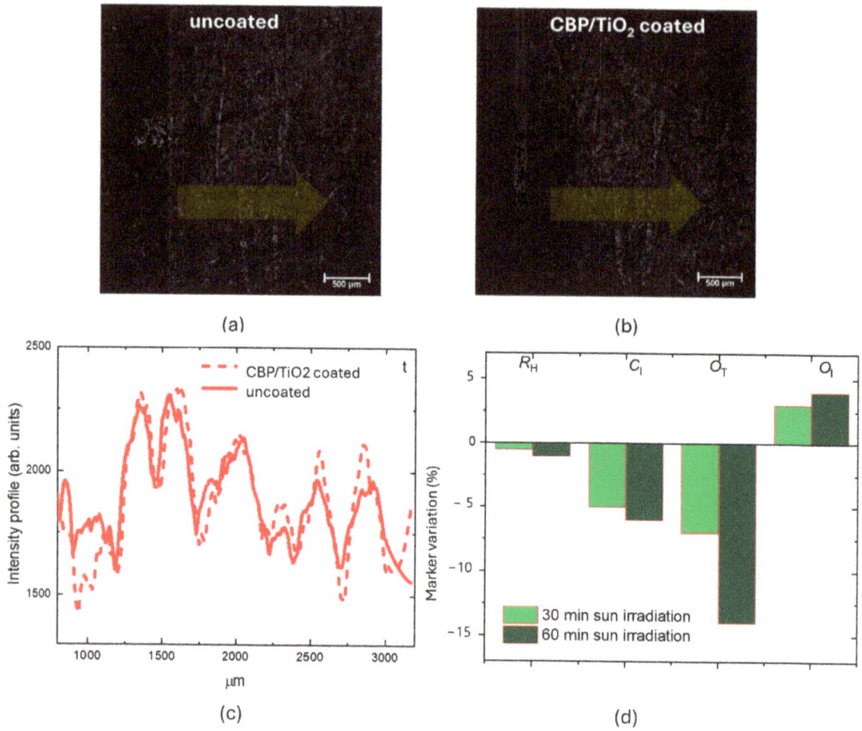

Figure 13. CLSM images obtained in reflection mode of the 2021 paper: (**a**) uncoated; (**b**) CBP/TiO_2-coated; (**c**) the intensity profiles measured along the yellow arrow; (**d**) marker variation percentage after 30 and 60 min of outdoor sun irradiation.

3. Discussion

The aim of the present study was to evaluate the effectiveness of cleaning and protective treatments using a CBP/TiO_2 composite gel activated by UV light, through a Raman spectroscopy-based diagnostic protocol.

Regarding the cleaning action, compared to the application of CBP alone, the addition of TiO_2 slightly improves the removal of oxidizing functional groups, while also ensuring biocidal activity. Moreover, by using CLS fitting of Raman spectra, we were able to monitor the effectiveness of gel manual removal.

To assess the ability of the CBP/TiO_2 coating to shield paper surfaces from light aging, we exposed various paper samples, coated and uncoated, to UVA radiation.

Our measurements show that, in the coated 2024 paper samples, hydrolysis is slowed down, and oxidizing agents are progressively removed through a self-cleaning process, while the protective layer is gradually depleted during exposition. In the case of 2021 paper naturally aged in the laboratory, the CBP/TiO_2 coating acts as paper consolidant (increase in R_H and C_I marker) counter acting the light induced oxidation. Applying a thicker layer of composite coating on 2021 paper, the same self-cleaning process observed for the 2024 modern paper occurred, even when the 2021 paper was exposed to outdoor sunlight.

Additionally, in all the experiments, the presence of the coating layer does not alter the original color of the paper and prevents light-induced yellowing, maintaining RGB values of paper.

4. Materials and Methods

4.1. Nanocomposite Hydrogel Preparation

Carbopol® Ultrez 10 (CAS No. 9003-01-4) from Lubrizol (Wickliffe, OH, USA) has a high level of purity. The following specifications were reported by the manufacturer: loss on drying $\leq 2\%$, heavy metals ≤ 10 ppm, and residual solvents $\leq 0.45\%$.

A 3.33 wt% solution of Carbopol® in water was used for the hydrogel matrix. Since Carbopol® reaches maximum viscosity at pH 7, a 20 wt% NaOH solution was employed as a neutralizing agent [25–27]. To achieve a homogeneous texture, TiO_2 nanoparticles were incorporated into the mixture, followed by sonication to ensure uniform dispersion. The resulting gel proves to be suitable for easy application on substrates. Furthermore, it holds long-term stability and retains its properties even at elevated temperatures, provided it is stored in a sealed container shielded from light.

The primary preparation conditions for the composite hydrogel are summarized in Table 1 and the resulting hydrogel structure is illustrated in Figure 14.

Table 1. Materials for hydrogel preparation.

Material	Structure	Function	Concentration or Volume
Carbopol® Ultrez 10	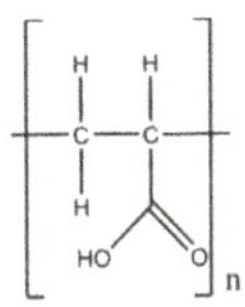	Polymer	3.33 wt%
Deionized water	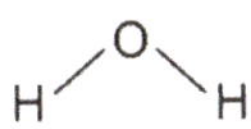	Solvent	15 mL
Sodium hydroxide	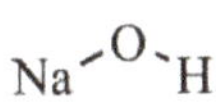	Neutralizing agent	260 µL
Titanium dioxide	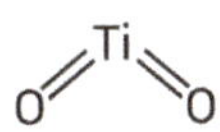	Photoactive element	800 ppm

Figure 14. Photo of hydrogel. On the left CBP gel (1–2–3) on the right CBP/TiO_2 gel (4–5–6).

Titania nanoparticles employed were synthesized via CO_2 laser-induced pyrolysis reactions in a titanium(IV) isopropoxide aerosol, following a previously established protocol [20]. This method ensures high chemical purity of the nanoparticles, as the reactor walls remain cold and non-reactive throughout the process. An estimation carried out through BET measurements shows a mean powder diameter of 20 nm.

When irradiated with photons of energy greater than its band gap, TiO_2 undergoes excitation and generates free electron–hole pairs:

$$TiO_2 + h\nu \rightarrow TiO_{2-} \quad (5)$$

These electron–hole pairs persist briefly before recombination, leaving the molecule in an excited state long enough to allow for them to interact with chemical species adsorbed on the NP surface. Organic contaminants are therefore removed by redox reactions [17,28].

4.2. Paper Samples

Paper samples of 100 cm^2 were cut from 2024 and 2021 laser printing paper. The 2021 sample was exposed for three years to ambient light. Additionally, we examined non printed areas of an ancient book (Brehm, Vita degli Animali, vol. 6, Torino 1896). The book was positioned on the sample holder.

4.3. Gel Application Procedure

The samples were divided into two subsamples and treated with CBP and CB10_TiO_2 composite gel. The gels were applied as a homogeneous layer using a spatula. In the case of the CB_{10}_TiO_2 composite gel, the sample was then placed under a UV light source (LED lamp) to activate the TiO_2 nanoparticles. After treatment, the gel was easily removed with the use of a spatula. Contact times for both cleaning procedures were set at 15 min and 1 h.

4.4. UV Source

The photocatalytic process of titania nanoparticles was activated with a high power UV LED source emitting a peak wavelength at 385 nm. The LED was used by mounting a collimation optics obtaining a closely collimated beam with a power of 120 mW on a beam area of 1450 mm^2. This UV LED was used also for light ageing process.

4.5. Raman Confocal Spectrometer

Raman spectra were acquired using a confocal micro-Raman spectrometer (XploRA Plus, Horiba, Lille, France SAS) with a 532 nm laser wavelength. The Raman signals were collected through a microscope equipped with 5×, 10×, 50×, and 100× objectives. Laser power was adjusted using neutral density filters. After preliminary studies, optimal conditions for both laser power and acquisition times were selected to achieve good signal-to-noise ratio while ensuring safe operating conditions for the samples. Raman spectra were recorded point-by-point across a selected area, following a predefined grid defining a 2D map and defining a 2D spectra array.

The fluorescence background was automatically subtracted from each Raman spectrum, for which the software also calculated the intensity, peak area, and peak width, which were used as contrast parameters. In the case of partially overlapping bands, the software performed spectral deconvolution, obtaining the spectrum as the best fit of a superposition of Gaussian curves. For each defined contrast parameter, a 2D Raman map was generated and associated with the optical image collected.

4.6. Confocal Laser Scanning Optical Microscope

Optical images were acquired using a confocal laser scanning microscope (CLSM) Nikon 80i-C1 (Amstelveen, The Netherlands), operating in both fluorescence and reflection modes [29]. In fluorescence mode, the samples were illuminated with a continuous 445 nm laser (nominal output power of 1.5 mW), and the spectrally integrated PL signal was detected by a system of two photomultiplier tubes, which separately and independently acquired signals in two distinct visible spectral ranges (red and green), selected using an optical filtering system: a long-pass filter at 560 nm for the red signal and a filter transmitting from 500 nm to 530 nm for the green signal. In reflection mode, the samples were illuminated using a 532 nm laser (nominal output power of 3 mW), and the reflected signal was detected by a PMT.

4.7. Analysis with ImageJ Software of Optical Microscope Images

The ImageJ processing used for RGB analysis of the optical images is based on the histogram elaboration for the separated color channels R, G, and B providing their value distributions. An identical region of interest was selected for all the samples in order to be able to correctly evaluate the results, with the same conditions, minimizing the effects of possible systematic non-uniformity of illumination.

5. Conclusions

Ancient books, artworks, and important documents preserved in archives are subject to chemical and biological deterioration processes, which can lead to irreversible damage.

Several hydrogel cleaning treatments are being studied to slow down these processes without altering the properties of the paper, while the existing literature suggests the potential of titanium dioxide nanoparticles for protecting artworks from light-induced degradation. In this study, we synthesized CBP/TiO_2 composite hydrogel and evaluated its effectiveness as cleaning and protective agent for paper artworks, employing a non-invasive diagnostic protocol based on Raman spectroscopy. Prolonged exposure to UVA radiation was used to simulate ageing and/or damaging effects.

It has been observed that papers coated with the composite gel layer exhibit better color stability, compared to uncoated ones, when exposed to UVA radiation; a lower content of oxidizing agents was observed as well as inhibition or elimination of hydrolysis process. Moreover, the application of CBP/TiO_2 coating does not alter the RGB values of the paper making this treatment of potential interest for cultural paper artifacts.

Author Contributions: Conceptualization, S.B. and F.B.; methodology, S.B. and F.B.; validation, S.B. and F.B.; data curation, S.B., F.B., J.R., R.D. and M.G.S.; writing—original draft preparation, S.B.; writing—review and editing, S.B., F.B. and J.R. All authors have read and agreed to the published version of the manuscript.

Funding: This research received no external funding.

Institutional Review Board Statement: Not applicable.

Informed Consent Statement: Not applicable.

Data Availability Statement: All data are available from the corresponding authors upon reasonable request.

Acknowledgments: S.B. and F.B. thank the IAEA Coordinated Research Project (CRP) F22082 "Development and Implementation of Cultural Heritage Preservation using Ionizing Radiation Technology".

Conflicts of Interest: The authors declare no conflicts of interest.

References

1. Manso, M.; Carvalho, M.L. Application of spectroscopic techniques for the study of paper documents: A survey. *Spectrochim. Acta Part B At. Spectrosc.* **2009**, *64*, 482–490. [CrossRef]
2. Bitossi, G.; Giorgi, R.; Mauro, M.; Salvadori, B. Spectroscopic techniques on cultural heritage conservation: A survey. *Appl. Spectrosc. Rev.* **2005**, *40*, 187–228. [CrossRef]

3. Chiriu, D.; Ricci, P.C.; Cappellini, G.; Carbonaro, C.M. Ancient and modern paper: Study on ageing and degradation process by means of portable NIR μ-Raman spectroscopy. *Microchem. J.* **2018**, *138*, 26–34. [CrossRef]
4. Librando, V.; Minniti, Z.; Lorusso, S. Ancient and modern paper characterization by FTIR and Micro-Raman Spectroscopy. *Conserv. Sci. Cult. Herit.* **2011**, *11*, 249–268.
5. Balakhnina, I.A.; Brandt, N.N.; Chikishev, A.Y.; Rebrikova, N.L. Raman microspectroscopy of old paper samples with foxing. *Appl. Spec.* **2018**, *68*, 495–500. [CrossRef]
6. Lojewska, J.; Miskowiec, P.; Lojesewski, T.; Proniewicz, L.M. Cellulose oxidative and hydrolytic degradation: In situ FTIR approach. *Polym. Degrad. Stabil.* **2005**, *88*, 512–520. [CrossRef]
7. Botti, S.; Bonfigli, F.; Nigro, V.; Rufoloni, A.; Vannozzi, A. Evaluating the Conservation State of Naturally Aged Paper with Raman and Luminescence Spectral Mapping: Toward a Non-Destructive Diagnostic Protocol. *Molecules* **2022**, *27*, 1712. [CrossRef] [PubMed]
8. Refugio Martínez, J.; Nieto-Villena, A.; De la Cruz-Mendoza, J.A.; Ortega-Zarzosa, G.; Lobo Guerrero, A. Monitoring the natural aging degradation of paper by fluorescence. *J. Cult. Herit.* **2017**, *26*, 22–27. [CrossRef]
9. Chiriu, D.; Ricci, P.C.; Cappellini, G.; Salis, M.; Loddo, G.; Carbonaro, C.M. Ageing of ancient paper: A kinetic model of cellulose degradation from Raman spectra. *J. Raman Spectrosc.* **2018**, *49*, 1802–1811. [CrossRef]
10. Teygeler, R. Preserving paper: Recent advances. In *Managing Preservation for Libraries and Archives;* Current Practice and Future Developments; Feather, J., Ed.; Ashgate: Aldershot, UK, 2004.
11. Mazzuca, C.; Micheli, L.; Carbone, M.; Basoli, F.; Cervelli, E.; Iannuccelli, S.; Sotgiu, S.; Palleschi, A. Gellan hydrogel as a powerful tool in paper cleaning process: A detailed Study. *J. Colloid Interface Sci.* **2014**, *416*, 205–211. [CrossRef]
12. De Filipo, G.; Palermo, A.M.; Tolmino, R.; Formoso, P.; Nicoletta, F.P. Gellan gum hybrid hydrogels for the cleaning of paper artworks contaminated with Aspergillus versicolor. *Cellulose* **2016**, *23*, 3265–3279. [CrossRef]
13. Botti, S.; Bonfigli, F.; D'Amato, R.; Rodesi, J.; Santonicola, M.G. Colorimetric Sensors Based on Poly(acrylic Acid)/TiO_2 Nanocomposite Hydrogels for Monitoring UV Radiation Exposure. *Gels* **2023**, *9*, 797. [CrossRef]
14. Talarico, F.; Caldi, C.; Valanzuela, M.; Zaccheo, C.; Zampa, A.; Nugari, M.P. Applicazione dei gel come supportanti nel restauro. *Boll. ICR* **2001**, *3*, 101–118.
15. De Filpo, G.; Mormile, S.; Nicoletta, F.P.; Chidichimo, G. Fast, self-supplied, all-solid photoelectrochromic film. *J. Power Sources* **2010**, *195*, 4365–4369. [CrossRef]
16. De Filpo, G.; Palermo, A.M.; Rachiele, F.; Nicoletta, F.P. Preventing fungal growth in wood by titanium dioxide nanoparticles. *Int. Biodeterior. Biodegrad.* **2013**, *85*, 217–222. [CrossRef]
17. De Filpo, G.; Palermo, A.M.; Munno, R.; Molinaro, L.; Formoso, P.; Nicoletta, F.P. Gellan gum/titanium dioxide nanoparticle hybrid hydrogels for the cleaning and disinfection of parchment. *Int. Biodeterior. Biodegrad.* **2015**, *103*, 51–58. [CrossRef]
18. Rantala, J.T.; Karkkäinen, A.H.O. Optical properties of spin-on deposited low temperature titanium oxide thin films. *Opt. Express* **2003**, *11*, 1406–1410. [CrossRef]
19. Afsharpour, M.; Rad, F.T.; Malekian, H. New cellulosic titanium dioxide nanocomposite as a protective coating for preserving paper-art-works. *J. Cult. Herit.* **2011**, *12*, 380–383. [CrossRef]
20. D'Amato, R.; Falconieri, M.; Gagliardi, S.; Popovici, E.; Serra, E.; Terranova, G.; Borsella, E. Synthesis of ceramic nanoparticles by laser pyrolysis: From research to applications. *J. Anal. Appl. Pyrolysis* **2013**, *104*, 461–469. [CrossRef]
21. Wiley, J.H.; Atalla, R.H. Band assignments in the Raman spectra of celluloses. *Carbohydr. Res.* **1987**, *160*, 113–129. [CrossRef]
22. Botti, S.; Di Lazzaro, P.; Flora, F.; Mezi, L.; Murra, D. Raman spectral mapping reveal molecular changes in cellulose aging induced by ultraviolet and extreme ultraviolet radiation. *Cellulose* **2024**, *31*, 749–758. [CrossRef]
23. Kacík, F.; Kacíková, D.; Jablonsky, M.; Katuscak, S. Cellulose degradation in newsprint paper ageing. *Polym. Deg. Stab.* **2009**, *94*, 1509–1514. [CrossRef]
24. Malesi, J.; Kolar, J.; Stirlic, M.; Kocar, D.; Fromageot, D.; Lemaire, J. Photoinduced degradation of cellulose. *Polym. Deg. Stab.* **2005**, *89*, 64–69. [CrossRef]
25. Lubrizol. Optimizing Performance of Carbopol®ETD 2020 and Ultrez 10 Polymers with Partial Neutralization of Polymer Dispersions. Technical Data Sheet TDS-243, Ed. Available online: https://www.lubrizol.com/-/media/Lubrizol/Health/TDS/TDS243 OptimizingPerformanceCarbopolETD2020Ultrez10PartialNeutralizationPolymerDispersions.pdf (accessed on 15 October 2007).
26. Lubrizol. Viscosity of Carbopol®Polymers in Aqueous Systems. Technical Data Sheet TDS-730, Ed. Available online: https://www.lubrizol.com/-/media/Lubrizol/Health/TDS/TDS-730_Viscosity_Carbopol_in_Aqueous-Systems.pdf (accessed on 13 August 2010).
27. Gutowski, I.A.; Lee, D.; de Bruyn, J.R.; Frisken, B.J. Scaling and mesostructure of Carbopol dispersions. *Rheol. Acta* **2012**, *51*, 441–450. [CrossRef]

28. Huang, C.; Chu, P.K. Recommended practices and benchmarking of foam electrodes in water splitting. *Trends Chem.* **2022**, *4*, 1065–1077. [CrossRef]
29. Diaspro, A. Confocal and Two-Photon Microscopy. In *Foundations, Applications and Advances*; Diaspro, A., Ed.; Wiley-Liss: New York, NY, USA, 2002.

molecules

MDPI

Article

Multi-Analytical and Non-Invasive Approach for Characterising Blackened Areas of Originally Blue Paints

Maria Labate [1], Maurizio Aceto [2,*], Giacomo Chiari [3], Simone Baiocco [4], Lorenza Operti [1] and Angelo Agostino [1,*]

1 Dipartimento di Chimica, Università degli Studi di Torino, Via P.Giuria 7, 10125 Torino, Italy; maria.labate@unito.it (M.L.); lorenza.operti@unito.it (L.O.)
2 Dipartimento per lo Sviluppo Sostenibile e la Transizione Ecologica (DISSTE), Università degli Studi del Piemonte Orientale "Amedeo Avogadro", Piazza S. Eusebio 5, 13100 Vercelli, Italy
3 Getty Conservation Institute (Retired), 1200 Getty Center Drive, Los Angeles, CA 90049, USA; gc.giacomochiari@gmail.com
4 Palazzo Madama—Museo Civico d'Arte Antica, Piazza Castello, 10122 Torino, Italy; simone.baiocco@fondazionetorinomusei.it
* Correspondence: maurizio.aceto@uniupo.it (M.A.); angelo.agostino@unito.it (A.A.)

Abstract: Azurite, a natural mineral pigment consisting of basic copper carbonate ($2CuCO_3{\cdot}Cu(OH)_2$), is one of the Middle Ages' most common blue pigments. Why paintings originally coated with azurite appear blackened today remains debated. Using a non-invasive multi-analytical approach, the study analysed several unexpectedly black-appearing details (objects such as books or clothing such as veils, robes, or mantles) in Antoine de Lonhy's works. The aim was to investigate if the black colour was due to intentional iconographic reasons, incautious restoration work, or painting deterioration. The analytical results displayed the presence of the blue pigment azurite, therefore, the expected original colour of various areas should be blue. To shed light on the discussion regarding the blackening, several other Renaissance paintings with similar black details were analysed, all from the same period and geographic area as de Lonhy's works and conserved under identical conditions. The reasons why the blackening takes place are still unclear. However, the combined use of X-ray fluorescence spectrometry (XRF), UV-visible diffuse reflectance spectrophotometry with optical fibres (FORS), portable X-ray diffraction (XRD), and the elemental mapping based on the XRF data revealed that these blackened areas were originally painted with azurite, suggesting they were once blue. This finding significantly changes the overall appreciation of these artworks.

Keywords: azurite; blue paintings; non-invasive analysis; FORS; XRF; XRD; elemental mapping

1. Introduction

A multidisciplinary study of artworks attributed to Antoine de Lonhy (ca. 1446–1490) [1] was facilitated by a 2021 exhibition dedicated to the artist, held in Palazzo Madama (Museo Civico d'Arte Antica di Torino). Antoine de Lonhy was an eclectic travelling painter previously known as Master of the Trinity of Turin. Trained in the extraordinary environment of the Flemish and French artists and culture in the middle of the 15th century in Bourgogne, the artist worked then in Toulouse and Barcelona, ending his career in the Duchy of Savoy, where he left a significant legacy. Different types of his artwork, such as manuscripts, frescoes, and panel paintings, were analysed through non-invasive techniques to define the artist's colour palette. Except for the illuminated manuscripts, in which the blue plays a role of particular relevance [2], the absence of this colour is immediately noticeable in all other artworks considered. In most cases, clothes, mantles, and the Virgin's veils appear unusually coloured in black. Other coeval paintings preserved at Palazzo Madama, for which darkening occurred, were also considered.

According to the literature, a large number of black areas in mediaeval paintings were originally blue [3] and made of azurite ($2CuCO_3{\cdot}Cu(OH)_2$), the most widespread mediaeval blue pigment. This is a valuable mineral from different localities of Europe [4–8] often used together with the more expensive ultramarine blue, obtained from Lapis Lazuli. As is well known, azurite is a low covering pigment, so it needs several coats of it and is always laid onto black (*veneda*, consisting of carbon black or vine black), red-brown (*morellone*, made of earths and ochres with carbon black) layers in the frescos, and different colour backgrounds were also used in panel paintings [9]. Pictorial layers with azurite are not the only ones to show blackening since similar behaviour has also been studied for Egyptian blue [10].

The blackened appearance of azurite paintings can be attributed to endogenous and exogenous causes, or other factors, i.e., due to the presence of other materials associated with it, such as binders or varnishes [3,11–21], to dust accumulation [22], or to underlying preparatory layer becoming visible because of the porosity of the azurite layers [3]. The darkening of azurite itself may depend on its preparation process and particle size [20,23–27] or on its conversion into other minerals due to interaction with the environment. The main black alteration products are tenorite [28–30], covellite [31], or copper carbamates [9]. The alteration can be induced by exposure to hydrogen sulphide gas and moisture [31–33] by heating treatment such as fires [34–39], or by the use of alkaline pH cleaning materials [12,39,40]. Many possible concurrent issues are discussed in the literature [41]. Still, no definitive answer is given about the darkening because of the uncertainty on the pigment purity, the binder employed, or the artistic technique adopted [42,43], the presence of a preparation layer and/or a varnish, and the artwork conservation context.

The paper involves ten paintings attributed to Antoine de Lonhy and twenty works painted by coeval artists. This study aims to characterise the blackened areas of various 15th- and 16th-century paintings using a totally non-invasive approach. The techniques adopted allow very fast, cost-effective, and easy measurements to be carried out in situ. Furthermore, possible identification of pigments in the blackened painted areas may provide important information for defining the original palette of the artworks, their conservation history, and useful indications for future restoration work (e.g., for defining pictorial integrations or for adopting the most suitable products in cleaning). To characterise possible blackening due to the alkalinity of restoration products, the behaviour of azurite in alkaline environments was tested by experiments conducted on azurite mock-ups using a restoration product of common historical practice.

2. Results

In the illuminated manuscripts attributed to Antoine de Lonhy (inv. 399), the blue colour plays a predominant role and, in contrast, it seems to be almost absent from several other artworks of the same epoch exhibited at Palazzo Madama. The historical blue pigments found in the in Book of Hours' palette are azurite and ultramarine [2]. The naked-eye impression of details of most of other works such as ropes, veils, mantles, and some objects like books is perplexing since they appear almost black. The surface looks flat, without the volume rendering of drapery which can instead be appreciated in the remaining non-blue parts of the paintings (Figure 1).

Stereomicroscope image acquisition was only possible on paintings of the Apostles predella (inv. 5) because of steric and handling reasons. Grains of a blue pigment were detected in green, white, and grey areas (Figure 2). Many blue pigment particles can also be found in dark parts. Some variability in pigment size is observed as well (Figure 3).

Several points in dark areas were analysed by XRF and, in most cases, copper was detected as shown in Table 1. In the 37 areas identified in 26 paintings, the copper was detected in 32 cases and in 5 cases it was not detected or detected in a concentration lower than 1%. These are God's mantle of 470/D, Madonna's veil of 428/D, the crow of 448/D, the robe of the bottom left figure of 460/D, and Saint Monica's veil of 787/D.

(**a**) (**b**)

Figure 1. The Virgin's mantle in Death of the Virgin (**a**) and Simon's tunic (inv. 5/C) in Apostles Predella (**b**) are good examples showing the recurring dark and rather flat coats under study.

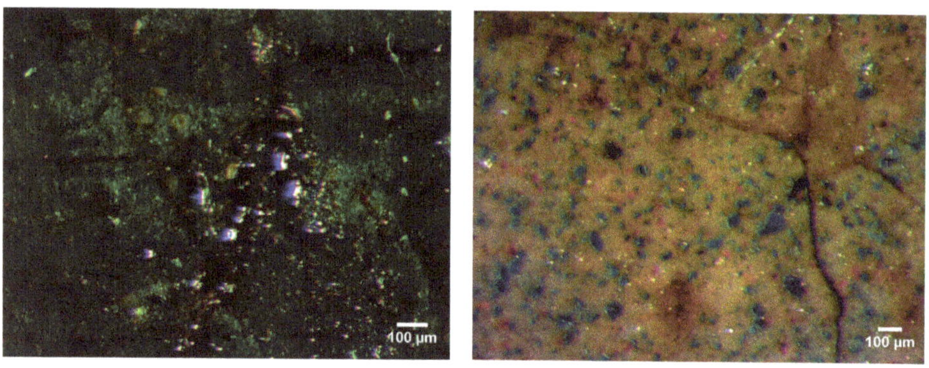

(**a**) (**b**)

Figure 2. Images acquired by stereo microscopy on painting details of apostle Matthew's robe (**a**) (inv. 5/A) and apostle Simon's mantle (**b**) (inv. 5/C). In these areas, blue pigment is present to improve the colour in the greens and whites.

XRD analysis was performed only on the Apostles' panels. The main phase recognisable in the diffraction pattern obtained on Simon's tunic (inv. 5/C) is azurite with a minor amount of covellite, carbon, calcium sulphate, and magnesium aluminium silicate mineral phases (Figure 4).

For steric reasons, the number of measurements for Duetto's XRD was not sufficient to generate maps of compounds using SmARTscan. The XRF measurements were on the contrary sufficient. We report here the maps of copper for Apostle Simon since this painting has the most extended regions of darkened blue (Figure 5a).

Table 1. FORS and XRF results of the measurement areas of paintings in which unintentional or probably deliberate darkening was observed. The 'unknown' reflectance spectrum type refers to that with the spectrum characteristics in Figure 6, with an increase in reflectance in the near-infrared (NIR) from about 850–900 nm.

Catalogue Code	Detail	FORS Results		XRF Results
		Num.	Attribution	Detection of Copper
Inv. 5/A	book	1	Carbon black	NA *
		2	Unknown	X
Inv. 5/B	Mantle	1	Unknown	X
Inv. 5/C	Tunic	1	Unknown	X
Inv. 5/D	Mantle	1	Unknown	X
470/D	God's mantle	1	Carbon black	
		2	Unknown	X
BBC	Madonna's veil	1	Unknown	X
	Mantle bottom center	2	Unknown	X
	Pillow	3	Unknown	X
418/D	Madonna's veil	1	Unknown	X
419/D	St. Anna's mantle	1	Unknown	X
	Madonna's mantle	2	Unknown	X
428/D	Madonna's veil	1	Carbon black	
431/D	Madonna's veil	1	Azurite	X
433/D	Madonna's veil	1	Unknown	X
434/D	Madonna's veil	1	Unknown	X
436/D	Jesus' tunic	1	Azurite	X
	Central fig.'s mantle	2	Unknown	X
441/D	S. Caterina sleeves	1	Azurite	X
442/D	Madonna's veil	1	Azurite, Unknown	X
448/D	Madonna's veil	1	Azurite	X
	Crow	2	Carbon black	
460/D	Bottom left fig.	1	Carbon black	
462/D	Madonna's veil	1	Azurite	X
476/D	Madonna's veil	1	Azurite	X
483/D	Madonna's veil	1	Azurite	X
	Madonna's veil (inner)	2	Unknown	X
489/D	Madonna's veil	1	Unknown	X
491/D	Madonna's veil	1	Unknown	X
512/D	Madonna's veil	1	NA *	X
516/D	Jesus' mantle	1	Azurite	X
	Maddalena's dress	2	Unknown	X
721/D	Madonna's veil	1	Azurite	X
		2	Unknown	X
774/D	Madonna's veil	1	Unknown	X
	Mantle bottom left	2	Unknown	X
787/D	S. Monica's veil	1	Carbon black	

* NA means not acquired.

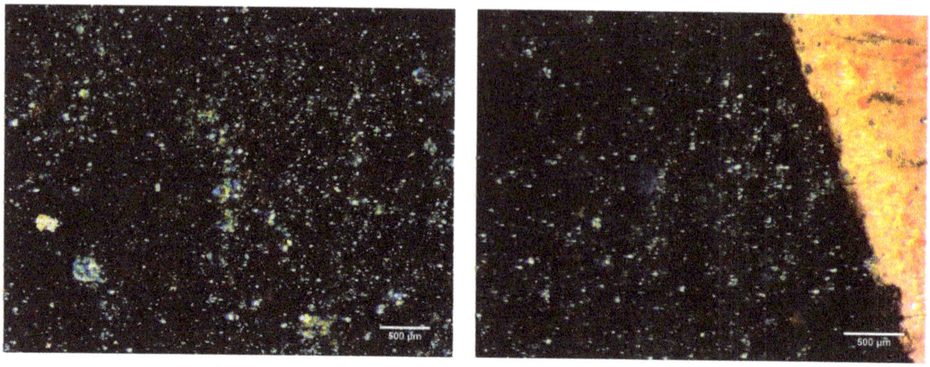

(**a**) (**b**)

Figure 3. Apostle Peter's mantle details (inv. 5/B) by stereo microscopy. The open texture of the blue coat in which pigment particles have fine (a)-to-coarse (b) size.

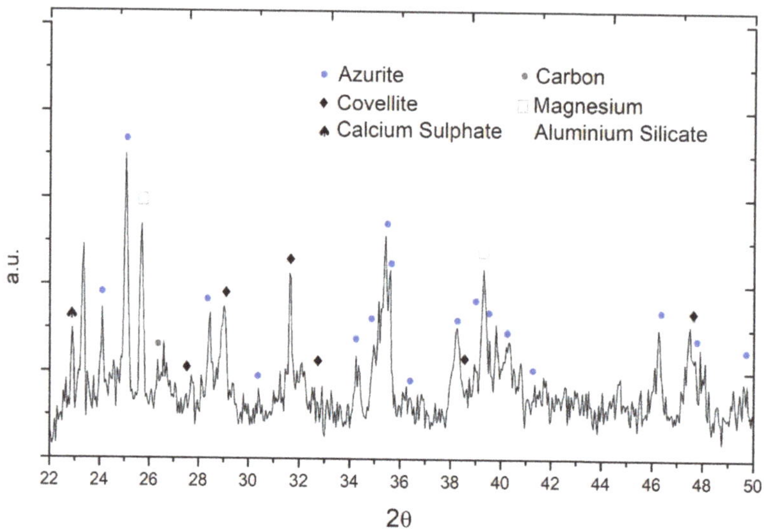

Figure 4. Diffraction pattern obtained from portable XRD analysis (Duetto) of Simon's tunic (inv. 5/C). Dominant azurite (PDF number: 00-011-0682) with covellite (PDF number: 00-006-0464), carbon (PDF number: 00-026-1076), calcium sulphate (PDF number: 00-026-0328) and magnesium aluminium silicate (PDF number: 00-027-0716).

Simon's rope is visibly darkened, almost black. The elemental map showing the distribution of copper perfectly fits the tunic, indicating that copper is equally distributed over the whole region. It may be interesting to notice that the pupils and the sword also contain minor quantities of copper, while they are grey in the visible image. The fact that copper is ubiquitous and homogeneously distributed on the dark part may be a further indication that the darkening and azurite are connected. This is confirmed by the XRD measurement that shows, in addition to the dominant azurite, the presence of covellite and carbon.

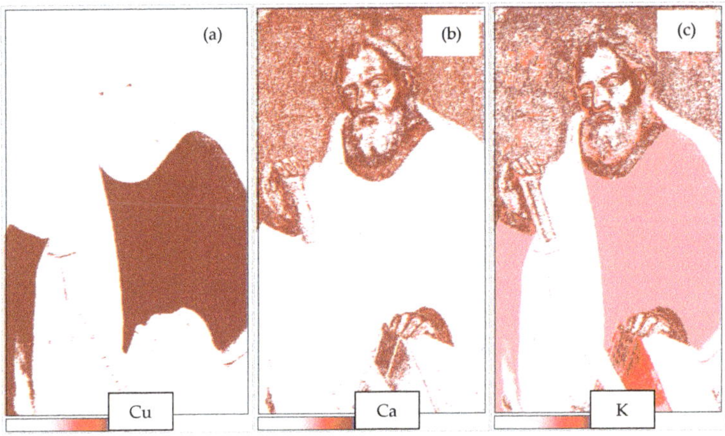

Figure 5. SmARTscan map of St. Simon (inv. 5/C) for copper (**a**), calcium (**b**), and potassium (**c**).

Calcium, shown in Figure 5b, may be in the form of calcite, as a pigment or as part of the priming. The map for potassium is also interesting (Figure 5c). The larger concentration of K is in the dark red parts, which is compatible with the hypothesis that it is part of the mordant of red lacquer. Considering the results of FORS analysis (Table 1), three types of reflectance spectra have been identified in dark areas, as shown in Figure 6:

- The first type of reflectance spectrum is characterised by low reflectance throughout the spectral range investigated. XRF's non-detection of characteristic elements of other black pigments suggests it may be carbon black.
- Another rather recurrent type of spectrum is observed, consisting of a steady absorbance throughout the visible range (Vis) with an increase in reflectance in the near-infrared (NIR) from about 850–900 nm. These kinds of spectra are defined in Table 1 as "unknown". XRF detected copper with well-defined signals in all corresponding areas while it was absent, or in irrelevant concentrations, in areas defined as carbon black.
- Despite the dark aspect, azurite is identified in some spectra by the typical reflectance minimum located at 640 nm [44]. This occurred in different measurements of Madonna's veils (431/D, 442/D, 448/D, 462/D, 476/D, 483/D, and 721/D), Jesus' tunic and mantle (436/D and 516/D) and Saint Caterina's sleeve (441/D).

However, in 442/D, 483/D, and 721/D the spectrum defined as "unknown", and the azurite spectrum, were both detected in different parts of the veils. In contrast, for two cases, both the reflectance spectrum attributable to carbon black and the "unknown" spectrum were detected (book of inv. 5A and God's mantle of 470/D).

The FORS spectra of the darkened areas were compared with the one measured on a tenorite (CuO) and covellite (CuS) samples. Furthermore, a comparison was made with reflectance spectra from azurite paintings treated with pure Contrad 2000 (Figure 7), the very alkaline restoration product described in Section 4.6.

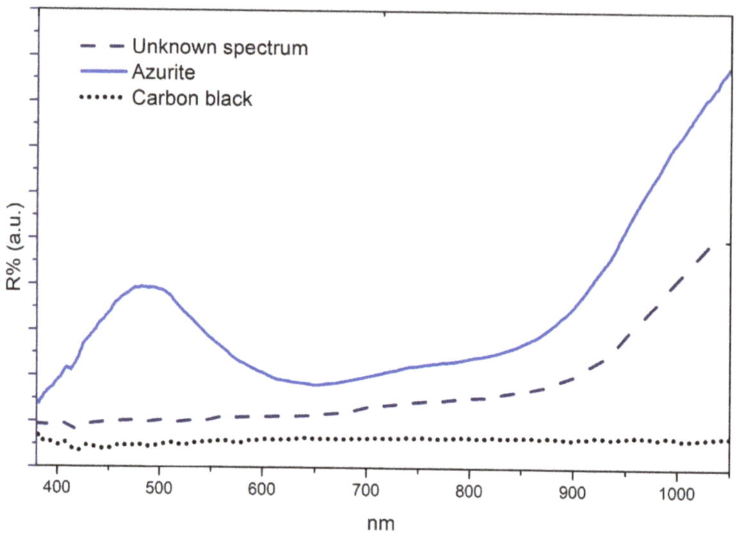

Figure 6. FORS spectra. Comparison among azurite, carbon black, and "unknown" reflectance spectra, acquired in correspondence with Simon's tunic (inv. 5/C).

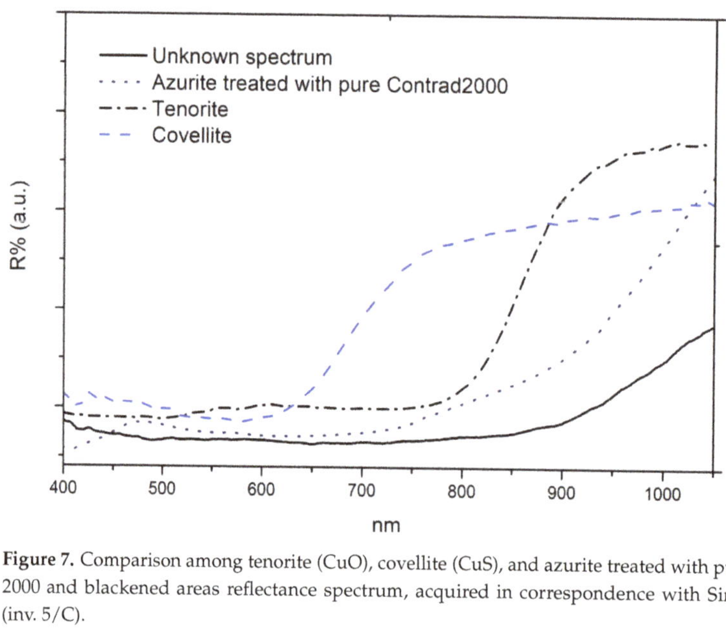

Figure 7. Comparison among tenorite (CuO), covellite (CuS), and azurite treated with pure Contrad 2000 and blackened areas reflectance spectrum, acquired in correspondence with Simon's tunic (inv. 5/C).

3. Discussion

The study of Antoine de Lonhy's colour palette applied to illuminated manuscripts (inv. 399) showed the presence of perfectly preserved azurite and ultramarine pigments, while a completely different *scenario* has been observed in the other panel paintings considered, by de Lonhy and other contemporaries.

Observing under the microscope the saint Apostles (inv. 5) a blue pigment is detected that was possibly intended for many purposes: to darken the green, to give a metallic look to some objects, such as Peter's key (inv. 5/B) or Simon's saw (inv. 5/C), and to give a colder tone to the white. The irregular shape and the heterogeneity of pigment particle size, appreciable under the microscope, points to its natural mineral origin, indicating that a restoration integration with synthetic fine material did not take place. The blue pigment particles observed in the dark backgrounds and traced back to azurite by XRD analysis do not refer to a coherent pictorial layering but rather to a residue that reveals underlying layers, suggesting that the original surface layers may have been removed.

Regarding dark areas of all the paintings considered and according to the results obtained from all the techniques adopted, in some cases, deliberate use of carbon black can be interpreted, but in other instances, the black, more likely, belonged to preparatory layers originally underlying the blue. Some objects and some robes such as the crow (448/D), the sky in a night scene (512/D) and the monastic robes (460/D and 787/D) are likely to be intentionally black. In the case of the Virgin's veil (428/D), the visible black refers to a layer made to accommodate overlying paintings that have been lost. In the cases of the Madonna's veil of the Death of the Virgin (BBC) and the God's veil of Trinity (470/D), the black portions are undoubtedly pictorial integrations due to the restoration work where, again, the blue layers originally in azurite had a conservation history that has compromised them, leaving only a residue detected with XRF and FORS analysis. In any case, the colour of the restoration intervention was certainly tuned to the colour of the surrounding original, suggesting that this was already darker than blue.

Although no longer visible as blue, in many cases, reflectance spectra ascribable to azurite were obtained, however, most measurement points revealed a spectrum not traceable to any pigment according to the reference spectra database present in the literature. The reflectance spectra were considered as "unknown" perhaps due to the incompleteness of the available databases but also because of the difficulty in characterising very dark pigments or mixtures by reflectance analysis. In the case of this study, all the reflectance spectra defined "unknown" have the same characteristics in the NIR spectral range, as defined in the results section. The optical microscope observation of blue pigments, the systematic detection of copper by XRF and, in cases where the use of XRD was possible also the identification of mineral phases, suggest that every darkened area in which "unknown" reflectance spectra were detected contained azurite. Achieving this through a totally noninvasive approach is of particular importance because it allows quick and relatively easy identification of azurite where it is not, or no longer, easily detectable. This was made possible through the use of an extended set of paintings traceable to a similar period of production but with different conservation histories.

The possibility that other copper-based pigments or alterations of azurite may have played a significant role was considered. A possible candidate is Kassel brown, a mixture of black and red copper oxides [45]. This pigment, however, has spectral characteristics not found in our observations [46].

The comparison among the reflectance spectra of our tenorite sample, tenorite sample, the azurite painting treated with pure Contrad C2000, and the blackened azurite from the original work (Figure 7) show that the spectral features of tenorite are quite different. The same can be said from a comparison with the covellite reflectance spectrum, available also in the literature [47,48]. Tenorite was compared although it is not detected with XRD because it may be present below the detection limit. Even if tenorite and covellite were present in low concentration, they do not mainly contribute to characterise the reflectance spectrum as observed in the areas investigated with both XRD and FORS (Simon's robe inv. 5/D) commented previously. On the other hand, the spectra obtained by applying Contrad 2000 on the azurite painting result in lowering the reflectance in the visible range, but not in the near-infrared range in a way comparable to the one shown in the blackened areas spectra.

The causes of the darkening remain somehow elusive, but the detection of azurite and darkened areas with azurite in the same region as it happens in 483/D (Figure 8) and 721/D, might suggest that the darkening is not always due to an altered painting but more likely to the purpose of making some specific areas darker.

Figure 8. Measurement points (red circles) on the Virgin's mantle in 483/D. FORS spectra acquired onto the light blue rope can be assigned to azurite while the dark inner part of the veil gives reflectance spectra of blackened azurite.

The darkened areas with azurite reflectance spectra could therefore be a result of the joint contribution of mixed carbon black and azurite to make some areas darker, while in some other cases the azurite layer has an open texture, due to the accidental partial loss of the pictorial layer, showing that the dark coating is an underlying painting layer laid as a background.

4. Materials and Methods

To define the materials of the darkened layers and to identify possible clues referable to any of the above-mentioned causes of the alteration process, a series of data were collected on dark areas of 30 paintings. A non-invasive and multi-analytical procedure was applied using portable instrumentation on the artworks listed in Table 2. Images of the works considered are available in Supplementary Materials. This type of approach was the only one possible since sampling was not permitted. Instrumental specifications are listed below.

FORS spectrophotometry and X-ray fluorescence spectrometry instruments have been described in detail in Agostino et al. [2]. The depth of information obtained by these analyses is different: in the first case, the surface is investigated (in a wavelength-dependent manner), and in the second case varying thicknesses could be analysed, depending on the excitation energy, the composition, and the density of the pictorial layers. According to various experiments, the thickness of hundreds of micrometres can be measured to detect information by reaching the preparatory layer of the painting. The portable XRD/XRF instrument DUETTO is well described by Chiari et al. [49]. The adopted analytical techniques do not allow for obtaining information about the binders, even though the investigated historical period includes the transition from the use of *tempera* to *tempera grassa* and oil techniques. There are no exhaustive reports of previous analyses of these artworks in the literature. Moreover, the documentation of the painting's conservation history is known and detailed in a few cases, but in most situations, previous restorations are poorly documented. For example, *Trinity* (Palazzo Madama no. 470/D) and *Death of the Virgins* (private collection) were both transferred from panel to canvas, for conservation reasons, without any description.

Table 2. List of the artworks analysed in this study with the code of the Museo Civico d'Arte Antica catalogue available online.

Title/Date	Type	Author	Catalogue Code
Sei apostoli/six apostles (Predella) 1465–1470	6 Panels (Originally panel)	Antoine de Lonhy	Inv. 5 A, B, C, D, E, F
Disputa di Gesù tra i dottori/ Christ among the doctors 1510	Panel	Defendente Ferrari	343/D
Libro d'Ore all'uso di Chalon-sur-Saône/ Book of Hours in use in Chalon-sur-Saône 1460	Illuminated manuscript	Antoine de Lonhy	inv. 399
Madonna col bambino e santi/ Madonna and Child with Saints 1435	Panel	Guglielmetto Fantini	418/D
Genealogia della Vergine/Holy Kinship 1501–1503	Panel	Gandolfino da Roreto	419/D
Pietà 1465	Mural painting	Collaborator of Antoine de Lonhy	428/D
Madonna col bambino e sant'Anna/ Madonna and Child with saint Anne 1510–1513	Panel	Giovanni Martino Spanzotti	431/D
Sposalizio della Vergine/ Marriage of the Virgin 1504	Panel	Defendente Ferrari	433/D
Sbarco di santa Maria Maddalena a Marsiglia/Arrival of saint Mary Magdalen in Marseille 1505	Panel	Defendente Ferrari	436/D
Santa Caterina/Saint Catherine end of XV century	Panel	Master of Santa Margherita of Crea Chapel	441/D
Incoronazione della Vergine e annunciazione/Coronation of the Virgin and Annunciation 1525–1530	Panel	Defendente Ferrari	442/D
Madonna in trono col bambino; San Giacomo Maggiore; San Giovanni Evangelista; San Giovanni Battista; San Tommaso d'Aquino; due donatori/Madonna enthroned with child; St. James the Greater; Saint John the Evangelist; St. John Baptis; Saint Thomas Aquinas; two donors. 1495	Panel	Macrino d'Alba	448/D

Table 2. *Cont.*

Title/Date	Type	Author	Catalogue Code
Madonna con bambino; Crocifissione; Sant'Orsola; Sant'Eulalia; San Giovanni Battista; San Giulio/Madonna with child; Crucifixion; Saint Ursula; Saint Eulalia; st. John Baptis; San Giulio 1525	Panel	Pietro Grammorseo	460/D
Presentazione di Gesù al tempio/Presentation of Jesus in the temple 1525–1530	Panel	Defendente Ferrari	462/D
Trinità/Trinity 1465–1470	canvas	Antoine de Lonhy	470/D
Madonna in trono col bambino e quattro angeli/Madonna enthroned with child and four angels 1475–1480	Panel	Giovanni Martino Spanzotti	476/D
Adorazione del bambino/Holy Child 1513	Panel	Gerolamo Giovenone	483/D
Crocifissione/Crucifixion mid XVI century	Panel	Gaudenzio Ferrari	489/D
Adorazione dei magi/Adoration of the Magi 1505	Panel	Gaudenzio Ferrari	491/D
Adorazione del bambino a lume di notte/Holy Child adoration in night light 1510	Panel	Defendente Ferrari	512/D
Noli me tangere 1515–1520	Panel	Sperindio Cagnoli	516/D
Madonna col bambino; Sant'Eusebio; Sant'Apollonia; Sant'Alberto Carmelitano; Santa Caterina; Donatori/Madonna and Child; Saint Eusebius; Saint Apollonia; Saint Albert the Carmelite; Saint Catherine; Donors 1519	Panel	Eusebio Ferrari	721/D
Stimmate di san Francesco; Madonna allattante il Bambino; la Trinità e Santi/Stigmata of Saint Francis; Madonna nursing the Child; the Trinity and Saints end of XIV century	Panel	Pietro d'Alba	774/D
S. Monica 1498	Panel	Maestro di Andriola de Barrachis	787/D
Morte della Vergine/Death of the Virgin 1470	Canvas (Originally panel)	Antoine de Lonhy	No code Balbo Bertone Collection (BBC)

4.1. Optical Microscopy (OM)

For OM a Leica MZ16 stereo microscope (Wetzlar, Germany) with 45° illumination was used, with Light source Leica Schott KL 1500 LCD. Microscope images were obtained exclusively from the Apostles panels (Palazzo Madama no. inv. 5/D) since their size permits the positioning of the areas of interest under the instrument.

4.2. X-Ray Fluorescence Spectrometry (XRF)

The elemental analyses were carried out by XRF. The instrumentation was a Thermo NITON spectrometer XL3T-900 GOLDD model (East Greenbush, NY, USA), equipped with an Ag anode, max. voltage 50 kV, max. current 100 μA and power 2 W, spot size 3 mm. The obtained spectra were processed with the commercial software BAxil v. 1.8 (Brightspec NV/SA, Belgium) derived from the academic software QXAS from IAEA.

4.3. UV-Visible-NIR Diffuse Reflectance Spectrophotometry with Optical Fibres (FORS)

FORS analysis was carried out using an Avantes AvaSpec-ULS2048XL-USB2 model spectrophotometer (Apeldoorn, The Netherlands) and AvaLight-HAL-S-IND tungsten halogen light source. The spectral range acquired is 375–1100 nm, spot size of 1.5 mm.

4.4. X-Ray Diffraction Analysis (XRD)

XRD analyses were performed using DUETTO, a portable instrumentation jointly developed by Examinart inXitu Inc. (Palo Alto, CA), and the Getty Conservation Institute (Los Angeles) designed specifically for non-invasive analysis of Cultural Heritage artefacts [44]. It consists of an X-ray source with copper target, voltage 30 kV, power 10 W and CCD detector. The analysis range is between 20° and 50° 2θ with step 0.05°, 2θ resolution from 0.15° to 0.5°. The ICDD (International Centre for Diffraction Data) diffractometric database JC-PDF2 was adopted for interpretation.

4.5. Elemental Maps Using the SmARTscan Procedure Based on XRF Measurements

The SmARTscan program [50] simulates the results of the macro-XRF scanners using a limited number of XRF measurements that are integrated with their position and colour. For each known point, six variables are available: X, Y, R, G, B, and the elemental concentration. All other unknown points miss the XRF results needed to generate the elemental map. This value is estimated from the shorter Euclidean distance from all the measured points in the 6-fold statistical space.

4.6. Mock-Up Samples for Comparison with the Originals and Conducting Cleaning Experimental Tests

For the Tenorite mock-up, copper(II) powder (Alfa Aesar, 44356) was used. The Covellite sample is a mineralogical specimen belonging to the collection of Museo Regionale di Scienze Naturali of Turin (code M/6913). Some painted samples with azurite (Azurite natural, standard, 0–120 μm, Kremer 10200) were made in egg yolk tempera (*Tempera grassa*). The coatings on the canvas or canvas board were treated with Contrad 2000 applied in varying dilutions (pure, 5% to 30%). To enhance the results, the product was tested in purity, a concentration theoretically not appropriate and not applied. For the same reasons in all tests, the product was applied but not rinsed off as would be required in its application. The making of the mock-ups was intentionally very simplified to test the interaction of azurite with a very basic material. For this reason, neither preparation nor paint was used, precisely to minimize the number of variables that play a role in any change in the pigment.

Contrad 2000 was a rather popular restoration product having an alkaline pH. Contrad consists of a complex emulsion in an aqueous base of anionic and non-ionic surfactants, stabilising agents, alkalis, detergents, and non-phosphate-based sequestering agents. Considering that pH is mentioned in the literature as one of the causes of blackening, Contrad 2000 was chosen because widely used in painting conservation in the past, to evaluate the effect on azurite of a product with an alkaline pH (pH > 13). Moreover, the paintings considered in this study may have been restored one or more times with alkaline products throughout their often poorly documented conservation history.

5. Conclusions

The present work defines a fast method for the detection of the pigment azurite even in those areas where the blue colour is no longer appreciable. Darkened areas on paintings were analysed by totally non-invasive techniques able to distinguish areas including an originally black pigment, from those presumed to be originally blue.

In some artworks, the observation of blue pigment grains by stereo microscope and the detection of copper by XRF point to the use of azurite, also confirmed by XRD, and azurite was shown to be the most abundant phase present. The FORS spectra show features comparable with very dark blue spectra and thanks to this study, they could refer to the painting layer in which azurite is present but no longer perceived as blue and as it was originally intended.

Since the works investigated have different conservation histories, various conditions may influence reflectance behaviour. The dark appearance of the painted areas could be attributable to several causes not inferable from the analyses adopted in this study: a surplus of darkened binder, the effect of the varnish, the possible presence of dark alteration products, and residual azurite with an open texture making the dark layer underneath visible or incautious cleaning. All this may be due, sometimes to conservative reasons, and at other times to deliberate artistic choices to achieve a darker rendering.

It is also possible to conclude that Antoine de Lonhy's artworks had to be much bluer than what we observe today as it happens for many other works taken into consideration. The pictorial surfaces now visible appear to observers as black and with no volume, but must originally have had pictorial layers that have now been lost.

Supplementary Materials: The following supporting information can be downloaded at: https://www.mdpi.com/article/10.3390/molecules29246043/s1, Figure S1. Sei apostoli/Six apostles, 1465–1470, Antoine de Lonhy (Museo Civico d'Arte Antica, Inv. 5); Figure S2. Sei apostoli/six apostles (panel A), 1465–1470, Antoine de Lonhy (Museo Civico d'Arte Antica, Inv. 5); Figure S3. Sei apostoli/six apostles (panel B), 1465–1470, Antoine de Lonhy (Museo Civico d'Arte Antica, Inv. 5); Figure S4. Sei apostoli/six apostles (panel C), 1465–1470, Antoine de Lonhy (Museo Civico d'Arte Antica, Inv. 5); Figure S5. Sei apostoli/six apostles (panel D), 1465–1470, Antoine de Lonhy (Museo Civico d'Arte Antica, Inv. 5); Figure S6. Sei apostoli/six apostles (panel E), 1465–1470, Antoine de Lonhy (Museo Civico d'Arte Antica, Inv. 5); Figure S7. Sei apostoli/six apostles (panel F), 1465–1470, Antoine de Lonhy (Museo Civico d'Arte Antica, Inv. 5); Figure S8. Libro d'Ore all'uso di Chalon-sur-Saône/Book of Hours in use in Chalon-sur- Saône, 1460, Antoine de Lonhy (Museo Civico d'Arte Antica, inv. 399, f.1r); Figure S9. Libro d'Ore all'uso di Chalon-sur-Saône/Book of Hours in use in Chalon-sur- Saône, 1460, Antoine de Lonhy (Museo Civico d'Arte Antica, inv. 399, f. 49v); Figure S10. Libro d'Ore all'uso di Chalon-sur-Saône/Book of Hours in use in Chalon-sur- Saône, 1460, Antoine de Lonhy (Museo Civico d'Arte Antica, inv. 399, f. 31v); Figure S11. Disputa di Gesù tra i dottori/Christ among the doctors, 1510, Defendente Ferrari (Museo Civico d'Arte Antica, 343/D); Figure S12. Madonna col bambino e santi/Madonna and Child with Saints, 1435, Guglielmetto Fantini (Museo Civico d'Arte Antica, 418/D); Figure S13. Genealogia della Vergine/Holy Kinship, 1501–1503, Gandolfino da Roreto (Museo Civico d'Arte Antica, 419/D); Figure S14. Pietà, 1465, collaborator of Antoine de Lonhy (Museo Civico d'Arte Antica, 428/D); Figure S15. Madonna col bambino e sant'Anna/Madonna and Child with saint Anne, 1510–1513, Giovanni Martino Spanzotti (Museo Civico d'Arte Antica, 431/D); Figure S16. Sposalizio della Vergine/Marriage of the Virgin, 1504, Defendente Ferrari (Museo Civico d'Arte Antica, 433/D); Figure S17. Sbarco di santa Maria Maddalena a Marsiglia/Arrival of saint Mary Magdalen in Marseille, 1505, Defendente Ferrari (Museo Civico d'Arte Antica, 436/D); Figure S18. Santa Caterina/Saint Catherine, end of XV century, Master of Santa Margherita of Crea Chapel (Museo Civico d'Arte Antica, 441/D); Figure S19. Incoronazione della Vergine e annunciazione/Coronation of the Virgin and Annunciation, 1525–1530, Defendente Ferrari (Museo Civico d'Arte Antica, 442/D); Figure S20. Madonna in trono col bambino; San Giacomo Maggiore; San Giovanni Evangelista; San Giovanni Battista; San Tommaso d'Aquino;

due donatori/Madonna enthroned with child; St. James the Greater; Saint John the Evangelist; St. John Baptis; Saint Thomas Aquinas; two donors., 1495, Macrino d'Alba (Museo Civico d'Arte Antica, 448/D); Figure S21. Madonna con bambino; Crocifissione; Sant'Orsola; Sant'Eulalia; San Giovanni Battista; San Giulio/Madonna with child; Crucifixion; Saint Ursula; Saint Eulalia; st. John Baptis; San Giulio, 1525, Pietro Grammorseo (Museo Civico d'Arte Antica, 460/D); Figure S22. Presentazione di Gesù al tempio/Presentation of Jesus in the temple, 1525–1530, Defendente Ferrari (Museo Civico d'Arte Antica, 462/D); Figure S23. Trinità/Trinity, 1465–1470, Antoine de Lonhy (Museo Civico d'Arte Antica, 470/D); Figure S24. Madonna in trono col bambino e quattro angeli/Madonna enthroned with child and four angels, 1475–1480, Giovanni Martino Spanzotti (Museo Civico d'Arte Antica, 476/D); Figure S25. Adorazione del bambino/Holy Child, 1513, Gerolamo Giovenone (Museo Civico d'Arte Antica, 483/D); Figure S26. Crocifissione/Crucifixion, mid XVI century, Gaudenzio Ferrari (Museo Civico d'Arte Antica, 489/D); Figure S27. Adorazione dei magi/Adoration of the Magi, 1505, Gaudenzio Ferrari (Museo Civico d'Arte Antica, 491/D); Figure S28. Adorazione del bambino a lume di notte/Holy Child adoration in night light, 1510, Defendente Ferrari (Museo Civico d'Arte Antica, 512/D); Figure S29. Noli me tangere, 1515–1520, Sperindio Cagnoli (Museo Civico d'Arte Antica, 516/D); Figure S30. Madonna col bambino; Sant'Eusebio; Sant'Apollonia; Sant'Alberto Carmelitano; Santa Caterina; Donatori/Madonna and Child; Saint Eusebius; Saint Apollonia; Saint Albert the Carmelite; Saint Catherine; Donors, 1519, Eusebio Ferrari (Museo Civico d'Arte Antica, 721/D); Figure S31. Stimmate di san Francesco; Madonna allattante il Bambino; la Trinità e Santi/Stigmata of Saint Francis; Madonna nursing the Child; the Trinity and Saints, end of XIV century, Pietro d'Alba (Museo Civico d'Arte Antica, 774/D); Figure S32. S. Monica, 1498, Maestro di Andriola de Barrachis (Museo Civico d'Arte Antica, 787/D); Figure S33. Morte della Vergine/Death of the Virgin, 1470, Antoine de Lonhy (Balbo Bertone Collection).

Author Contributions: Conceptualization, M.L., M.A. and A.A.; data curation, M.L., M.A. and A.A.; formal analysis, M.L., M.A., G.C. and A.A.; investigation, M.L., M.A., G.C. and A.A.; methodology, M.L., M.A. and A.A.; supervision, M.A. and A.A.; validation, M.A., L.O. and A.A.; writing—original draft, M.L.; writing—review and editing, M.A., G.C., S.B., L.O. and A.A. All authors have read and agreed to the published version of the manuscript.

Funding: This research received no external funding.

Institutional Review Board Statement: Not applicable.

Informed Consent Statement: Not applicable.

Data Availability Statement: The data presented in this study are available on request from the corresponding authors.

Acknowledgments: The authors would like to thank Simonetta Castronovo and Museo Civico di Arte Antica of Turin for the availability to analyse the artworks. Special thanks also to Erica Bittarello and Massimiliano Senesi of the Museo Regionale di Scienze Naturali of Turin for making some mineralogic samples available for the study. Sincere gratitude to Vincenzo Basiglio of the Gabbantichità company for his contribution to the cleaning experimental test. Special thanks to Amalia Nani for her valuable suggestions.

Conflicts of Interest: The authors declare no conflicts of interest.

References

1. Aceto, M.; Angelo, A.; Maria, L. I Materiali Pittorici di Antoine de Lonhy: Informazioni della diagnostica. In *Il Rinascimento europeo di Antoine de Lonhy*; Baiocco, S., Natali, V., Eds.; Sagep Editori Srl: Genova, Italy, 2021; pp. 230–235.
2. Agostino, A.; Pellizzi, E.; Aceto, M.; Castronovo, S.; Saroni, G.; Gulmini, M. On the Hierarchical Use of Colourants in a 15th Century Book of Hours. *Heritage* **2021**, *4*, 1786–1806. [CrossRef]
3. Thompson, D.V. *The Materials and Techniques of Medieval Painting*; Dover Publications: New York, NY, USA, 1956.
4. Gettens, R.J.; Fitzhugh, E.W. Azurite and Blue Verditer. *Stud. Conserv.* **1966**, *11*, 54–61. [CrossRef]
5. Scott, D.A. *Copper and Bronze in Art: Corrosion, Colorants, Conservation*; Getty Trust Publications: Los Angeles, CA, USA, 2002.
6. Dik, J. *Scientific Analysis of Historical Paint and the Implications for Art History and Art Conservation. The Case Studies of Naples Yellow and Discoloured Smalt.*; University of Amsterdam: Amsterdam, The Netherlands, 2003.
7. Aru, M.; Burgio, L.; Rumsey, M.S. Mineral Impurities in Azurite Pigments: Artistic or Natural Selection? *J. Raman Spectrosc.* **2014**, *45*, 1013–1018. [CrossRef]

8. Capriotti, S.; Medeghini, L.; Mignardi, S.; Petrelli, M.; Botticelli, M. The Blue Road: Provenance Study of Azurite Samples from Historical Locations through the Analysis of Minor and Trace Elements. *Heliyon* **2023**, *9*, e19099. [CrossRef]
9. Lalli, C.G.; Innocenti, F. Appunti Sulle Caratteristiche Chimico-Fisiche Dell'Azzurrite e Del Lapislazzuli. *OPD Restauro* **2014**, *26*, 78–82.
10. Daniels, V.; Stacey, R.; Middleton, A. The Blackening of Paint Containing Egyptian Blue. *Stud. Conserv.* **2004**, *49*, 217–230. [CrossRef]
11. Ruhemann, H. Criteria for Distinguishing Additions from Original Paint. *Stud. Conserv.* **1958**, *3*, 145–161. [CrossRef]
12. Hoenigher, C. The Identification of Blue Pigments in Early Sienese Paintings by Color Infrared Photography. *J. Am. Inst. Fo Conserv.* **1991**, *30*, 115–124. [CrossRef]
13. van Eikema Hommes, M.H. *Discoloration in Renaissance and Baroque Oil Paintings. Instructions for Painters, Theoretical Concepts, and Scientific Data*; University of Amsterdam: Amsterdam, The Netherlands, 2002.
14. d'Acapito, F.; Maurizio, C.; Brunetti, B.; Cartechini, L.; Sgamellotti, A.; Quartieri, S.; Arletti, R. Archaeometric Studies at the GILDA Beamline at the European Synchrotron Radiation Facility. *Il Nuovo Cimento C* **2007**, *30*, 1–9. [CrossRef]
15. Lluveras, A.; Boularand, S.; Andreotti, A.; Vendrell-Saz, M. Degradation of Azurite in Mural Paintings: Distribution of Copper Carbonate, Chlorides and Oxalates by SRFTIR. *Appl. Phys. A* **2010**, *99*, 363–375. [CrossRef]
16. Gil, M.; Carvalho, M.L.; Longelin, S.; Ribeiro, I.; Valadas, S.; Mirão, J.; Candeias, A.E. Blue Pigment Colors from Wall Painting Churches in Danger (Portugal 15th to 18th Century): Identification, Diagnosis, and Color Evaluation. *Appl. Spectrosc.* **2011**, *65*, 782–789. [CrossRef]
17. Duce, C.; Bramanti, E.; Ghezzi, L.; Bernazzani, L.; Bonaduce, I.; Colombini, M.P.; Spepi, A.; Biagi, S.; Tine, M.R. Interactions between Inorganic Pigments and Proteinaceous Binders in Reference Paint Reconstructions. *Dalton Trans.* **2013**, *42*, 5975–5984. [CrossRef] [PubMed]
18. Stoner, J.H.; Rushfield, R.A. *Conservation of Easel Paintings*, 2nd ed.; Routledge: New York, NY, USA, 2021.
19. Švarcová, S.; Hradil, D.; Hradilová, J.; Čermáková, Z. Pigments—Copper-Based Greens and Blues. *Archaeol. Anthropol. Sci.* **2021**, *13*, 190. [CrossRef]
20. Cavaleri, T.; Pelosi, C.; Giustetto, R.; Andreotti, A.; Bonaduce, I.; Calabrò, G.; Caliri, C.; Colantonio, C.; Manchinu, P.; Legnaioli, S.; et al. The Northern-Italy Renaissance in a Panel by Defendente Ferrari: A Complete Study with a Multi-Analytical Investigation. *J. Archaeol. Sci. Rep.* **2022**, *46*, 103669. [CrossRef]
21. Pottasch, C.; van Loon, A.; Delaney, J.K.; Dooley, K.A. The Materials and Techniques of The Lamentation of Christ (ca. 1460, Mauritshuis), Attributed to Rogier van Der Weyden and Studio: Combining MA-XRF, Reflectance Imaging Spectroscopy and Paint Cross-Section Analysis. *X-Ray Spectrom.* **2023**, *52*, 463–485. [CrossRef]
22. de Viguerie, L.; Pladevall, N.O.; Lotz, H.; Freni, V.; Fauquet, N.; Mestre, M.; Walter, P.; Verdaguer, M. Mapping Pigments and Binders in 15th Century Gothic Works of Art Using a Combination of Visible and near Infrared Hyperspectral Imaging. *Microchem. J.* **2020**, *155*, 104674. [CrossRef]
23. C., A.H. Ancient Pigments. *Nature* **1914**, *92*, 646. [CrossRef]
24. Price, M. A Renaissance of Color: Particle Separation and Preparation of Azurite for Use in Oil Painting. *Leonardo* **2000**, *33*, 281–288. [CrossRef]
25. Rosado, T.; Gil, M.; Caldeira, A.T.; Martins, M.d.R.; Dias, C.B.; Carvalho, L.; Mirão, J.; Candeias, A.E. Material Characterization and Biodegradation Assessment of Mural Paintings: Renaissance Frescoes from Santo Aleixo Church, Southern Portugal. *Int. J. Archit. Herit.* **2014**, *8*, 835–852. [CrossRef]
26. Cardell, C.; Herrera, A.; Guerra, I.; Navas, N.; Rodríguez Simón, L.; Elert, K. Pigment-Size Effect on the Physico-Chemical Behavior of Azurite-Tempera Dosimeters upon Natural and Accelerated Photo Aging. *Dyes Pigments* **2017**, *141*, 53–65. [CrossRef]
27. Brocchieri, J.; de Viguerie, L.; Sabbarese, C.; Boyer, M. Combination of Noninvasive Imaging Techniques to Characterize Pigments in Buddhist Thangka Paintings. *X-Ray Spectrom.* **2021**, *50*, 320–331. [CrossRef]
28. Gutscher, D.; Mühlethaler, B.; Portmann, A.; Reller, A. Conversion of Azurite into Tenorite. *Stud. Conserv.* **1989**, *34*, 117–122. [CrossRef]
29. Mattei, E.; de Vivo, G.; De Santis, A.; Gaetani, C.; Pelosi, C.; Santamaria, U. Raman Spectroscopic Analysis of Azurite Blackening. *J. Raman Spectrosc.* **2008**, *39*, 302–306. [CrossRef]
30. Kotulanová, E.; Švarcová, S.; Hradil, D.; Bezdička, P.; Schweigstillová, J.; Grygar, T. Wall Painting Damage by Salts: Causes and Mechanisms | The Institute of Rock Structure and Mechanics of the Czech Academy of Sciences. *Acta Res. Rep.* **2009**, *18*, 27–31.
31. Coccato, A.; Moens, L.; Vandenabeele, P. On the Stability of Mediaeval Inorganic Pigments: A Literature Review of the Effect of Climate, Material Selection, Biological Activity, Analysis and Conservation Treatments. *Herit. Sci.* **2017**, *5*, 12. [CrossRef]
32. (Odlyha, M.; Cohen, N.S.; Foster, G.M.; West, R.H. Dosimetry of Paintings: Determination of the Degree of Chemical Change in Museum Exposed Test Paintings (Azurite Tempera) by Thermal and Spectroscopic Analysis. *Thermochim. Acta* **2000**, *365*, 53–63. [CrossRef]
33. Smith, G.D.; Clark, R.J.H. The Role of H2S in Pigment Blackening. *J. Cult. Herit.* **2002**, *3*, 101–105. [CrossRef]
34. Simpson, D.R.; Fisher, R.; Libsch, K. Thermal Stability of Azurite and Malachite. *Am. Mineral.* **1964**, *49*, 1111–1114.
35. Rickerby, S. Heat Alterations to Pigments Painted in the Fresco Technique. *Conservator* **1991**, *15*, 39–44. [CrossRef]
36. Frost, R.; Ding, Z.; Kloprogge, J.; Martens, W. Thermal Stability of Azurite and Malachite in Relation to the Formation of Mediaeval Glass and Glazes. *Thermochim. Acta* **2002**, *390*, 133–144. [CrossRef]

37. Sotiropoulou, S.; Daniilia, S.; Miliani, C.; Rosi, F.; Cartechini, L.; Papanikola-Bakirtzis, D. Microanalytical Investigation of Degradation Issues in Byzantine Wall Paintings. *Appl. Phys. A* **2008**, *92*, 143–150. [CrossRef]
38. Doménech-Carbó, M.T.; Edwards, H.G.M.; Doménech-Carbó, A.; del Hoyo-Meléndez, J.M.; de la Cruz-Cañizares, J. An Authentication Case Study: Antonio Palomino versus Vicente Guillo Paintings in the Vaulted Ceiling of the Sant Joan Del Mercat Church (Valencia, Spain). *J. Raman Spectrosc.* **2012**, *43*, 1250–1259. [CrossRef]
39. Perusini, G. *Il restauro dei dipinti e delle sculture lignee. Storia, teorie e tecniche*; Del Bianco Editore: Udine, Italy, 2004.
40. de Vivo, G.S. Studio Sull'alterazione Dell'azzurrite in Tenorite Indotta da Sostanze Tradizionalmente Impiegate per la Pulitura di Dipinti Murali. Master's Thesis, Istituto Centrale per il Restauro (ICR), Roma, Italy, 2006.
41. Saunders, D.; Kirby, J. The Effect of Relative Humidity on Artists' Pigments. *Natl. Gallery Tech. Bull.* **2004**, *25*, 62–72.
42. Anselmi, E. Corso Sulla Manutenzione di Dipinti Murali, Mosaici, Stucchi. Part I—Module 3; ICR: Rome, Italy, 1978; pp. 25–40.
43. Punis, R.; Zoleo, A. Cu(II)-Binder Complexes in Azurite and Malachite Pictorial Mixtures: An EPR Study. *Microchem. J.* **2024**, *200*, 110303. [CrossRef]
44. Aceto, M.; Agostino, A.; Fenoglio, G.; Idone, A.; Gulmini, M.; Picollo, M.; Ricciardi, P.; Delaney, J.K. Characterisation of Colourants on Illuminated Manuscripts by Portable Fibre Optic UV-Visible-NIR Reflectance Spectrophotometry. *Anal. Methods* **2014**, *6*, 1488–1500. [CrossRef]
45. Krekel, C.; Haller, U.; Burmester, A. Artists' Pigments Reconsidered: Does Modern Science Match the Historical Context? *Stud. Conserv.* **2006**, *51*, 244–248. [CrossRef]
46. The IFAC-CNR Spectroscopic Measurement Database. 2024. Available online: https://spectradb.ifac.cnr.it/index.php (accessed on 10 October 2024).
47. Fang, J.; Xuan, Y. Investigation of Optical Absorption and Photothermal Conversion Characteristics of Binary CuO/ZnO Nanofluids. *RSC Adv.* **2017**, *7*, 56023–56033. [CrossRef]
48. Fuentes, R.; Luarte, D.; Sandoval, C.; Myakalwar, A.K.; Yáñez, J.; Sbarbaro, D. Data Fusion of Laser Induced Breakdown Spectroscopy and Diffuse Reflectance for Improved Analysis of Mineral Species in Copper Concentrates. *Miner. Eng.* **2021**, *173*, 107193. [CrossRef]
49. Chiari, G.; Sarrazin, P.; Heginbotham, A. Non-Conventional Applications of a Noninvasive Portable X-Ray Diffraction/Fluorescence Instrument. *Appl. Phys. A* **2016**, *122*, 990. [CrossRef]
50. Martin-Ramos, J.D.; Chiari, G. SmART_scan: A Method to Produce Composition Maps Using Any Elemental, Molecular and Image Data. *J. Cult. Herit.* **2019**, *39*, 260–269. [CrossRef]

molecules

MDPI

Article

First Insights into the Formation of Metal Soaps in Alkyd-Based Paints: A Proof-of-Concept Investigation Using FTIR Spectroscopy

Tommaso Poli [1,*], Michael P. Haaf [2], Anna Piccirillo [3,*], Abby P. Costa [2], Rebecca L. Craig [2] and Federica Pozzi [3]

1 Dipartimento di Chimica, Università di Torino, Via Giuria 7, 10125 Turin, Italy
2 Department of Chemistry, Ithaca College, 953 Danby Road, Ithaca, NY 14850, USA; mhaaf@ithaca.edu (M.P.H.); acosta2@ithaca.edu (A.P.C.); rcraig2@ithaca.edu (R.L.C.)
3 Centro per la Conservazione ed il Restauro dei Beni Culturali "La Venaria Reale", Via XX Settembre 18, 10078 Venaria Reale, Turin, Italy; federica.pozzi@ccrvenaria.it
* Correspondence: tommaso.poli@unito.it (T.P.); anna.piccirillo@ccrvenaria.it (A.P.); Tel.: +39-011-4993017-x217 (T.P.); +39-011-4993024-x224 (A.P.)

Abstract: The formation of metal soaps or carboxylates in oil paintings is a widely studied topic. Expanding upon the extant research on the subject, this work investigates the formation of metal soaps as pertaining to alkyd-based media. Especially popular in modern and contemporary art, alkyd paints are complex commercial formulations mainly containing a polyester backbone obtained by the condensation of glycerol and phthalic acids, where the third free alcoholic function is esterified with a blend of saturated and unsaturated fatty acids. The formulation may also contain cross-linking agents, dries, and catalysts. Compared to drying oils, alkyd systems have different stability and mobility, likely prompting different kinetics in the formation of metal soaps. This work explores the formation of metal carboxylates in mock-up paints prepared as mixtures of an alkyd binder with potassium hydroxide and three commonly used pigments (smalt, zinc white, and lead white) at ambient conditions and upon artificial aging. As a second step, samples from two contemporary works by Italian artists Franco Garelli and Luciano Minguzzi were investigated for comparison. The results confirm the formation of metal soaps in both mock-up and real paints, paving the way for future research, with significant implications especially for the conservation of modern and contemporary artworks.

Keywords: alkyd paints; metal soaps; cultural heritage

1. Introduction

The formation of metal soaps, or carboxylates, in oil paintings is a well-known issue that, in the past few decades, has been studied thoroughly by numerous research groups worldwide [1,2]. In modern paints, metal soaps such as zinc or aluminum stearates might be added deliberately to commercial formulations as plasticizers, favoring the dispersion of pigments within the organic medium [3,4]. However, these compounds can also form from the reaction of free fatty acids with cations from pigments or additives as a result of paint aging and degradation [5]. Indeed, oil paint films typically undergo oxidative processes that promote triglyceride cross-linking on the double bonds of unsaturated fatty acids, enabling oil drying [6–8]. Other possible degradation phenomena include, for instance, the hydrolysis of the molecules' ester bond [9]. The presence of humidity mostly in the form of water, impurities, and pollutants within the paint layers might favor the release of acidic functions both from the free saturated fatty acids and their cross-linked counterparts. These free acidic functions are thus available to react with pigment cations, yielding the formation of free carboxylates, or metal soaps, along with more complex ionomeric structures [2,10,11]. The presence of metal soaps in a paint film does not constitute an issue per se, as demonstrated by the fact that most commercial oil paint formulations contain zinc or aluminum stearates as dispersive agents to prevent the separation of the pigment/medium system during the paint's shelf life. Nevertheless, it has been assessed that the newly formed carboxylate products can migrate, aggregate, and form protrusions within the paint layers, thus compromising their physical and mechanical properties while also altering the surface's aesthetical appearance [1,12–14]. The most common pigments involved in these types of reactions are zinc white, lead white, smalt, and copper-based pigments (the latter are involved to a more limited extent due to their lower reactivity) [1,2,15]. The reactivity of metal soaps typically leads to even more complex interactions at the interface between the paint layers and the uppermost varnish or finishing coating, involving ion exchange processes with terpenic acids from natural resins such as colophony, sandarac, dammar, and copal [16–18].

Building on the extant research on oil paints, this work investigates the formation of metal soaps as pertaining to alkyd-based media. Widely used by world-renowned modern and contemporary artists such as Picasso, Rauschenberg, and Pollock [19–22], alkyd paints consist of complex commercial formulations mainly containing a polyester backbone obtained by the condensation of 1,2,3-propanetriol (glycerol) and phthalic acids (typically orthophthalic acid), where the third free alcoholic function is usually esterified with a blend of C16-18 saturated and unsaturated fatty acids depending on the exact formulation (Figure 1) [23–28].

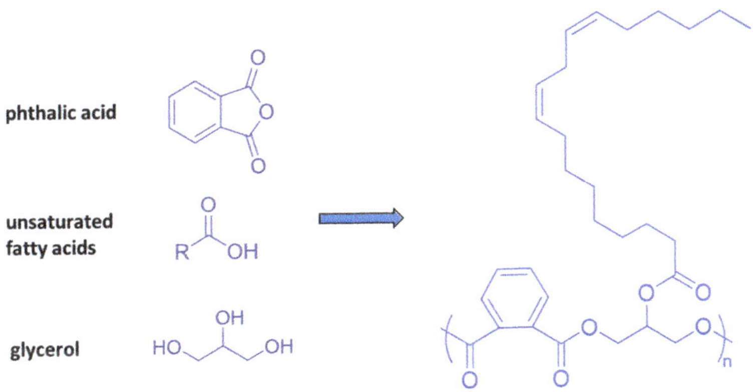

Figure 1. Main components and structure of a generic alkyd resin, where R is typically an unsaturated C16-18 fatty acid.

Commercial formulations may significantly vary and contain different additives (e.g., nitrocellulose), modifiers (e.g., styrene, divinylbenzene), cross-linking agents (e.g., pentaerythritol, sorbitol, maleic acid), and driers or catalysts (often cobalt carboxylates) [23–25]. Alkyd-based paints can also be modified with urea/formaldehyde, phenol/formaldehyde, and polyurethane resins to obtain enamels or metal coatings [26,27]. The fatty acid content typically ranges from <40% (short oil) to 70% *w*/*w* (long oil) [27,28]: as a result, a generic alkyd paint formulation may contain on average from 30 to 70% *w*/*w* of fatty acids condensed on the glycerol in mid-position. Based on these features, it is reasonable to hypothesize that alkyd-based paints may undergo similar hydrolysis processes as those described for drying oil, leading to the formation of metal soaps. Compared to drying oil, however, alkyd systems are characterized by different stability and mobility, likely prompting different kinetics in the formation of metal soaps. While the formation of metal soaps in oil paintings and its implications for the conservation of artworks has been the subject of extensive research, only one article to date has focused on alkyd-based paints to the best of our knowledge [29]. The author of that study reported the loss of gloss in zinc white-containing alkyd gloss paints, which was attributed to the presence of protuberances resulting from the formation of zinc soaps, with the hazing effect being worst in the case of alkyd resins with high saturated and oleic acid ester content.

In this context, the present work aims to fill the knowledge gaps on the formation of metal carboxylates in paints that contain alkyd media under controlled aging conditions. In the first phase of this project, a series of mock-up paints were prepared in the laboratory by mixing three commonly used, oil-reactive pigments—namely smalt (a cobalt-colored potassium glass), zinc white (zinc oxide, ZnO), and lead white (basic lead carbonate, $Pb(CO)_3{\cdot}Pb(OH)_2$)—with an alkyd binder. The selection of materials was based on the available data on pigments that had previously shown high reactivity with oil media, as well as on the frequent occurrence of their cations in modern commercial formulations and substrates: for instance, lead oxides and zinc stearate are typically found as anticorrosive and dispersive agents [3–5,30], while zinc is used for galvanized steel. Additionally, zinc oxide is often present in titanium white formulations [31]. Potassium hydroxide (KOH) was also included in the experimentation as a reference since it is a strong base, able to promote the hydrolysis of the polyester backbone fatty acids and is prone to reacting with the newly formed acidic functions. Moreover, potassium is also relevant to the reaction with smalt, as it is the cation involved in the formation of carboxylates. The possibility of metal soap formation in these mock-up paints at ambient conditions and upon artificial aging was assessed by means of transmission FTIR spectroscopy, by detecting the appearance and growth of the -COO-Metal carbonyl band within the same spectral region relevant to drying oils (approximately 1520–1590 cm^{-1}) [1,2,4]. The systematic study of mock-up paints was followed by the investigation of samples from two contemporary artworks, namely *Costruzione* (1970) by Franco Garelli and *Uomini* (1971) by Luciano Minguzzi, which both underwent scientific analysis and conservation treatment at the Centro per la Conservazione ed il Restauro dei Beni Culturali (CCR) "La Venaria Reale", just outside of Turin, Italy. The results of this proof-of-concept work confirmed that the formation of metal soaps occurred both in mock-up and real paints in all cases examined.

2. Results

As a preliminary step, FTIR analysis was conducted on the bare alkyd binder reference to confirm the absence of any possible carboxylates (e.g., those containing cobalt, aluminum, and zinc cations), which are often added to the formulation either as driers, catalysts, or dispersive agents [3,5,32]. The results showed that the selected reagent did not contain any unwanted or interfering substances. Sections 2.1 and 2.2, below, illustrate the main results obtained from a systematic study of the alkyd-containing mock-up paints and from the comparative analysis of two works by Italian artists Garelli and Minguzzi.

2.1. *Alkyd-Containing Mock-Up Paints*

The next four paragraphs discuss data obtained from the FTIR investigation of alkyd-containing potassium hydroxide, smalt, zinc white, and lead white mock-up paints. The selection of spectra shown in the following sections, demonstrating the formation of metal soaps, was collected from mock-up paints prepared with a 1:1 *w*/*w* binder/pigment ratio. Carboxylate formation was also verified for paints with a 2:1 *w*/*w* binder/pigment ratio.

2.1.1. Potassium Hydroxide Mock-Up

Analysis of the potassium hydroxide mock-up highlighted the formation of potassium carboxylates, identified by a strong band at 1560 cm^{-1} (Figure 2), which is consistent with data from potassium-containing drying oil paints.

The reaction takes place instantaneously, as the above-mentioned distinctive band was detected immediately after paint mixing. The obtained spectrum also shows a decrease in the C=O ester signal at 1736 cm^{-1}, along with a significant decrease in the alkyd backbone bands in the 1000–1300 cm^{-1} region. This observation likely indicates that the potassium hydroxide started to promote the hydrolysis of the polyester chain, resulting in the formation of an OH function, characterized by a broad band centered at 3300 cm^{-1} and other oligomeric oxidated structures.

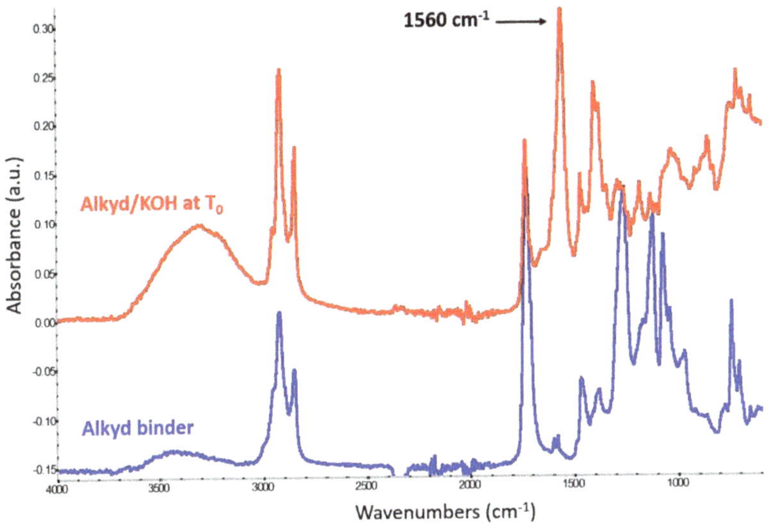

Figure 2. A reference spectrum of the bare alkyd binder (blue trace) compared with the spectrum obtained from the alkyd-containing potassium hydroxide mock-up paint at T_0 (red trace). The latter shows the formation of potassium carboxylates, identified by a strong band at 1560 cm^{-1}.

2.1.2. Smalt Mock-Up Paint

As expected, the reaction of the alkyd medium with smalt is slower compared to potassium hydroxide, yet the appearance of a band at 1560 cm^{-1} (Figure 3), related to the formation of potassium carboxylates as in the previous case, is clearly noticeable after 40 cycles of artificial aging.

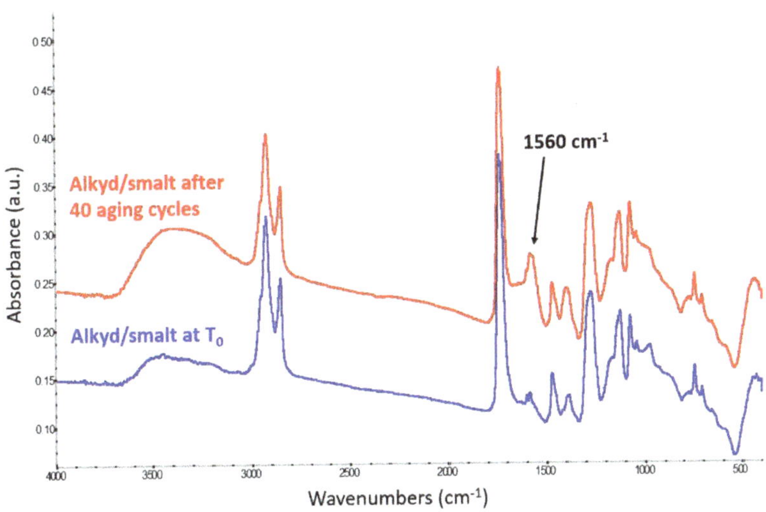

Figure 3. Spectrum of the alkyd-containing smalt mock-up paint at T_0 (blue trace) compared with the spectrum obtained from the same mixture after 40 aging cycles (red trace). The latter shows the formation of potassium carboxylates, identified by a strong band at 1560 cm^{-1}.

This suggests that the alkyd medium might interact with smalt in a similar way as drying oil. Potassium is used as a finer and melter in the preparation of smalt. In this case, there are no signs of other ongoing degradation patterns, such as a decrease in the carbonyl band or the appearance of new infrared signals. It is worth pointing out that, nine months after mock-up paint preparation, the band assigned to metal soap formation was also detected in the mock-up paints that were not subjected to artificial aging but rather were kept in the laboratory at ambient conditions.

2.1.3. Zinc White Mock-Up Paint

A carboxylate-related band, centered at 1575 cm^{-1}, was detected in the FTIR spectra obtained from the zinc white mock-up paint, too (Figure 4). In particular, the signal became clearly detectable after 40 aging cycles, around the same wavenumbers as those of zinc soaps from drying oil paints.

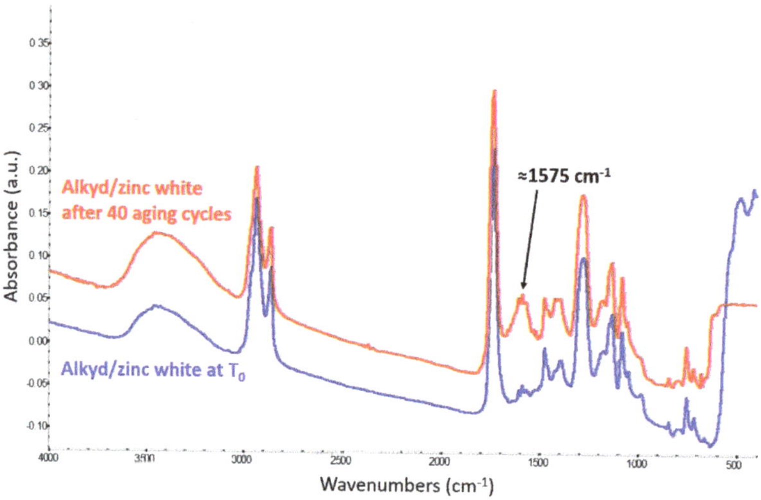

Figure 4. Spectrum of the alkyd-containing zinc white mock-up paint at T_0 (blue trace) compared with the spectrum obtained from the same mixture after 40 aging cycles (red trace). The latter shows the formation of zinc carboxylates, identified by a strong band at ≈1575 cm^{-1}.

This band is broader than the one originating from potassium hydroxide likely due to the simultaneous presence of mono- and di-carboxylated zinc salts in the sample analyzed and to an overall greater structural inhomogeneity. As in the case of smalt, nine months after mock-up paint preparation, an FTIR band related to metal soap formation was also noticed in the mock-up paints kept at ambient conditions, while the spectra show no evidence of the degradation of the alkyd structure.

2.1.4. Lead White Mock-Up Paint

In the alkyd/lead white mock-up paint, the formation of lead carboxylates was barely noticeable even after 40 aging cycles (Figure 5), though the band's observed location (1543 cm^{-1}) corresponds exactly to that detected for lead soaps originating from lead-containing drying oil paints. It is not clear if such a slow reaction might be due to the actual reactivity of lead ions with the alkyd medium or to the type of lead white used in our experiment, in which the relative proportion of cerussite ($Pb(CO)_3$) and hydrocerussite ($Pb(CO)_3 \cdot Pb(OH)_2$) is unknown. The ratio between these two components may vary significantly in lead white paint formulations. The spectra collected show no signals of decay in the alkyd structure and, in this case, metal soap formation was not detected in the mock-up paints kept at ambient conditions even after nine months of aging.

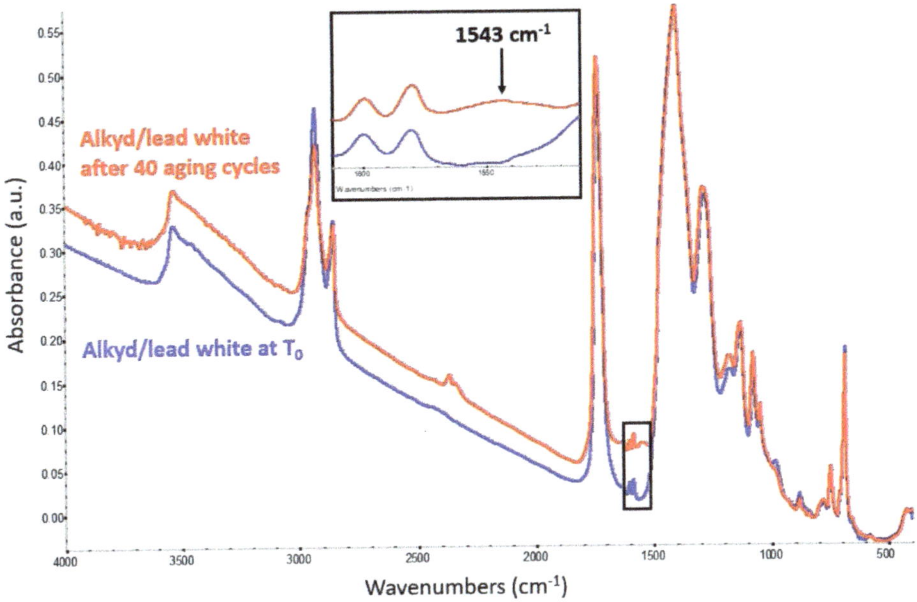

Figure 5. Spectrum of the alkyd-containing lead white mock-up paint at T_0 (blue trace) compared with the spectrum obtained from the same mixture after 40 aging cycles (red trace). The latter shows the formation of lead carboxylates, identified by a weak band at 1543 cm^{-1}.

2.2. *Contemporary Artworks*

After carrying out systematic work on the mock-up paints, confirming the occurrence of metal soaps in the presence of alkyd binders, FTIR data from previous studies conducted on two contemporary artifacts featuring alkyd-based paints were reviewed to assess the possible presence of carboxylates. The main results are shown below.

2.2.1. Franco Garelli, *Costruzione* (1970)

An exponent of the Informal movement, Franco Garelli (1909–1973) was an Italian physician, painter, sculptor, and ceramist who was presented to the CoBra group by Asger Jorn and later followed the Japanese Gutai experience. *Costruzione* (1970) is representative of the artistic and cultural history of the city of Torre Pellice, Turin, and is one of the most relevant artifacts featured in the Filippo Scroppo Civic Gallery (Figure 6). During conservation treatment at the CCR "La Venaria Reale", a few multi-layered samples were removed from the work and mounted as cross sections for micro-invasive analysis to gain insight into the materials and techniques used. Microscopic observation of a sample collected from the painted decoration of a side metal bar highlighted a sequence of layers of different colors, consisting of mixtures of pigments with binders of various nature (acrylic, vinylic, and alkyd-based). This complex stratigraphy might be related to deliberate compositional changes on the artist's part and/or to the reuse of elements from other artifacts.

Investigations by means of scanning electron microscopy with energy-dispersive X-ray spectroscopy (SEM/EDS) showed that the violet layer, indicated by an arrow in Figure 6, consists of a mixture of zinc white and Prussian blue. Transmission FTIR analysis of a loose scraping from the same layer confirmed the presence of Prussian blue with an alkyd binder. Aside from the pigment's characteristic C≡N stretching at 2091 cm^{-1}, spectra from this material, all featuring a strong band at 1575 cm^{-1} attributed to the formation of zinc carboxylates, show a close correspondence in terms of overall shape and wavenumbers with data acquired from the alkyd-containing zinc white mock-up paint that was prepared in the laboratory and artificially aged (Figure 7).

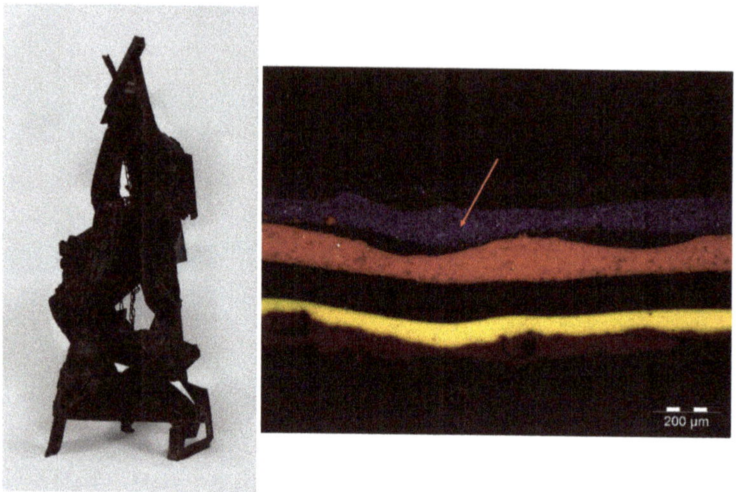

Figure 6. (**Left**) Franco Garelli, *Costruzione* (1970). (**Right**) Visible light image of a cross section removed from the painted decoration of a side metal bar. The arrow indicates a violet paint layer that contains zinc white and Prussian blue.

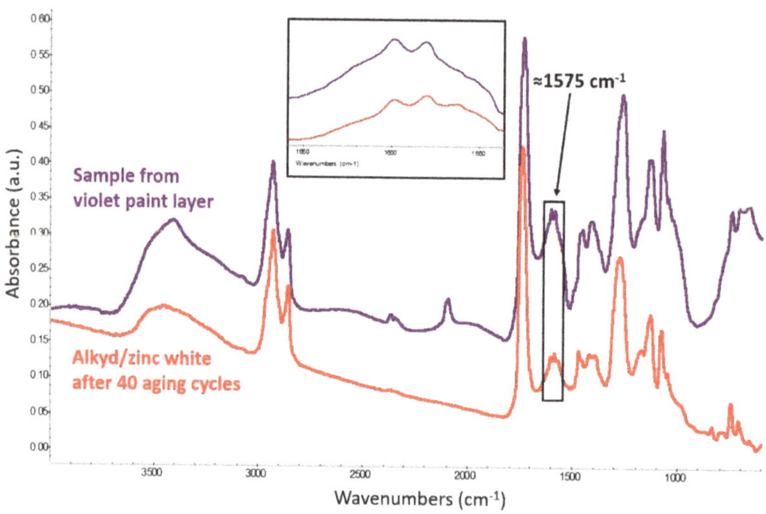

Figure 7. Spectrum of a violet layer in Garelli's work (violet trace) compared with the spectrum obtained from the alkyd-containing zinc white mock-up paint after 40 aging cycles (red trace). Both show the formation of zinc carboxylates, identified by a strong band at ≈1575 cm^{-1}.

Other cross sections from the same artwork, including a sample removed from a fastening element near the chain, display a thin, orange layer of minium (lead oxide, $2PbO \cdot PbO_2$) applied directly onto the metal, below the paint layers, likely serving an anticorrosive purpose (Figure 8).

Transmission FTIR spectra of a loose scraping from this lead-containing layer revealed the presence of an alkyd binder, along with polyvinyl acetate (PVA) likely from an adjacent layer (with main signals at 1244 and 1026 cm^{-1}). The 1543 cm^{-1} band, ascribable to the formation of lead carboxylates, appears significantly stronger in this artwork's sample compared to the aged alkyd/lead white mock-up paint (Figure 9). It is worth noticing that, albeit forming the same type of carboxylate, the lead cations involved in the reaction in this case do not originate from the lead white pigment, but from lead oxide, possibly leading to different reactivity.

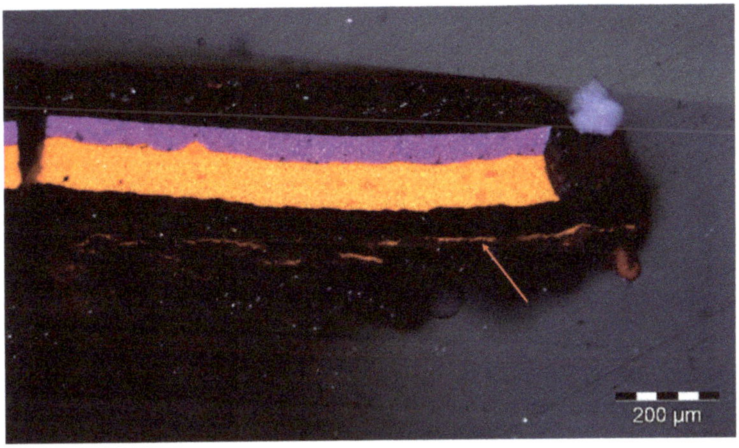

Figure 8. Visible light image of a cross section removed from a fastening element near the chain on Garelli's work. The arrow indicates a thin, orange layer of minium applied directly onto the metal, below the paint layers, likely serving an anticorrosive purpose.

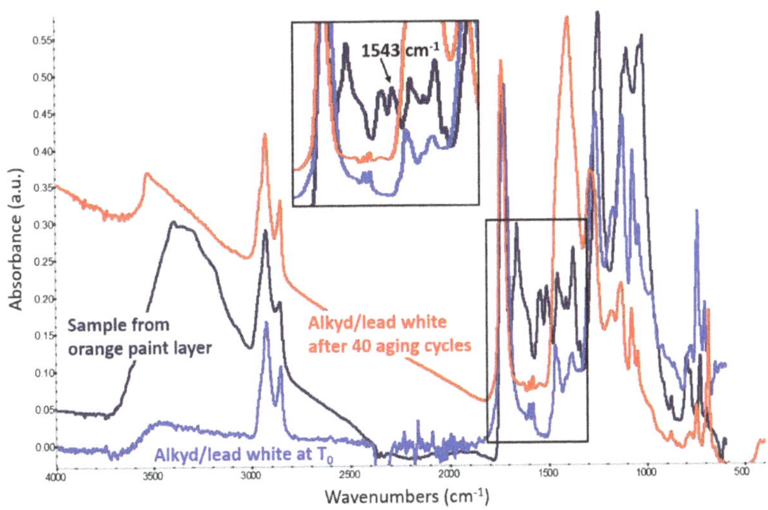

Figure 9. Spectrum of a minium layer sample from Garelli's work (dark blue trace) compared with spectra obtained from the alkyd-containing lead white mock-up paint at T_0 (light blue trace) and after 40 aging cycles (red trace). The first and latter show the formation of lead carboxylates, identified by a band at 1543 cm^{-1} that appears stronger in the case of the artwork's sample.

2.2.2. Luciano Minguzzi, *Uomini* (1971)

Luciano Minguzzi (1911–2004) was an Italian sculptor, medallist, and engraver who was selected to design the fifth door of the Milan Cathedral and is currently featured at the Peggy Guggenheim collection in Venice. *Uomini* (1971) is a large-scale installation made of metal alloys and resins, normally located in Saint-Vincent, Aosta Valley, which recently underwent conservation treatment at the CCR "La Venaria Reale" (Figure 10).

Figure 10. (**Left**) Luciano Minguzzi, *Uomini* (1971). (**Right**) Detail of the sampling site. The arrow indicates the exact location subjected to sampling.

After non-invasive XRF investigations of selected spots of interest on the artwork's surface, a sample of degraded white paint removed from the proper right edge of one of the panels was characterized with transmission FTIR spectroscopy. The analysis highlighted the presence of zinc white dispersed in an alkyd resin. As in the case of Garelli's violet paint layer, the FTIR spectrum shows a distinctive band centered at 1575 cm^{-1}, ascribable to the formation of zinc carboxylates (Figure 11).

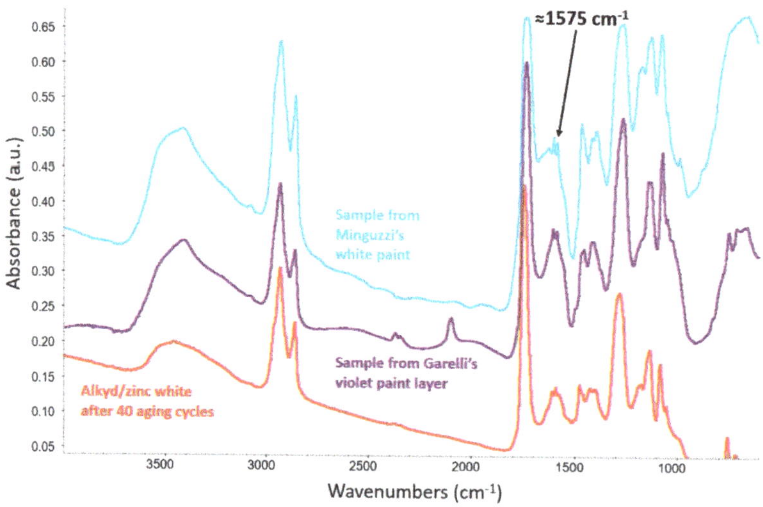

Figure 11. Spectrum of a white paint sample from Minguzzi's work (light blue trace) compared with spectra obtained from Garelli's violet layer (violet trace) and the alkyd-containing zinc white mock-up paint after 40 aging cycles (red trace). All spectra show the formation of zinc carboxylates, identified by a strong band at ≈1575 cm^{-1}.

3. Discussion

As expected, the systematic work on mock-up paints illustrated herein highlighted a similar reactivity in alkyd-based media to that of drying oils due to an abundance of fatty acids in their formulations (with a theoretical minimum of 30% *w*/*w*). In the FTIR spectra, the metal carboxylates originating from the reaction of an alkyd binder with the chosen pigments gave rise to absorption bands at the same wavenumbers as metal soaps from drying oil. The choice of cations in this work aimed to verify the formation of metal soaps in alkyd-based paints under controlled aging conditions, starting from pigments that are known to be the most reactive in drying oil media. While smalt and lead white are not present in modern alkyd-based paint formulations, their cations are commonly available in other materials that are relevant to cultural heritage research: for instance, potassium can be found in stones, plasters, and mortars employed in wall painting, while lead compounds are often used as anticorrosive layers, as also reported in the Garelli case study.

At this stage, a direct comparison of the kinetics of metal soap formation is challenging because the reactivity of each material strongly depends on the cation's availability, i.e., on the compound in which it is found and its chemical environment. On a qualitative level, based on a preliminary comparison of the relative intensities of the observed carboxylate bands, we detected a greater reactivity for zinc white compared to smalt, and a greater reactivity of both these pigments compared to lead white. However, it must be taken into consideration that the seemingly lower reactivity of lead might also depend on the availability of the lead cations within the specific material used, in which the exact hydrocerussite/cerussite ratio is unknown. Additionally, materials containing the same cation, such as lead white and minium, might prompt different scenarios, even though the formation of carboxylates was verified at the same wavenumbers: in other words, the formation of lead carboxylates gives rise to a band at 1543 cm^{-1} when lead cations originate from both a lead white pigment and from an anticorrosive layer of minium.

Compared to drying oil, the kinetics of metal soap formation in alkyd-based paints are expected to be slower, likely due to the lower number of fatty acids present and to the overall lower mobility of the system under study, in which fatty acids are not attached to glycerol to form triglycerides but to a polymeric polyester backbone.

Scientific analysis revealed an exact correspondence between metal carboxylates found in laboratory mock-up paints and samples from actual artifacts. This confirms that the same type of metal carboxylates observed in the mock-up paints also form in real artwork, even though the latter are likely to contain commercial alkyd binders or enamels whose formulations might be more complex and diverse than the media used here.

The inhomogeneity of alkyd-based paints might pose a challenge to the conservation of artworks, as modern and contemporary artists often do not pursue pure alkyd media, but rather make extensive use of lacquer, spray paints, or enamels originally formulated for different purposes (including, but not limited to, metal coating and surface finishing). These products can be heavily modified, for example, by adding melamine–formaldehyde resins, typically yielding significantly more reticulated structures than alkyd paint media. Moreover, the system's cross-linking rate and "stiffness" might deeply affect the hydrolizability of the fatty acids, their mobility, and their ability to interact with cations. As a matter of fact, no signs of migration or aggregation of the newly formed carboxylates were detected in the mock-up paints or contemporary artwork samples, likely due to their relatively short lifespan so far and possibly to the system's lower mobility. On the other hand, one must consider that works featuring alkyd-based paints are often significantly more recent than oil paintings, which means that the onset of issues in alkyd-containing artwork might be yet to occur.

4. Materials and Methods

4.1. Preparation of Mock-Up Paints

As virtually no literature exists on the formation of metal soaps in alkyd-based paints, the choice of pigments for this proof-of-concept study was based on the available data on pigments that had previously shown high reactivity with oil media. With this consideration in mind, mock-up paints were prepared as mixtures of an alkyd binder with potassium hydroxide and three commonly used pigments (smalt, zinc white, and lead white). The alkyd binder, described by the manufacturing company as "Médium Alkyde Qualitè Beaux Arts-Satinè", was purchased from Lefranc Bourgeois (Paris, France). Smalt, zinc white, and lead white were supplied by Kremer Pigmente (Aichstetten, Germany), while potassium hydroxide was supplied by Merck (Boston, MA, USA).

Two sets of four mock-up paints were prepared with two different binder/pigment ratios (1:1 and 2:1 *w/w*). The resulting paints were applied onto silicon float zone polished windows (13 mm diameter × 1 mm thickness) purchased from Crystran Ltd. (Dorset, UK). In addition, a mock-up sample with the bare alkyd resin and no pigments was prepared as a reference.

Each mock-up was prepared in three replicas to ensure experiment reproducibility. Three mock-up replicas for each cation were kept for nine months at ambient conditions without direct exposure to sunlight (22 °C and 55% relative humidity). Three mock-up replicas for each cation were also subjected to artificial aging, undergoing wet (3 days, 22 °C, and 95% relative humidity) and dry (3 days in the oven, 65 °C, and 20% relative humidity) alternated cycles for up to nine months, totaling 40 aging cycles. The initial FTIR measurements were conducted one week after paint film curing; after that, analyses were carried out once a day for the first week and once a week for the rest of the aging timeframe.

4.2. FTIR Measurements

FTIR analysis was performed with a Bruker Vertex 70 FTIR spectrometer (Billerica, MA, USA) equipped with a DTGS detector, coupled with a Bruker Hyperion 3000 infrared microscope with a mercury cadmium telluride (MCT) detector. Mock-up paints applied onto silicon float zone polished windows were analyzed as is using the system's optical bench, while samples from the contemporary works were examined upon compression in a diamond cell through a 15× objective. Data were collected in transmission mode, in the 4000–400 cm^{-1} spectral range, at a spectral resolution of 4 cm^{-1}, as the sum of 64 scans.

5. Conclusions

While the formation of metal soaps in oil paintings has been the subject of extensive research to date, virtually no data exist in the literature on this phenomenon as pertaining to alkyd-based artists' paints. In this work, the possibility of metal carboxylate formation in the presence of alkyd media was assessed for the first time through a systematic investigation of mock-up paints by means of transmission FTIR spectroscopy. A series of mock-up paints were prepared by mixing a commercial alkyd binder with potassium hydroxide and three widely used pigments (i.e., smalt, zinc white, and lead white). The resulting paints were applied onto silicon wafers and subjected to natural and artificial aging. Measurements were taken at regular intervals, after curing the paint film, over a period of nine months. After testing the mock-ups, samples from two contemporary installations by Italian artists Franco Garelli and Luciano Minguzzi were also investigated for comparison. The results confirmed the formation of metal carboxylates in both mock-up paints and artwork samples, revealing different reactivity for the materials selected that strongly depends on the cations' availability within their own chemical environment. Remarkably, the data collected highlight an exact correspondence between metal carboxylates found in laboratory mock-ups and samples from actual artifacts, even though the latter typically feature commercial alkyd binders with more complex formulations than the media used here. Overall, the work presented herein provides readers with preliminary, yet essential insights on a topic of special interest in the field of cultural heritage, paving the way for future research with potentially significant implications, especially from the point of view of the conservation of modern and contemporary artworks.

In the future, this proof-of-concept study involving mock-up paints will be extended to a selection of ready-to-use commercial alkyd-based tube paints, spray paints, and enamels. In addition to evaluating the kinetics of metal carboxylate formation in a series of materials of interest, a thorough investigation of the paint film's physical and mechanical properties, behavior, and changes thereof will be conducted. Future research on this topic might also include an in-depth characterization of the newly formed carboxylates so as to understand the impact of metal soap formation on the behavior of alkyd-based paints and predict possible conservation issues arising from these degradation processes. At this research stage, it is unclear whether the metal soaps formed within a work's paint layers in the presence of alkyd binders may undergo migration and/or aggregation as is observed in the case of drying oils: this aspect will be investigated using imaging techniques. A detailed comparison between the kinetics of metal soap formation in oil-based and alkyd-based paints will be also performed.

From an analytical standpoint, FTIR spectroscopy is currently the technique of choice for the detection and characterization of metal carboxylates in paints. This is thanks to the availability of benchtop and portable instrumentation as part of the basic laboratory infrastructure, not only in highly specialized research institutions and universities, but also in most museums and conservation centers. More work needs to be done to identify additional, reasonably accessible instrumental techniques that could be employed along with FTIR in this type of studies: this might be especially useful in the case of complex mixtures, where the distinctive infrared carboxylate-related bands are sometimes obscured or disguised by the overlapping signals of other materials.

Author Contributions: Conceptualization, T.P., A.P. and F.P.; methodology, T.P., A.P. and F.P.; investigation, T.P., A.P., M.P.H., A.P.C. and R.L.C.; data curation, T.P., A.P. and M.P.H.; writing—original draft preparation, T.P.; writing—review and editing, F.P., A.P. and M.P.H.; supervision, F.P. All authors have read and agreed to the published version of the manuscript.

Funding: This research received no external funding.

Institutional Review Board Statement: Not applicable.

Informed Consent Statement: Not applicable.

Data Availability Statement: All data are either displayed in this article or available from the corresponding authors upon reasonable request.

Acknowledgments: The authors would like to acknowledge Chiara Andrea Lombardi, Università degli Studi di Milano and La Sapienza Università di Roma, Italy, who occasionally assisted in the collection of FTIR measurements from the mock-up paints.

Conflicts of Interest: The authors declare no conflicts of interest.

References

1. Casadio, F.; Keune, K.; Noble, P.; Van Loon, A.; Hendriks, E.; Centeno, S.A.; Osmond, G. (Eds.) *Metal Soaps in Art: Conservation and Research*; Springer International Publishing: Cham, Switzerland, 2019.
2. Izzo, F.C.; Kratter, M.; Nevin, A.; Zendri, E. A critical review on the analysis of metal soaps in oil paintings. *ChemistryOpen* **2021**, *10*, 904–921. [CrossRef] [PubMed]
3. van den Berg, K.J.; Burnstock, A.; Schilling, M. Notes on metal soap extenders in modern oil paints: History, use, degradation, and analysis. In *Metal Soaps in Art: Conservation and Research*; Springer International Publishing: Cham, Switzerland, 2019; pp. 329–342.
4. Izzo, F.C. 20th Century Artists' Oil Paints: A Chemical-Physical Survey. Ph.D. Thesis, Università Ca' Foscari Venezia, Venice, Italy, 2011.
5. Izzo, F.C.; van den Berg, K.J.; van Keulen, H.; Ferriani, B.; Zendri, E. Modern oil paints–formulations, organic additives and degradation: Some case studies. In *Issues in Contemporary Oil Paint*; Springer International Publishing: Cham, Switzerland, 2014; pp. 75–104.
6. Pizzimenti, S.; Bernazzani, L.; Tinè, M.R.; Treil, V.; Duce, C.; Bonaduce, I. Oxidation and cross-linking in the curing of air-drying artists' oil paints. *ACS Appl. Polym. Mater.* **2021**, *3*, 1912–1922. [CrossRef]
7. Bonaduce, I.; Duce, C.; Lluveras-Tenorio, A.; Lee, J.; Ormsby, B.; Burnstock, A.; van den Berg, K.J. Conservation issues of modern oil paintings: A molecular model on paint curing. *Acc. Chem. Res.* **2019**, *52*, 3397–3406. [CrossRef] [PubMed]
8. van den Berg, J.D.J.; van den Berg, K.J.; Boon, J.J. Chemical changes in curing and ageing oil paints. In Proceedings of the ICOM-CC 12th Triennial Meeting, Lyon, France, 29 August–3 September 1999; Volume 1.
9. Meilunas, R.J.; Bentsen, J.G.; Steinberg, A. Analysis of aged paint binders by FTIR spectroscopy. *Stud. Conserv.* **1990**, *35*, 33–51. [CrossRef]
10. Hermans, J.J.; Keune, K.; Van Loon, A.; Iedema, P.D. Toward a complete molecular model for the formation of metal soaps in oil paints. In *Metal Soaps in Art: Conservation and Research*; Springer International Publishing: Cham, Switzerland, 2019; pp. 47–67.

11. Hermans, J.J.; Keune, K.; Van Loon, A.; Corkery, R.W.; Iedema, P.D. Ionomer-like structure in mature oil paint binding media. *RSC Adv.* **2016**, *6*, 93363–93369. [CrossRef]
12. Cotte, M.; Checroun, E.; De Nolf, W.; Taniguchi, Y.; De Viguerie, L.; Burghammer, M.; Walter, P.; Rivard, C.; Salomé, M.; Koen, J.; et al. Lead soaps in paintings: Friends or foes? *Stud. Conserv.* **2017**, *62*, 2–23. [CrossRef]
13. Noble, P. A brief history of metal soaps in paintings from a conservation perspective. In *Metal Soaps in Art: Conservation and Research*; Springer International Publishing: Cham, Switzerland, 2019; pp. 1–22.
14. Ortiz Miranda, A.S.; Kronkright, D.; Walton, M. The influence of commercial primed canvases in the manifestation of metal soaps protrusions in Georgia O'Keeffe's oil paintings. *Herit. Sci.* **2020**, *8*, 107. [CrossRef]
15. Robinet, L.; Corbeil, M.C. The characterization of metal soaps. *Stud. Conserv.* **2003**, *48*, 23–40. [CrossRef]
16. Poli, T.; Piccirillo, A.; Zoccali, A.; Conti, C.; Nervo, M.; Chiantore, O. The role of zinc white pigment on the degradation of shellac resin in artworks. *Polym. Degrad. Stabil.* **2014**, *102*, 138–144. [CrossRef]
17. Poli, T.; Piccirillo, A.; Nervo, M.; Chiantore, O. Interactions of natural resins and pigments in works of art. *J. Colloid Interf. Sci.* **2017**, *503*, 1–9. [CrossRef]
18. Poli, T.; Chiantore, O.; Diana, E.; Piccirillo, A. Drying oil and natural varnishes in paintings: A competition in the metal soap formation. *Coatings* **2021**, *11*, 171. [CrossRef]
19. Cappitelli, F.; Koussiaki, F. THM-GCMS and FTIR for the investigation of paints in Picasso's Still Life, Weeping Woman and Nude Woman in a Red Armchair from the Tate Collection, London. *J. Anal. Appl. Pyrol.* **2006**, *75*, 200–204. [CrossRef]
20. Dooley, K.A.; Coddington, J.; Krueger, J.; Conover, D.M.; Loew, M.; Delaney, J.K. Standoff chemical imaging finds evidence for Jackson Pollock's selective use of alkyd and oil binding media in a famous 'drip' painting. *Anal. Methods* **2017**, *9*, 28–37. [CrossRef]
21. Lake, S.; Ordonez, E.; Schilling, M. A technical investigation of paints used by Jackson Pollock in his drip or poured paintings. *Stud. Conserv.* **2004**, *49* (Suppl. 2), 137–141. [CrossRef]
22. Rosi, F.; Miliani, C.; Delaney, J.K.; Dooley, K.; Stringari, L.; Subelyte, G.; Pensabene Buemi, L. Jackson Pollock's drip paintings: Tracing the introduction of alkyds through non-invasive analysis of mid-1940s paintings. In *Science and Art*; Royal Society of Chemistry: London, UK, 2020.
23. Schilling, M.R.; Keeney, J.; Learner, T. Characterization of alkyd paint media by gas chromatography-mass spectrometry. *Stud. Conserv.* **2004**, *49* (Suppl. 2), 197–201. [CrossRef]
24. La Nasa, J.; Degano, I.; Modugno, F.; Colombini, M.P. Alkyd paints in art: Characterization using integrated mass spectrometry. *Anal. Chim. Acta* **2013**, *797*, 64–80. [CrossRef]
25. Ploeger, R.; Scalarone, D.; Chiantore, O. The characterization of commercial artists' alkyd paints. *J. Cult. Herit.* **2008**, *9*, 412–419. [CrossRef]
26. Lin, K.F. Alkyd resins. In *Kirk-Othmer Encyclopedia of Chemical Technology*; John Wiley & Sons: Hoboken, NJ, USA, 2000.
27. Gan, S.N.; Tan, B.Y. FTIR studies of the curing reactions of palm oil alkyd–melamine enamels. *J. Appl. Polym. Sci.* **2001**, *80*, 2309–2315. [CrossRef]
28. Honzicek, J. Curing of air-drying paints: A critical review. *Ind. Eng. Chem. Res.* **2019**, *58*, 12485–12505. [CrossRef]
29. Bell, S.H. Controlling loss of gloss. *Paint. Varn. Prod.* **1970**, *60*, 55–60.
30. Veleva, L. Protective coatings and inorganic anti-corrosion pigments. In *Paint and Coating Testing Manual*; ASTM International: West Conshohocken, PA, USA, 2012; pp. 282–299.
31. Palladino, N.; Occelli, M.; Wallez, G.; Coquinot, Y.; Lemasson, Q.; Pichon, L.; Stankic, S.; Etgens, V.; Salvant, J. An analytical survey of zinc white historical and modern artists' materials. *Herit. Sci.* **2024**, *12*, 47. [CrossRef]
32. Tanase, S.; Bouwman, E.; Reedijk, J. Role of additives in cobalt-mediated oxidative crosslinking of alkyd resins. *Appl. Catal. A. Gen.* **2004**, *259*, 101–107. [CrossRef]

molecules MDPI

Article

Non-Invasive Characterisation of Bromoil Prints by External Reflection FTIR Spectroscopy

Raquel Freixas-Jambert [1,*], Cristina Ruiz-Recasens [1], Alejandra Nieto-Villena [2] and Marta Oriola-Folch [1]

1 Research Group of Conservation of Cultural Heritage, Art and Conservation Department, Fine Arts Faculty, University of Barcelona, Pau Gargallo, 4, 08028 Barcelona, Spain
2 Department of Art History, Faculty of Geography and History, University of Valencia, Avda. de Blasco Ibáñez, 28, 46010 Valencia, Spain
* Correspondence: rfreixas@outlook.es

Abstract: The bromoil process, developed in 1907, was a photographic technique highly esteemed by pictorialist photographers for its capacity for image manipulation, which aligns its expressiveness with traditional pictorial techniques. Despite the artistic and technical value of bromoils and their prevalence in renowned collections, there is little research on their composition and structural characteristics. This study uses non-invasive external reflection FTIR spectroscopy to characterise 16 bromoil prints dating from the 1920s to 2010. FTIR spectroscopy allowed the discrimination of key components such as cellulose, gelatine, pigments, and oils, elucidating the structural and compositional complexity of bromoil prints. The study presents new perspectives on the traditionally recognised structure of bromoil prints, based on FTIR spectroscopy results along with evidence from microscopic examination, particularly regarding the role of certain strata in their identification. These results suggest a need to reconsider the understanding and characterisation of bromoil prints in relation to the currently available literature. This research also proposes a measurement model adapted to the studied samples and addresses the advantages and limitations found in the different types of equipment used. It thus proposes a key methodology for the identification and provides a source for the physicochemical studies of photographic processes

Keywords: bromoil process; photography; oil positives/prints; FTIR spectroscopy; barite layer; pigment process; gelatine; greasy ink; reflection mode

1. Introduction

Bromoil is one of the so-called photographic pigment processes. It was first developed in 1907 as a refinement of Rawlins's oil process (invented in 1904), and it represents the culmination of decades of research and experimentation aimed at identifying a definitive noble process. The bromoil process was particularly embraced by pictorialist photographers due to its high capacity for manipulation and intervention during the processing of the image, as well as its unique visual characteristics and proximity to the pictorial aesthetic of the images it produced [1–5].

The final image, although photographic, is materialised from a greasy ink similar to the one used in lithography. It is formed on a silver gelatine developing-out paper, usually bromide paper, from which the metallic silver has been removed by bathing the paper in different chemical solutions during the production of the bromoil. This reaction hardens the gelatine and renders it hydrophobic to a degree that is proportional to the quantity of silver present in the image. This results in the formation of a gelatine matrix that can be moistened and inked in accordance with its degree of hardening. The outcome is a pigment image comprising greasy ink that faithfully reproduces the original silver image. Furthermore, the author has the option of modifying and adding changes to the image during the inking process and subsequent retouching with remarkable ease, a capability that is unique to this historical photographic process.

Due to the intricate nature of the production process and the singularity of the resulting object, bromoil prints are often considered to possess remarkable artistic and technical value. They are frequently included in a multitude of private and public collections, held within archives, museums, and cultural institutions. Nevertheless, there is a notable absence of scientific studies and research focused on these photographic processes.

To gain a comprehensive understanding of the composition and technical aspects of oil processes, it is essential to seek out primary sources from the era of the greatest production of these photographs, which spans the first decades of the 20th century [6,7]. Despite the aforementioned references describing the materials utilised in these processes, it remains unclear which compounds constitute these photographs, a factor of significant importance for their conservation. Potential variations between authors, residual substances not eliminated by the baths, by-products generated during the manufacturing process, and alterations due to the passage of time, all contribute to this uncertainty. Secondary sources of information on these processes are also relatively scarce and mainly focus on the production aspects. Moreover, most of them are compilations of data from primary sources.

A review of the literature on the characterisation of historical photographic processes through morphological study reveals a notable scarcity of references to oil processes. Among the sources consulted, the Graphics Atlas [8] stands out as the most comprehensive on this topic. Together with Pavão's manual [9], these constitute the sole direct sources on the morphological identification of bromoil prints. Moreover, the limitations of morphological characterisation preclude the determination of composition and deterioration processes, as well as the distinction between photographic processes that look similar and are easily mistaken for one another, such as silver gelatine DOP or dichromate gum and its derivatives [8].

It is therefore necessary to develop a more objective research methodology that relies on advanced identification techniques in order to obtain detailed and specific data on the composition and interaction between materials and their environment [10]. FTIR spectroscopy has yielded optimal results in the characterisation of photographic materials. Some studies have made significant advances in the possibilities of studying photographs and paintings by using external reflection FTIR spectroscopy (ER-FTIR), a non-destructive and non-invasive method [11,12]. However, the majority of existing FTIR studies focus on silver processes, with only a few examining alternative processes [10,13–19]. A recent study has applied instrumental analysis techniques to the study of bromoil prints [20]. Nevertheless, it does not address the numerous inconsistencies identified in the existing literature on the structure and composition of bromoil prints, which remain unresolved.

For example, some authors include bromoil positives in the category of two-layer photographic processes [9], thereby indicating that these do not contain the barite layer that is typical of photographic papers. However, other reference sources [8] explicitly state that the barite layer is a key in the identification of bromoil prints.

In light of the aforementioned considerations, this study seeks to examine the potential of FTIR spectroscopy in external reflection mode as a non-destructive characterisation tool for bromoil prints. The applicability of this technique offers significant results in the analysis of cultural and artistic heritage [21], and it offers some important advantages such as the portability of some equipment [22,23], the ability to explore both organic and inorganic compounds [24,25], the possibility of discerning between the different layers of a sample [11,26], and the capacity to obtain valuable information without damaging the sample [14,27]. This approach allows us to expand our knowledge of the composition and structure of bromoil prints and to determine the information that can be obtained by reflection FTIR spectroscopy for their characterisation. In order to optimise the use of FTIR spectroscopy and obtain valuable results for the photographic process under study, the aim is to describe the methods and define the parameters to be used, as well as to evaluate the advantages and limitations offered by both desktop and portable spectrometers.

2. Results and Discussion

2.1. Results

The results demonstrate that the FTIR spectroscopy in external reflection mode is a valuable technique for the study and characterisation of bromoil prints. The methodology employed to carry out the analysis, along with the selection of the areas to be analysed based on different image densities, has provided information about the materials present in the works and their stratigraphic position. This allowed for the characterisation of the layered structure of the studied print, based on the spectral response of the materials and the location of the measurement points.

FTIR analysis has made it possible to identify the main compounds present in each of the layers that make up the bromoil prints: cellulose, as the main component of the paper support; the barite layer, when this is present in the work immediately above the paper; the gelatine, which is present in the next layer; and, in the most superficial layer, the oils and certain pigments of the image-forming layer.

Each of these compounds has been detected with a greater or lesser degree in the resulting FTIR spectrum, depending on the proportion in which it is present at the point analysed and/or on the presence or absence of other upper layers, which may make its presence more evident or, on the contrary, make its identification more difficult.

In accordance with the above, after this study, we can affirm that, in bromoil prints, the intensity of the FTIR response of the surface layer of the greasy ink is directly proportional to the density of the image, while it is inversely proportional in the case of gelatine, barite, and cellulose. Taking this aspect into account, the comparative study of the zones of different density (Dmax, Dmid, and Dmin) allows us to characterise the stratigraphic structure of the copy.

2.1.1. Cellulose

The first layer of this structure is the paper support. The analysis of the back of the bromoil samples allows us to characterise it without interference from the other layers superimposed on the front to form the photographic print. The backs of all the bromoil samples selected in the study gave a similar response in all cases, corresponding to the spectral characteristics of cellulose, the main component of the paper, and with a high response in FTIR. For example, in Figure 1, which shows the spectrum obtained from the analysis of the back of the *Janini_026* bromoil print, the broad and strong band of the OH group at 3460 cm^{-1}, and the absorptions in the region between 2970 and 2860 cm^{-1} approx., which presents a broad band centred at 2901 cm^{-1}, are related to the asymmetric and symmetric stretching of the methyl (CH_3) and methylene (CH_2) groups [28,29]. The bands located at 1433, 1375, 1338, and 1323 cm^{-1} are assigned to CH_2 symmetric bending and CH bending vibrations [11,29,30]. A band appears at 1642 cm^{-1}, corresponding to the combined C=O and H-O-H (intramolecular water) vibrations [28]. Finally, the region from 1180 to 1040 cm^{-1} corresponds to the stretching vibrations of the CO and CC bonds in the cellulose chain [11]. The small peak appearing near 1000 cm^{-1} could also be attributed to CO stretching vibrations, while the absorption peak observed at 899 cm^{-1} appears to correspond to the C-O-C bond deformation [11,31], which is assigned to the amorphous region in cellulose [29].

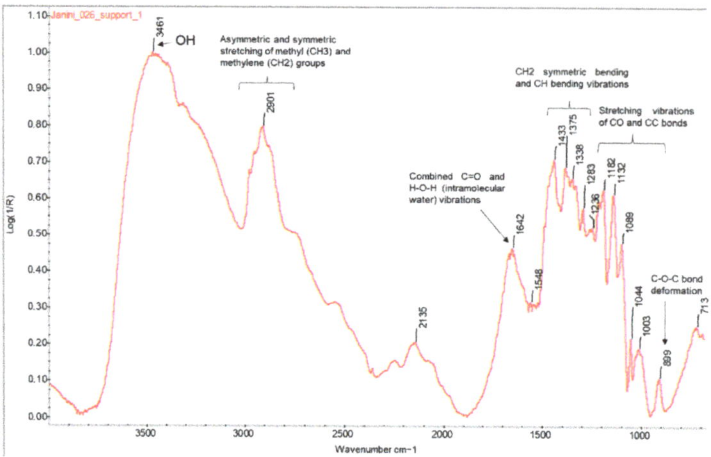

Figure 1. FTIR spectrum in external reflection mode of the reverse side of the primary support of the *Janini_026* bromoil sample obtained with the Nicolet™ iN10 MX spectrometer (ThermoFisher Scientific, Waltham, MA, USA) of the Scientific and Thechnological Centers at the University of Barcelona (CCiTUB) (Barcelona, Spain). A clear spectrum of cellulose is identified from the assignment of the main bands.

From the front, the cellulose is hardly visible at most points in the Dmin and edge regions due to the strong response of the proteinaceous material forming the photographic binder layer on the paper support. A large part of the spectral bands of the cellulose are covered by the amide bands of the protein binder (~1690–1470 cm^{-1}) at the fingerprint region (see Figure 2, blue spectral line). Only the absorptions below 1130–700 cm^{-1} could be faintly intuited, although their interpretation is doubtful, and this information is not sufficient to identify or corroborate the presence of cellulose. Furthermore, it is also not possible to distinguish the bands of the OH, methyl (CH3), and methylene (CH2) groups of cellulose located in the upper range of the spectrum (approximately 3400–2860 cm^{-1}) due to the spectral distortions that appear in the bands of this range. Nevertheless, the objective is not to corroborate the presence of cellulose in the sample, as this is plausible through general observation of the sample and demonstrable by analysing the back of the copies (see Figure 1), but rather to study how the upper layers, such as the protein binder layer, affect its detection from the front, even in areas of minimal image density. Therefore, it confirms that there are indeed other layers located between the paper support and the material forming the image (corresponding to the Dmid and Dmax areas). These data are very useful when characterising photographic prints because, when correlated with observations from microscopic study, they allow differentiation between photographic copies with one and those with two or three layers (the former without photographic binder) and, thus, between the different photographic processes that characterise them. Moreover, as discussed in the results discussion, the detection of protein material covering the absorptions of cellulose in the Dmin areas is an indication that discards processes like gum bichromate or carbon.

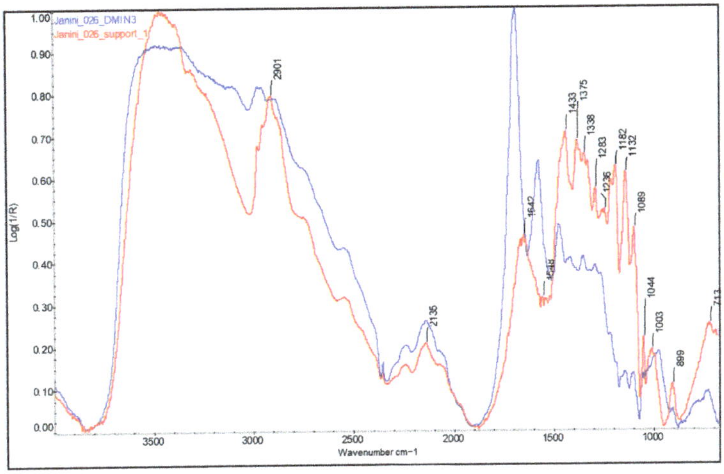

Figure 2. FTIR spectrum of the Dmin (see blue line) and paper support of the bromoil sample *Janini_026*, obtained using the Nicolet™ iN10 MX desktop spectrometer (ThermoFisher Scientific, Waltham, MA, USA) at the CCiTUB (Barcelona, Spain). The bands of the protein material (binder) cover a major part of the cellulose spectral signature. Only the absorptions below 1130–700 cm^{-1} could be faintly intuited, although their interpretation is doubtful.

2.1.2. Gelatine

Measurements in the areas corresponding to the non-image margins and the areas of minimum image density allow the IR signals of the protein binder to be detected with a minimum of spectral interference from other materials present in the samples, particularly the oil-based ink in the image. The spectra obtained in these areas are quite sharp and show intense bands due to the fact that the binder layer is the first surface layer in these areas of the bromoil prints (regardless of any loose and dotted pigment particles). In Figure 3, the two sharp and intense bands of amide I (carbonyl absorption) and II (combination of C-N and N-H vibrations) at ~1693 and ~1573 cm^{-1}, respectively, are clearly distinguishable [28,32]. The CH bending vibration occurring in this spectrum near 1469 cm^{-1} is also clearly visible [32]. The bands corresponding to amides A and B, which usually appear between 3300 and 3100–3030 cm^{-1} [33], are not well distinguishable due to the distortion in this range.

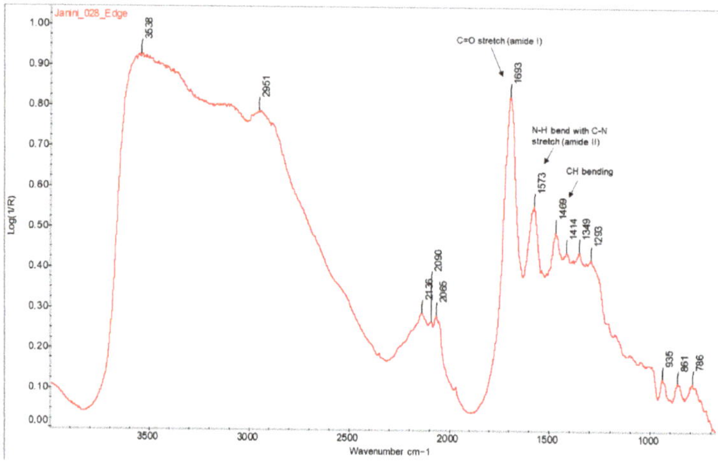

Figure 3. FTIR spectrum in the reflection mode of the area corresponding to the non-image edge (absence of ink) of the bromoil sample *Janini_028*. The bands of amides I, II, and III are clearly visible in the region ~1690–1460 cm^{-1} corresponding to the protein material. Spectrum recorded on a Nicolet™ iN10 MX desktop spectrometer (ThermoFisher Scientific, Waltham, MA, USA).

The results demonstrate the effectiveness of FTIR spectroscopy in identifying the proteinaceous binder in bromoil prints but also the possibility of distinguishing it from albumen, which is not present in these works, although it is present in prints made using other photographic processes. Figure 4 clearly shows the three specific bands of gelatine located at approximately 1472, 1419, and 1348 cm^{-1}, which gradually decrease in intensity, with the first being much more noticeable, but all three show a significant response (even with interference from the spectral bands of other compounds such as cellulose), unlike albumen, which has bands located at 1451 and 1399 cm^{-1} of very similar intensity [10].

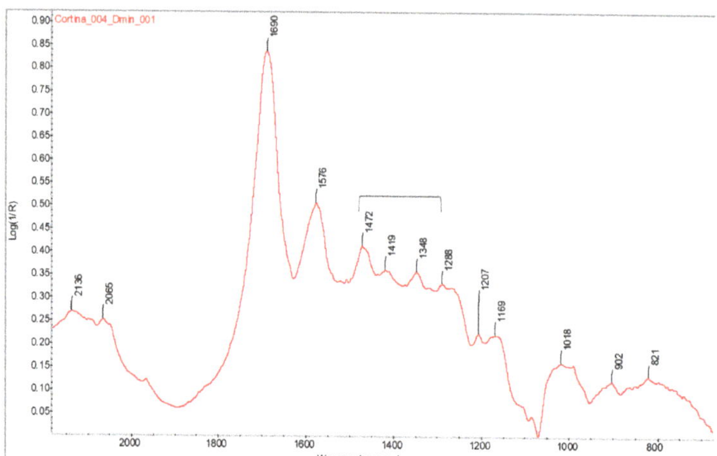

Figure 4. Detail of the FTIR spectrum in the reflection mode of the Dmin region of a photograph of a bromoil print (*Cortina_004*), showing the typical pattern of the gelatine photographic binder. The spectrum was obtained using a Nicolet™ iN10 MX desktop spectrometer (ThermoFisher Scientific, Waltham, MA, USA).

In this study, the spectral interpretation of the photographic binders of the copies did not reveal any relevant differences that would indicate a change in the period or authorship of the bromoil prints analysed, not even between the oldest (first decades of the 20th century) and the most recent (2009–2010) (see Figure 5). It should be noted, however, that although the bromoil samples analysed come from three different collections, each with specific storage conditions, no obvious conservation problems or samples in an advanced state of degradation were found. The only plausible difference is the appearance of combination bands related to barium sulphate from the barite layer in the sample *Cortina_001*, located beneath the gelatine layer (see Figure 5, red spectral line).

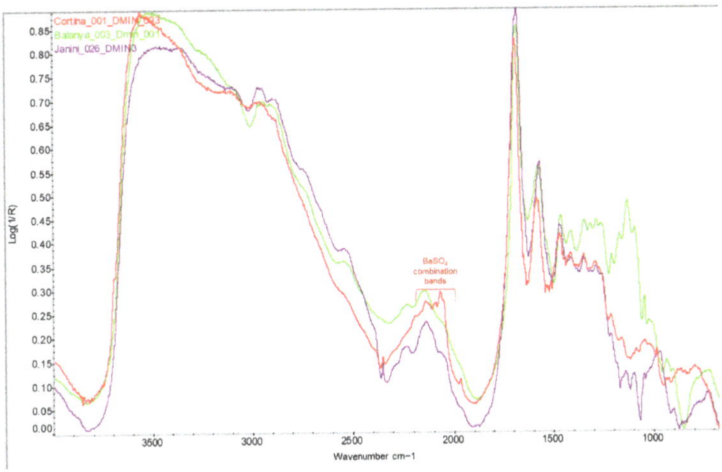

Figure 5. FTIR reflectance spectra taken through the Dmin areas of the bromoil prints *Janini_026* (first half of the 20th century) (purple), *Cortina_001* (1940s—late 1950s) (red), and *Balanya_003* (2009–2010) (green). Despite the difference in the relative intensities of the spectral bands, which could not be related to the age of the print, there is no significant difference in the bands of the protein material between the three prints. In the sample *Cortina_001*, it is also possible to distinguish the combination bands of barium sulphate (red). Spectra acquired using the Nicolet™ iN10 MX desktop spectrometer (ThermoFisher Scientific, Waltham, MA, USA).

As mentioned above, the protein binder in the bromoil samples is detected by FTIR in a way that is inversely proportional to the image density, confirming that this analytical technique can be used to characterise the material as coming from a layer below the image, i.e., the photographic support. This means that areas of lower image density (Dmin) show a higher intensity of gelatine bands, while, in areas of maximum density (Dmax), these bands are significantly reduced. This is an important point, as it allows bromoil positives to be distinguished from other very similar photographic positives, such as DOP (chemical development) prints or carbon prints. In DOP positives, the gelatine binder is detected with the same intensity over the entire surface, as it is the most superficial layer of the copy (except for possible coatings), since it contains the silver particles of the image inside (emulsion) [10]. In carbon prints, on the other hand, the pattern is the opposite; the amide bands I and II of the gelatine are more clearly detected in the areas of maximum density of the image, where there is a greater thickness of pigmented gelatine [34].

For example, looking at Figure 6, broad shoulders between 1715 and 1745 cm^{-1} begin to form in the areas of medium density (Dmid) as the presence of ink, and therefore, the drying oil increases. When the ink concentration is at its maximum exponent, this back is defined as a broad band or the merging of two bands, at ~1715 cm^{-1} and ~1745 cm^{-1}, in the areas of maximum density (Dmax) (see Figures 6–8). In parallel with the appearance of these bands related to the linseed oil present in the ink, specifically the carbonyl group band at 1750–1740 cm^{-1} [32] and a small peak at around 1715 cm^{-1} related to the level of polymerisation of the oil and suggesting the formation of acidic carboxylic products [35], there is a decrease in the sharpness and shape of the bands related to the protein material and, in some cases, also in their relative intensity, around 1690, 1572, and 1460–1470 cm^{-1} (see Figures 6–8). Along with the appearance of the shoulder at 1746 cm^{-1} related to the linseed oil present in the ink, the characteristic band of the triple CN bond of Prussian blue also appears at Dmax in the sample *Cortina_004* (see Figure 8, red spectral line).

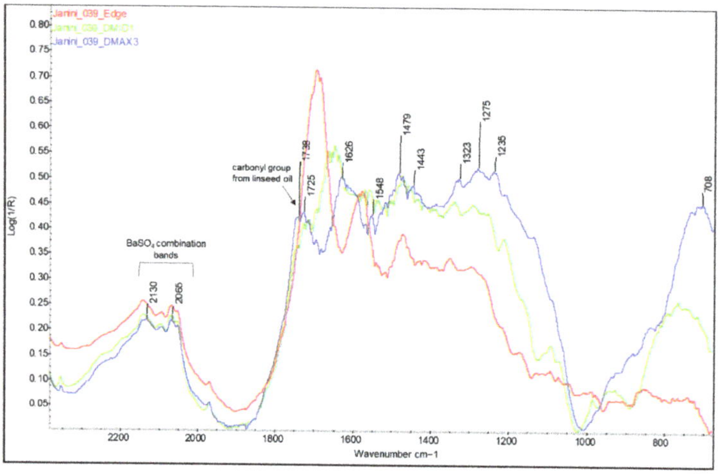

Figure 6. Comparison of the FTIR spectra corresponding to the non-inked edge (red) in the Dmid (green) and Dmax (blue) regions of the bromoil print *Janini_039*. As the image density increases, and with it the concentration of oil-based ink, the amide bands of the protein lose definition and become obscured by the overlapping of the typical bands of oils in this range of the spectrum. In the spectrum, it is also possible to observe the combination bands of barium sulphate from the barite layer. Spectra obtained using a Nicolet™ iN10 MX desktop spectrometer (ThermoFisher Scientific, Waltham, MA, USA).

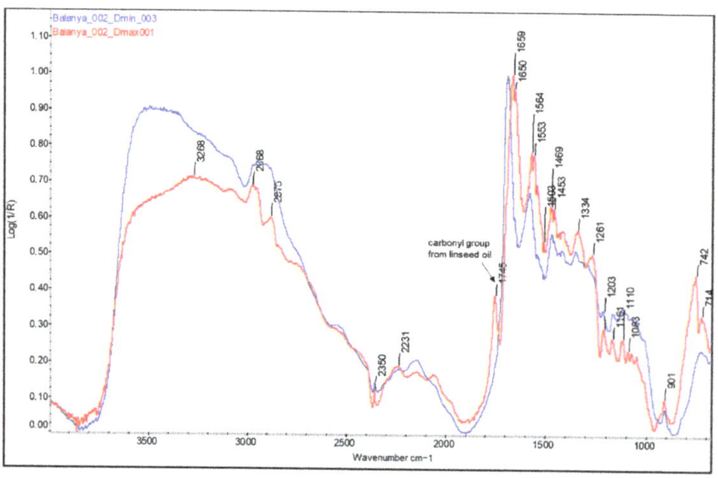

Figure 7. FTIR spectra of the Dmin (blue) and Dmax (red) regions of the bromoil sample *Balanya_002*, showing the loss of definition of the binder bands in the Dmax region of the image. The spectra were acquired using a Nicolet™ iN10 MX desktop spectrometer (ThermoFisher Scientific, Waltham, MA, USA).

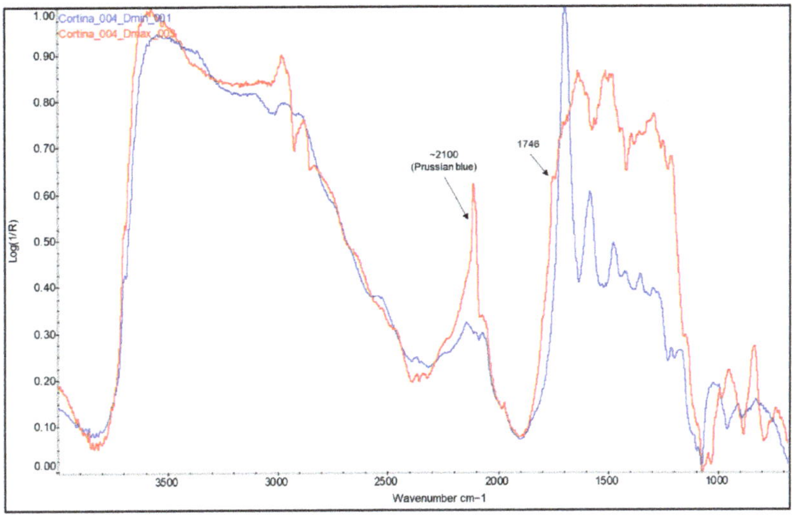

Figure 8. FTIR spectra of the Dmin (blue) and Dmax (red) regions of the bromoil sample *Cortina_004*. The amide bands of the protein are barely recognisable in the spectrum obtained in the Dmax area. Along with the appearance of the shoulder at 1746 cm^{-1} related to the linseed oil present in the ink, the characteristic band of the triple CN bond of Prussian blue also appears at Dmax. The spectra were acquired using a Nicolet™ iN10 MX desktop spectrometer (ThermoFisher Scientific, Waltham, MA, USA).

It is therefore shown that this behaviour is directly related to the presence of the surface layer of greasy ink that forms the image of the bromoil print. In other words, it is mainly associated with the interference of a layer on top, the oily ink, which is interposed between the gelatine substrate and the light beam of the spectrometer, so that the response of the binder is less strong, affecting, in some cases, the relative intensity and sharpness of the protein bands. We can therefore confirm that the variability in the relative intensities and definition of the bands gives us information about the stratigraphic position of the different layers detected in the FTIR spectrum.

2.1.3. Barite

In contrast to the spectral overlap generated between the protein bands, which hide the IR signals of the cellulose of the paper layer, the results show that the presence of barium sulphate, the main component of the barite layer, does not generate any significant interference for the detection of the other layers of the photographic support (gelatine and cellulose) (see Figure 9).

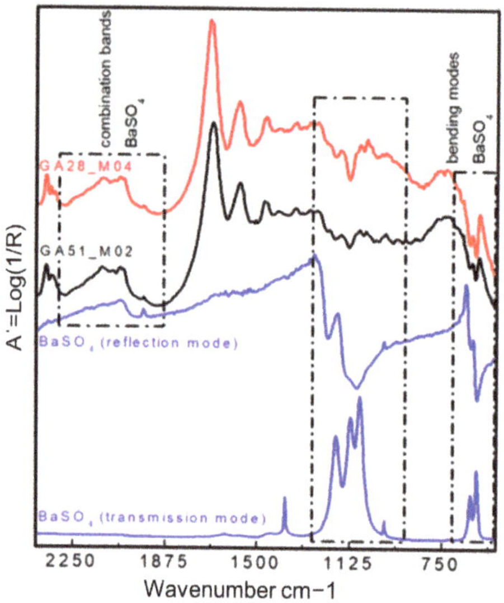

Figure 9. Comparing transmission mode and reflection mode spectra recorded from a barium sulphate reference compound—blue spectral lines—with the spectra obtained in the areas of minimum density (Dmin) of two original bromoil samples from Pla Janini (*Janini_028* and *Janini_051*). The characteristic IR bands of barium sulphate and their corresponding assignments are highlighted by black boxes in the range of 2250 to 1870 cm^{-1} (combination bands) and the v_4 bending vibration of the $SO2^{-4}$ between 700 and 550 cm^{-1}. Spectra obtained with the Alpha mobile spectrometer (Bruker Corporation, Billerica, MA, USA).

In fact, the identification of barite in bromoil prints by FTIR presents certain difficulties, resulting from the superposition of the stretching modes located in the range of 1280 to 970 cm^{-1} (indicator generally used for the detection of the sulphate anion) with bands of different materials present in the sample, such as cellulose, gelatine, and, in some cases, silicate compounds such as kaolinite. The second problem, as stated by Miliani et al. [24], is that the bands located between 1200 and 1000 cm^{-1}, typical of the asymmetric stretching v_3 of the tetrahedral anion $SO_4{}^{2-}$ present in barite, are not observed in the reflection mode due to its high absorption index k, by which all the v_3 bands of the sulphate compounds are affected by the so-called Reststrahlen effect. As a result, they appear in the spectrum as inverted bands [24], which are difficult to recognise when they are due to the different materials that make up bromoil prints and which can sometimes cause confusion when the experimental spectra are interpreted, especially if no comparison with another spectrometer or technique (as transmission mode) is available [36].

However, in the bromoils analysed with the Alpha mobile spectrometer (Bruker Corporation, Billerica, MA, USA), the combination bands in the range of 2250 to 1870 cm^{-1}, together with the v_4 bending vibration of the $SO_2{}^{-4}$ between 700 and 550 cm^{-1}, can be read as an indication of barium sulphate, especially when clearly detected in the areas of minimum density or at the margins without images of the combination bands (see Figure 9).

This information was very relevant in confirming the presence of the barite layer in some of the samples where the previous microscopic examination had not given clear results (see Figures 10 and 11), especially in the case of very thin barite layers. For example, in the case of samples *Janini_028* and *Janini_026*, the analysis with the mobile FTIR spectrometer has allowed us to confirm the presence or absence of the barite layer, information that was intuitively present in these samples at the morphological level (see Figures 10 and 11). On the one hand, the measurements carried out in the Dmin regions of sample *Janini_028* (see Figure 12, red spectral line) clearly show the inverted bands corresponding to the bending modes of barium sulphate at about 615 and 640 cm^{-1}, as well as the combination bands located between 2500 and 1870 cm^{-1}. For sample *Janini_026*, however, it is confirmed that there is indeed no barite layer, since the IR bands associated with barium sulphate are not identified (see Figure 12, blue spectral line).

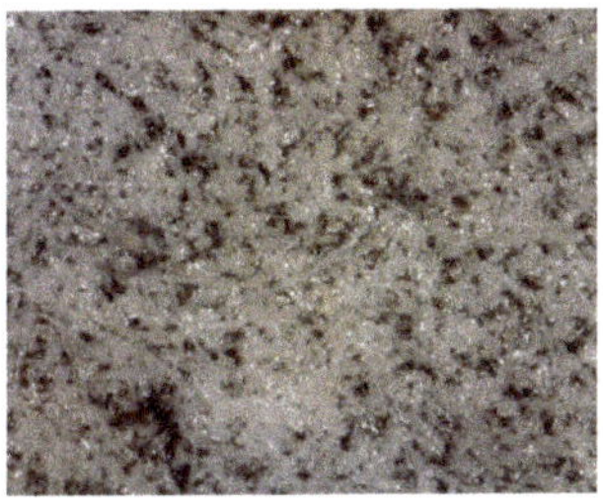

Figure 10. Janini_026_Dmin_1 (5) 180× (Digital Surface Microscope).

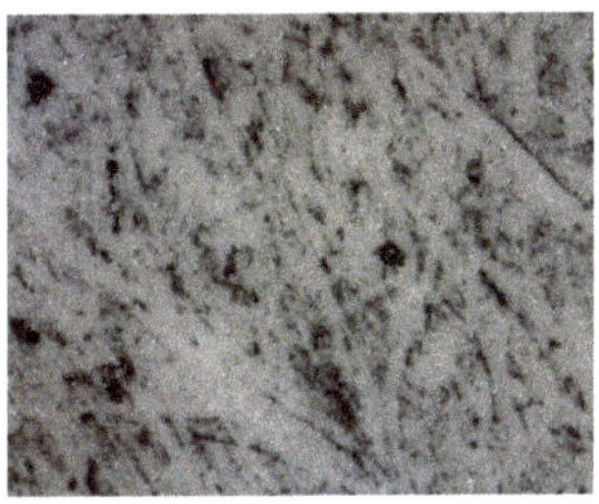

Figure 11. Janini_028_Dmin_2 (2) 180× (Digital Surface Microscope).

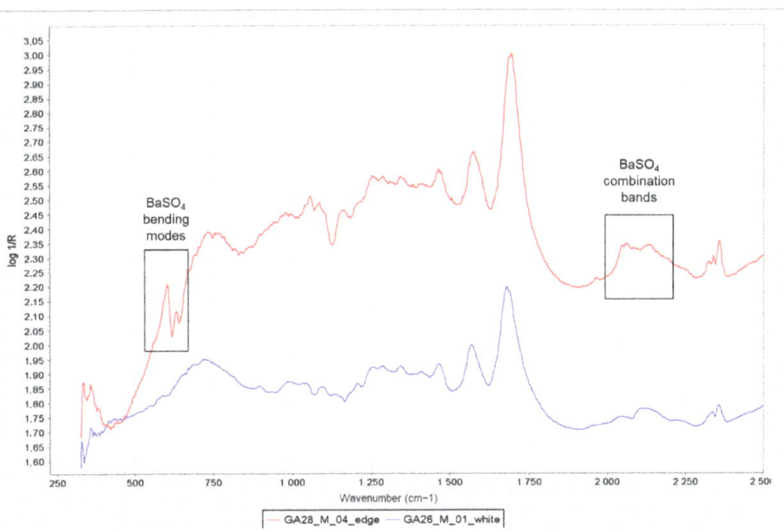

Figure 12. A comparison of two-point FTIR spectra corresponding to the Dmin region of a barite-coated bromoil sample, *Janini_028* (red), and a barite-free sample, *Janini_026* (blue). The combination bands of barium sulphate are visible in the 2500–1870 cm^{-1} region in the *Janini_028* sample (red), as are the two strong inverted bands at 640 and 617 cm^{-1} corresponding to the bending modes; neither of these bands is visible in the *Janini_026* sample (blue). Spectra obtained with the Alpha mobile spectrometer (Bruker Corporation, Billerica, MA, USA).

In the spectra obtained with the mobile spectrometer, it is again observed how, as the image density increases, the relative intensity of the described bands corresponding to barium sulphate decreases in Dmid and Dmax (see Figure 13).

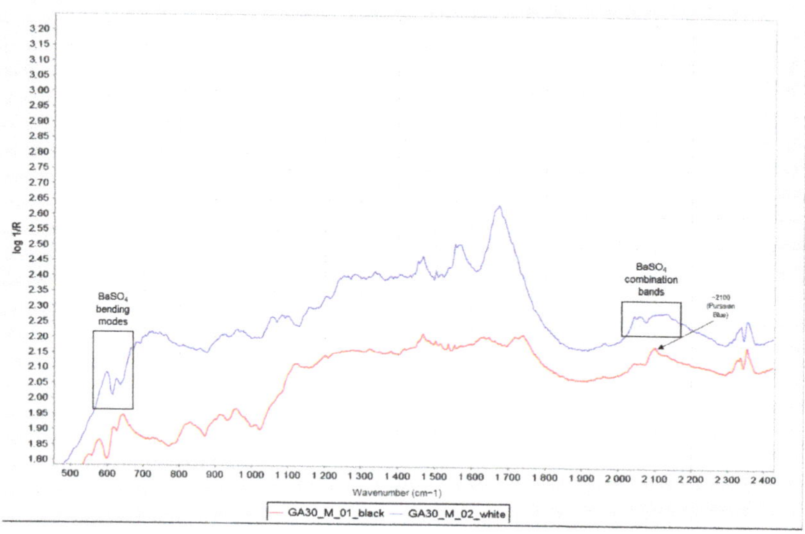

Figure 13. Spectra corresponding to the Dmin (blue) and Dmax (red) regions of the bromoil sample *Janini_030* with a barite layer. It can be seen how the relative intensity of the barium sulphate bands in Dmax decreases with respect to the Dmin region and the typical band of Prussian blue appears at 2100 cm^{-1}. Spectra obtained with the Alpha mobile spectrometer (Bruker Corporation, Billerica, MA, USA).

Thus, according to the results presented, it is possible to identify IR bands characteristic of barium sulphate in bromoil photographic prints by means of FTIR spectroscopy in the reflection mode, which are particularly evident in the areas of minimum density of the image, while they may be hidden or have lower relative intensities in the dense areas of the image. Specifically, by correlating the data previously obtained with the optical microscope and the digital surface microscope with the results of the mobile FTIR (applied only to the Pla Janini samples), it was possible to determine that three of the eight historical samples of Joaquim Pla Janini, namely *Janini_018, 020,* and *026*, showed no signs of barite.

On the other hand, it was very complex or uncertain to confirm the presence of the barite layer using the Nicolet iN10 MX desktop spectrometer (ThermoFisher Scientific, Waltham, MA, USA), since the spectral range of the instrument does not allow the inverted bands at 640 and 610 cm^{-1} to be displayed, and therefore, the identification can only be made using the combination bands at 2500–1870 cm^{-1} (see Figure 14). These data were confirmed by comparison with the FTIR results from the mobile instrument.

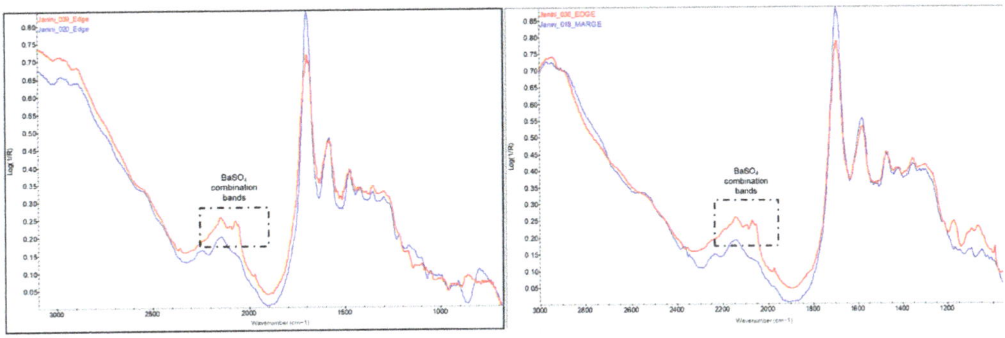

Figure 14. On the right, a comparison of two-point FTIR spectra corresponding to the Dmin region of a barite-coated bromoil sample (*Janini_030*, red) with one without barite (*Janini_018*, blue). On the left, comparison of two points corresponding to the Dmin area of a barite-coated bromoil sample (*Janini_039*, red) with one without barite (*Janini_020*, blue). The combination bands of barium sulphate in the range of 2500 to 1870 cm^{-1} are evident in samples *030* and *039* (with barite), whereas they are absent in samples *018* and *020* (without barite). Spectra obtained using a Nicolet™ iN10 MX desktop spectrometer (ThermoFisher Scientific, Waltham, MA, USA).

Despite the limitations of the desktop equipment for the characterisation of barite, this system has allowed the detection, in some samples, of inorganic compounds that can be attributed to the paper support, such as small traces of kaolinite (3685–3620 cm^{-1}). The similar response of this compound in the spectra of the back, the front in Dmin, and the areas of the margin without ink shows that it is not part of the layers found only on the front of the copy but is intrinsic to the paper (see Figure 15).

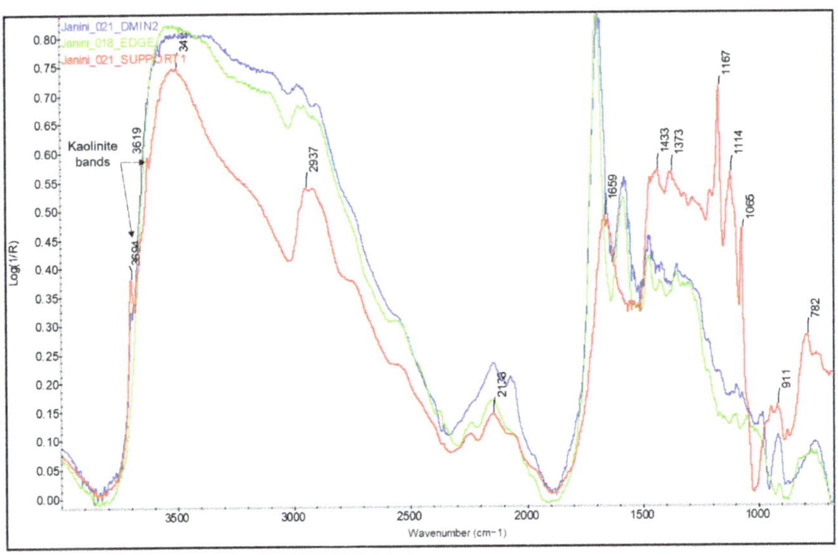

Figure 15. FTIR spectra of the edge region (Dmin) of the historic barite-free bromoil sample *Janini_018* (green), the Dmin region (blue), and the back of the primary support of sample *Janini_021* (red), which appears to contain a barite layer. The typical bands associated with kaolinite at 3685 and 3620 cm^{-1} from the primary support of sample *Janini_021* are observed, but there is no significant difference in the detection of cellulose from the back side between the two samples. Spectra obtained using the Nicolet™ iN10 MX desktop spectrometer (ThermoFisher Scientific, Waltham, MA, USA).

2.1.4. Oil-Based Ink

Contrary to the other compounds presented, and as mentioned at the beginning of this section, this study confirms that the oil-based ink is the only layer of a bromoil print whose components show a response in FTIR that is directly proportional to the density of the image.

Using FTIR in the external reflection mode it is possible to confirm the presence of drying oils in the bromoil prints, mainly in the Dmax areas, thus confirming the oily nature of the ink with one of the two spectrometers used in the study, although the desktop equipment has allowed a more precise identification of the bands associated with the oils and their level of polymerisation.

The carbonyl band (1630–1750 cm^{-1}), which is sharply defined at 1750–1740 cm^{-1} by the ester group (see Figure 16), is the most characteristic absorption band of oils [37]. This band is a clear characteristic of oil because it is the only natural organic material of the five main categories (proteins, oils, vegetable and insect resins and waxes) that has an intense carbonyl band in this region. The problem is that when mixed with certain pigments, the carbonyl band can be shifted to lower wavenumbers [32]. Other characteristic bands of oils are the aliphatic C-H bands that appear at about 1464, 1378, and 723 cm^{-1} and the C-O bands at 1240, 1165, and 1100 cm^{-1} [32,37], but in the case of compositionally complex samples, such as bromoil prints, these bands are overlapped by other compounds present in the prints, such as the protein binder, cellulose or barium sulphate, etc., making it difficult to confirm the presence of drying oils in this region (see Figure 16).

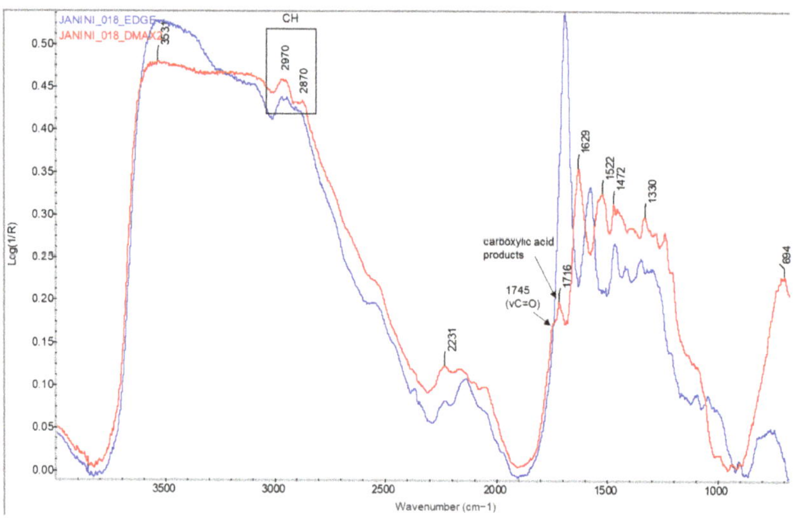

Figure 16. FTIR spectra of the edge area (blue) and Dmax (red) of bromoil sample *Janini_018*. Spectral changes related to the appearance of absorption bands characteristic of the drying oil present in the greasy ink can be seen at Dmax, due to the appearance of the carbonyl band at 1745 cm^{-1} and the formation of carboxylic acid products at 1716 cm^{-1}, related to the polymerised state of the oil. Changes are also observed in the CH region of the oil at 2970 and 2870 cm^{-1}. Spectra obtained using the Nicolet™ iN10 MX desktop spectrometer (ThermoFisher Scientific, Waltham, MA, USA).

In the typical oil spectrum, the CH_2 stretching bands are also prominent in the upper wavenumber region of the spectrum. The olefinic C-H stretching band at 3010 cm^{-1} is attributed exclusively to non-conjugated and symmetrically disubstituted cis double bonds [35], while the methylene bands are located at 2926 and 2855 cm^{-1} [32]. The visualisation of these bands is also often compromised by the superposition of bands from different compounds or spectral distortions, as will be shown below. For this reason, the detection of drying oil in bromoil samples has been mainly based on the carbonyl bands appearing in the 1630–1750 cm^{-1} region.

On the other hand, it is important to note that in order to associate the presence of esters with the image-forming material, and thus to attribute them to a print with greasy inks, these esters must be detected with a relative intensity that is directly proportional to the density of the image, since the amount of ink, and therefore the concentration of drying oil, is also proportional. Otherwise, the presence of esters in a photographic print could be circumstantial (e.g., possible retouching) and not due to the specific photographic process. In other words, the intensity of the bands corresponding to the oils is greater in the areas of greater image density (Dmax) and may be imperceptible in the lighter areas of the image (Dmin), even if they contain a certain amount of ink. This is because, in these Dmin areas, the compounds of the lower layers appear between the ink-free areas, so that they are detected clearly and with greater intensity than in the areas where they are covered by the ink layer (Dmax). It is in these Dmid and Dmax areas that the typical oil bands begin to be distinguished.

For example, Figure 16 shows the Dmin and Dmax spectra of a Dmin region of the *Janini_018* sample. The Dmin or edge spectrum (blue spectral line) consists of the typical bands of the protein binder and some coming from the lower cellulose. Comparing the Dmax spectrum (red spectral line), a new rather broad band appears centred at 1716 cm^{-1}, consisting of a strong shoulder at 1745 cm^{-1} related to the carbonyl band characteristic of oils. The peak protruding from the centre of the band at 1716 cm^{-1}, according to the studies carried out by Meilunas et al. [35], suggests the formation of carboxylic acid products, components associated with oxypolymerised oily materials. Meilunas et al. [35] also point out that the absorption peak centred at 1776 cm^{-1} is useful for identifying aged oil samples, since other materials, including pigments, are generally transparent in this region of the spectrum. Other authors, such as Lazzary and Chiantore [38] and Derrick et al. [32], who have studied the structural and spectral changes in linseed oil during drying and aging, also point to a broadening of the carbonyl absorption (1850–1650 cm^{-1}) and the formation of additional ketones, esters, and acid carbonyls (1750–1700 cm^{-1}). In the present sample, the back at 1776 cm^{-1} of the polymerised oils mentioned in the references is not very prominent (see Figure 16), although it may be hidden under the broad carbonyl band, which appears as a prominent back at 1745 cm^{-1}. It should be noted that in the spectra of the reference literature [35], this back at 1770–1780 cm^{-1} is also not pronounced but rather weak, similar to the experimental spectrum of sample *Janini_018* (see Figure 16).

This pattern is clearly repeated in all the bromoil prints examined, although the clarity of the oil absorption bands obviously varies according to the point, area, and sample analysed.

As mentioned above, it is difficult to distinguish between the aliphatic C-H and C-O bands in the 1460–700 cm^{-1} region due to the bands of other compounds present in the sample, such as gelatine and cellulose. Although these materials are found in lower layers and their absorptions are clearly influenced by the presence of a thick layer of oily ink, their bands eventually predominate and are completely mixed with those of the oil in this region of the spectrum. It should also be noted that other materials may be involved, such as the pigment itself, additives, fillers, etc. Finally, in the region between 3000 and 2850 cm^{-1}, it is difficult to distinguish significant absorptions that may be associated with the drying oil, such as the prominent hydroxyl/hydroperoxide band between 3600 and 3400 cm^{-1} [35,38], mainly due to the confluence of bands from several compounds. However, a smooth curve is observed in this region, as well as a slight spectral shift at 2969 and 2870 cm^{-1} between the spectrum of the edge region (see Figure 16, blue spectral line) and the Dmax region (see Figure 16, red spectral line), bands that may be associated with the C-H stretching modes of the oil.

A very similar analytical profile was obtained in the samples to the bromoil of Josep Cortina Molist. In the figure below (see Figure 17), the spectrum of the Dmin region (blue spectral line) and Dmax region (red spectral line) corresponding to sample *Cortina_002* are shown. In the spectrum of the Dmax region, the carbonyl band of the oil appears centred at 1746 cm^{-1} and a slight shoulder seems to start to form at ~1780 cm^{-1}. In this case the band at 1715 cm^{-1} is not clearly distinguishable, probably because the bands in this region appear very close together and practically merge with the absorption bands of the gelatine, although a small peak can be seen that stands out near 1715 cm^{-1}. The fact that the carbonyl band appears perfectly centred at 1746 cm^{-1} and the shoulder at 1715 cm^{-1} of the carboxylic acids is not so present indicates that there is perhaps a lower level of polymerisation of the oil compared to the *Janini_018* sample seen previously. Again, the appearance or definition of bands at 2972 and 2876 cm^{-1} can be seen, which can be related to the CH vibrational modes of the oil. However, in this sample, the appearance of oxidation products with hydroxyl characteristics (3600–3000 cm^{-1}) seems to be better distinguished.

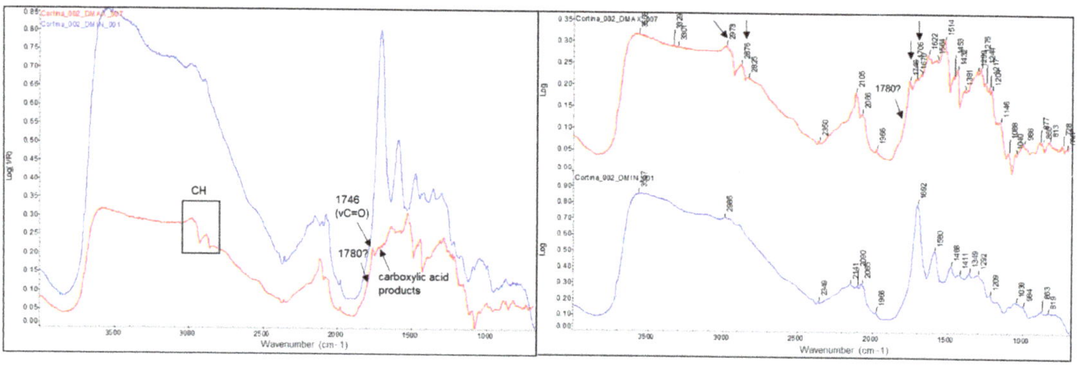

Figure 17. FTIR spectra of the Dmin (blue) and Dmax (red) regions of the bromoil sample *Cortina_002*. The carbonyl absorption bands at 1746 cm^{-1} and the CH bands of the oil at 2972 and 2876 cm^{-1} can be seen. Spectra obtained using a Nicolet™ iN10 MX desktop spectrometer (ThermoFisher Scientific, Waltham, MA, USA).

Finally, it is interesting to compare the spectra of the original historical samples from the mid-20th century with those of the more recent bromoil samples by Jaume Balanyà. Despite the fact that the latter are almost 14 years old, the spectral differences due to the polymerisation, oxidation, and aging reactions are obvious when compared with a sample that is almost 90 years old, such as the *Janini_018* print. If we look at the Dmax spectrum of the *Balanya_003* sample (see Figure 18), we can see some small changes, especially in the region between 1790 and 1710 cm^{-1}. In this case, the typical carbonyl band of drying oils appears more angular and perfectly defined, centred at 1749 cm^{-1}. Next, associated with the formation of carboxylic acid products, is a very small peak at 1717 cm^{-1}. However, this is almost imperceptible, especially when compared to the sharp peak of the *Janini_018* Dmax spectrum (see Figure 16) or the broader peak at 1715 cm^{-1} of the *Cortina_002* sample (see Figure 17, red spectral line), which is consistent with a lower oxidative polymerisation of the oil compared to the historical samples. There is also a small peak at 1775 cm^{-1}. In addition, the presence of oxidation products with hydroxyl characteristics (3600–3000 cm^{-1}) is not identified, unlike the older samples where, although they are mixed with bands of other compounds due to the position of the absorptions, it is possible to perfectly guess the curvature of the spectral line between 3600 and 3400 cm^{-1}, which on the other hand does not appear in the Dmax spectra of the contemporary samples (see Figure 18, red spectral line).

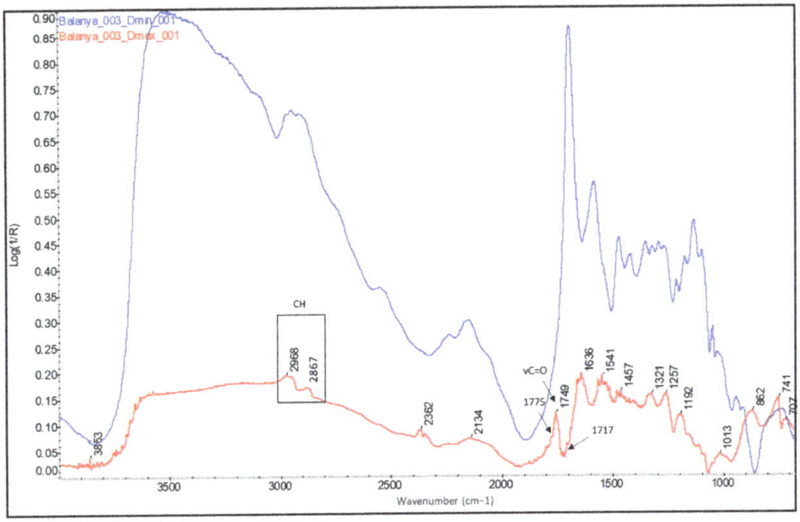

Figure 18. Dmin (blue) and Dmax (red) area spectra of the contemporary bromoil sample *Balanya_003*. The Dmax area spectrum clearly shows the carbonyl band of the oil at 1748 cm^{-1}, accompanied by a small shoulder at 1775 cm^{-1} and a small peak at 1717 cm^{-1}, absorptions associated with oxypolymerised lipid materials. C-H stretching bands are also detected at 2968 and 2867 cm^{-1}. Spectra acquired on a Nicolet™ iN10 MX desktop spectrometer (ThermoFisher Scientific, Waltham, MA, USA).

2.1.5. Pigments

Finally, reflection-mode FTIR spectroscopy has proved useful in detecting certain pigments present in the sample images, particularly using Bruker Corporation's mobile Alpha instrument (Billerica, MA, USA), thus allowing the print to be characterised as a pigmented photographic print.

The main pigments detected were Prussian blue or ferric ferrocyanide ($Fe_4[Fe(CN)_6]_3$) and earth pigments composed of iron oxides, detectable thanks to the Fe-O and Fe-OH vibrations that make them identifiable with FTIR.

Prussian blue has been detected in almost all the samples, except for the three contemporary bromoil samples by Jaume Balanyà, which show a warm black image, and the historical samples *Janini_018* and *Janini_039*, both of which were produced with a clearly orange-brown pigment. In the great majority of samples where the Prussian blue pigment has been detected, it is present as a secondary pigment in the black ink, i.e., as an additive to give more intensity and contrast with the warm tone of the black pigment. In other samples with more bluish or greenish tones, such as *Cortina_003* and *Janini_028*, it has been visualised microscopically as a single colour within the palette that forms the images (see Figures 19 and 20).

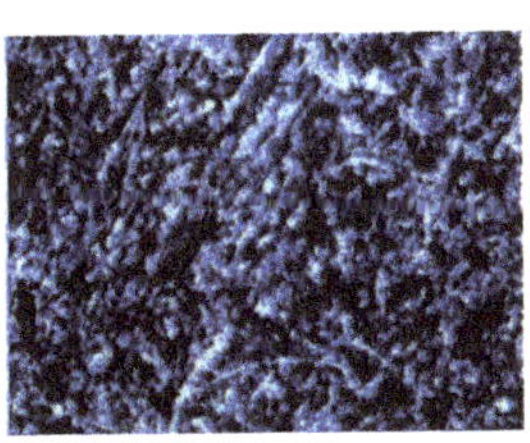

Figure 19. Cortina_003_Dmax 1 (6) 180× (digital surface microscope).

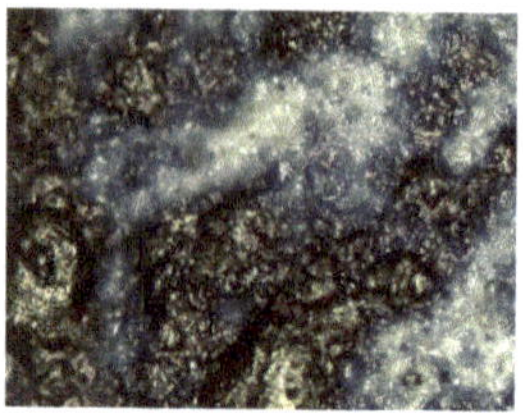

Figure 20. Cortina_003_Dmax 100× (optical microscope).

Using FTIR spectroscopy, Prussian blue is an easily identifiable pigment because it shows a strong absorption band at around 2100 cm^{-1}, due to the triple bond CN (see Figure 21). It also usually shows a band at 500 cm^{-1} and two weak bands at ~608 and ~1420 cm^{-1} [39], but these are difficult to distinguish in the recorded spectra due to the superposition of bands from other compounds in these regions of the spectrum.

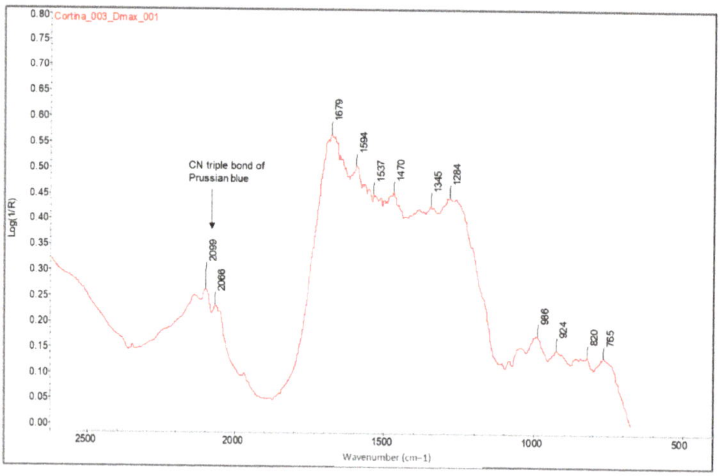

Figure 21. Spectrum of the Dmax region of the bromoil sample *Cortina_003*, where the characteristic band of the CN triple bond of Prussian blue is identified, in agreement with the blue-tinted image of the copy as observed in the microscope images. Spectrum obtained using a Nicolet™ iN10 MX desktop spectrometer (ThermoFisher Scientific, Waltham, MA, USA).

Further FTIR spectra are shown below, taken from the areas of maximum density in the image of samples where ferric ferrocyanide has also been detected thanks to the clearly visible absorption peak at ~2100 cm^{-1} (see Figure 22). In addition, sample *Janini_028* contains a higher amount of this pigment, which can be deduced from the more intense signals it shows (see Figure 22, left, red spectral line).

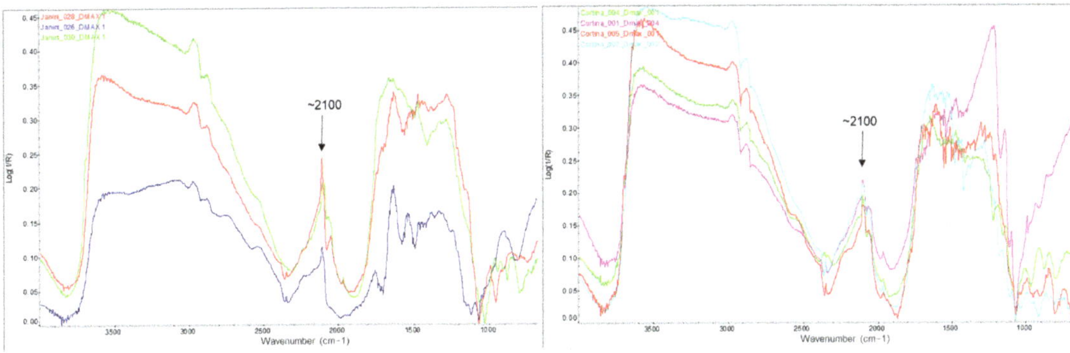

Figure 22. FTIR spectra of the Dmax regions of the bromoil samples *Janini_026* (blue), *028* (red), and *030* (green) on the left and *Cortina_001* (pink), *002* (blue), *004* (green), and *005* (red) on the right, where the strong absorption band at ~2100 cm^{-1} of ferric ferrocyanide (Prussian blue) is clearly visible. Spectra obtained using a Nicolet™ iN10 MX desktop spectrometer (ThermoFisher Scientific, Waltham, MA, USA).

Iron oxide pigments have been found mainly in association with those bromoil images that have a distinctly umber tone, although the chromatic range varies from more yellowish light brown tones to darker or reddish tones. This type of pigment was found in samples *Janini_018*, *020*, *021*, *039*, and *051*. The problem with umber or brown pigments is that, although FTIR spectroscopy has been used to verify the presence of one or more iron-oxide-based pigments in these images, it has not been possible to identify the specific pigments of which they are composed by FTIR spectroscopy, which would require a combination of other analytical techniques, such as Raman spectroscopy or X-ray fluorescence. This study has therefore been able to demonstrate the potential but also the limitations of FTIR in identifying pigments.

FTIR spectroscopy allows the detection of iron oxide pigments from two absorption bands located at the edge of the spectral range corresponding to low wavenumbers (~530 and ~440–430 cm^{-1}) (see Figures 23 and 24). However, these bands are not characteristic enough to identify the pigment in question [40], and in addition, the infrared bands of ferric oxides and hydroxides are outside the accessible range of some instruments [24], such as the Nicolet iN10 MX desktop spectrometer (ThermoFisher Scientific, Waltham, MA, USA). Therefore, iron oxides were only identified in the samples analysed with the mobile instrument Alpha (Bruker Corporation, Billerica, MA, USA). On the other hand, when using the reflection mode, these bands are distorted by specular (surface) reflection, producing inverted bands or residual peaks located between 600 and 400 cm^{-1} [40] (see Figure 23). In the case of the bromoil samples analysed, these bands were more specifically located at wave numbers 530 and 450–440 cm^{-1} (see Figures 23 and 24).

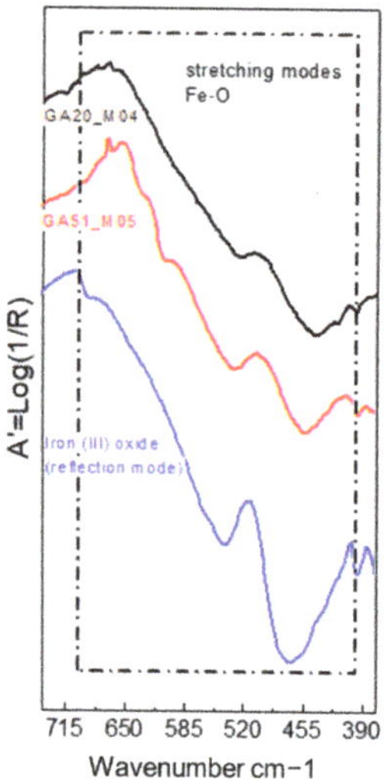

Figure 23. Using reflection mode FTIR spectroscopy, iron (III) oxides are identified from two inverted bands in the 670–400 cm^{-1} range corresponding to the Fe-O stretching modes (see comparison of reflection mode spectra recorded for the corresponding reference compound (blue line) and for two of the samples analysed (black and red). Spectra obtained with the Alpha mobile spectrometer (Bruker Corporation, Billerica, MA, USA).

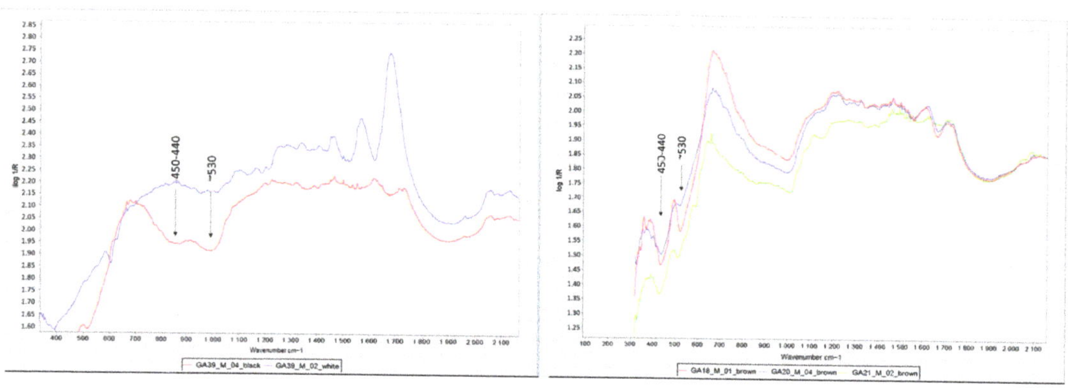

Figure 24. Location of the inverted bands of the stretching modes (Fe-O) of the ferric oxides corresponding to the umber pigments detected in the bromoil images of the *Janini_039, 018, 020,* and *021* samples. The FTIR spectrum on the left shows a comparison between a Dmin (in blue) and Dmax (in red) area; note that the bands located at 530 and 450–440 cm^{-1} appear only in the Dmax area. Spectra obtained with the Alpha mobile spectrometer (Bruker Corporation, Billerica, MA, USA).

In the case of both iron oxide pigments and Prussian blue, the characteristic IR absorption bands appear parallel to the increase in image density, i.e., in relation to the areas of maximum image density. This confirms that these are compounds involved in image formation. The presence of pigment particles with these characteristics was also previously confirmed by microscopy (see Figures 25 and 26).

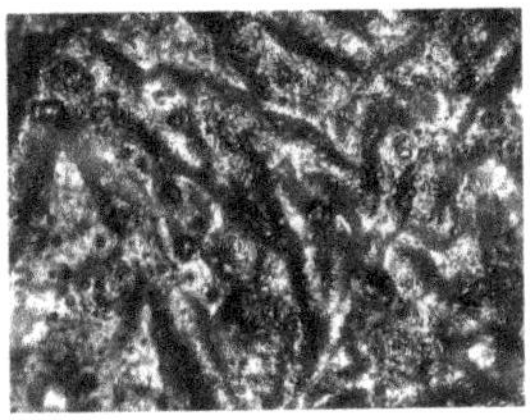

Figure 25. Janini_020_Dmax_100× (optical microscope).

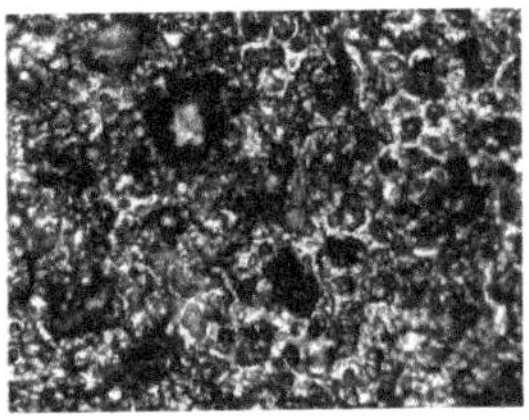

Figure 26. Janini_051_Dmax_100× (optical microscope).

2.2. Discussion

In the current study, we have succeeded for the first time in characterising the main components of all the layers of bromoil prints via FTIR spectroscopy. Given the inherent heritage value of these works, the methodology developed is non-destructive and non-invasive.

The paper proposes a study methodology through which FTIR spectroscopy in external reflection mode emerges as a highly valuable technique for identifying bromoils prints, characterising them, gaining a deep understanding of their composition and structure, and even detecting mechanisms of alteration and/or aging of their components.

As a result of applying this methodology, the possibilities and limits of the application of FTIR in the study of bromoils prints are unveiled, considering the capabilities of two types of spectrometers, one desktop and one mobile. The results of the study display the compounds that can be detected, as well as the conditions under which these, when detected in a photographic print, can be considered indicators or evidence for the identification of bromoils prints.

The basis of the proposed methodology lies in the identification system introduced by James Reilly in 1986 [41], which classifies historic photographic prints, specifically from the 19th century, into copies of one, two, or three stratigraphic layers depending on how the fibres of the paper support are observed from the front through different image areas. With the purpose of employing some of Reilly's methods for the study and identification of photographic copies, we have adapted to the bromoil prints analysis methodology two fundamental issues: on one hand, discerning measurement areas based on image density and, on the other, interpreting the results of the spectra based on the stratigraphic structure of the sample. Adopting these criteria allows us to demonstrate the potential to recognise the qualities of the different compounds present in the copy and to relate them to their stratigraphic position, thus establishing a deeper interpretation of the composition and structure of the bromoil prints.

The combination of these two pieces of information—composition and structure of the layers—is indispensable not only for understanding the original but also for its identification as a bromoil print.

Thus, the method used has proven to be very effective in obtaining optimal results, so it can be said that it significantly contributes to the methodological issues for future studies of bromoil copies—and quite likely of fatty positives in general—using spectroscopic techniques.

The results have demonstrated that all those materials inherent to the image, that is, the oils and the pigments that make up the oil-based ink, have been detected in intensity proportional to the image density, mainly in the Dmid and Dmax areas. In contrast, all those compounds associated with the original photographic support, such as the protein binder, the barite layer, and the cellulosic substrate, which are located in layers beneath the greasy ink layer, have been detected with inverse proportion to the image density. This is due to the greasy ink acting literally as a physical layer that, as it increases in concentration in the areas of medium and maximum image density, covers and hides the FTIR signals coming from the lower substrates. This aspect is characteristic of oil-based ink copies and should, therefore, be considered as a key to the identification of these works. This is crucial to avoid incorrect identifications that may result from erroneously attributing the spot detection of oily materials on the surface of a photograph made by another procedure—such as possible retouches where paints containing drying oils are used—to the presence of a greasy ink layer that forms the image, characteristic of a bromoil copy.

Thus, when analysing a bromoil copy, measurements made in areas of maximum image density (Dmax) mainly allow detecting and recognising the materials that form the image, such as pigments and drying oils that make up the oil-based ink, while measurements in areas of minimum image density (Dmin) detect more intensely and clearly the characteristic bands of the materials coming from underlying layers (barite layer, photographic binder, and paper support), as these areas of the image are not hidden under the greasy ink layer, as happens in Dmax. In the case of cellulose, it remains almost hidden even in the Dmin areas due to the overlap of the characteristic bands of the protein binder. If possible, it is also advisable to perform measurements at points of medium image density (Dmid) to study more deeply the changes generated between image areas. Finally, if the sample allows, it is recommended to conduct tests on the back of the primary support to rule out materials specific to the paper support that may interfere in the detection of materials of interest when analysing the sample from the front.

This study demonstrates that FTIR spectroscopy in external reflection mode is a technique capable of characterising the major components of each of the layers that constitute bromoils and their location in the work: the cellulose of the paper, the barite layer, the gelatine layer, the drying oils, and certain pigments of the image-forming layer. Thus, the results obtained both from close examination under the microscope and FTIR spectroscopy allowed us to propose a clear representation of the stratigraphic structure of the bromoil copies (see Figure 27).

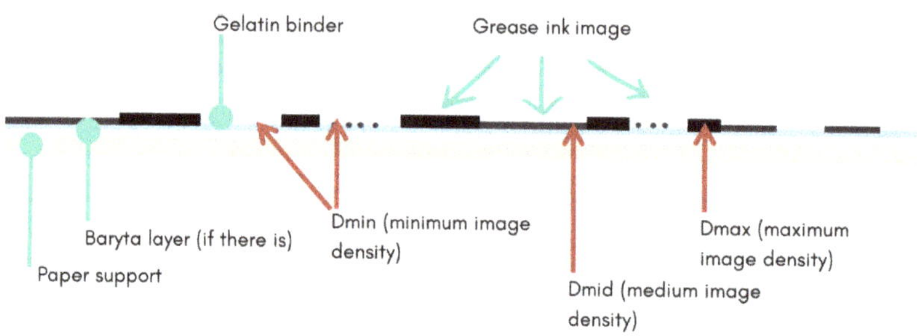

Figure 27. Graphic representation of the stratigraphic structure of a bromoil print. At least three layers are distinguished: the paper substrate, the barite layer (if present), the gelatine binder, and the layer of greasy ink that forms the image. The areas of Dmin correspond to the areas with the minimum ink content, the Dmid areas contain a medium density of the image, and the Dmax areas correspond to the blacks or points of maximum ink density.

On one hand, the results of our study demonstrate that in bromoil, the image-forming material, namely the greasy ink, is not located inside the gelatine, as is the case in most non-fatty photographic processes with a protein binder. Instead, it is situated in a top layer, in line with other reference sources [8]. Therefore, the gelatine cannot be considered a colloid that suspends the particles forming the image, whether silver or pigments, as in other non-fatty and silver pigmentary photographic processes, but rather a layer that functions similarly to a matrix at the time of image creation, when the oil-based ink is manually incorporated into the copy, adhering to it based on the degree of wetness of the gelatine, inversely proportional to the degree of hardening suffered during the prior bleaching process, according to the silver concentration of the original bromide [7,42–44]. From these results, we propose to reconsider the use of the terms *colloid* or *emulsion* when talking about the gelatine layer in bromoils (terms widely used for pigmentary and silver copies). Considering the function this protein layer has in bromoils, it is suggested to better understand it as a *matrix* or *intermediary layer.*

This conceptual and terminological issue holds pivotal relevance in the characterisation of bromoils through instrumental analysis as, as shown by the results of this study, the location of the ink layer as an independent stratum that overlays the rest of the layers directly affects how the other materials of the work inherent to the photographic support are detected, namely, those situated in layers beneath the ink, constituting this phenomenon a key identification factor of bromoil copies.

On the other hand, this study reveals the existence of bromoil prints without a barite layer, which cannot be considered exceptional cases. Of the eight historic samples from Joaquim Pla Janini, analysed with the Alpha mobile spectrometer (Bruker Corporation, Billerica, MA, USA), three prints without a barite layer were detected, coinciding with the results of the microscopic study. This indicates that the presence of barite is not an indicator for the identification of the bromoil technique. Moreover, the microscopic examination also did not provide evidence of a barite layer in the three contemporary bromoil prints, information that was confirmed by the testimony of the author himself, who provided samples and the exact brand of the photographic paper used, Kentmere® Document Art (Balanyà, personal communication, June 2019). These observations help clarify the current contradictions that appear in the literature on this topic where, on one hand, bromoil copies are included within the category of two-layer photographic processes, alongside carbon and gum dichromate prints [9], overlooking the fact that many bromoil prints contain a barite layer. However, in other references [8], the presence of this barite layer is considered one of the keys to the identification of bromoil prints. The use of FTIR spectroscopy for the analysis of heritage collections of bromoil prints has served to deepen the recognition of the compounds forming their structure, especially in terms of detecting the barite layer, which may be often absent. This allows us to suggest that the presence or absence of the barite layer can no longer be considered an identification key for bromoil copies. This study contributes to the generation of new knowledge of bromoil prints verified through the study of a heterogeneous collection of samples by means of FTIR analysis.

Therefore, it is important to consider that the characterisation of bromoil prints using the visual recognition system by layers must consider the possibility that these works have four layers in cases where they include the barite layer and three layers when using a paper without this intermediate stratum. The classification of bromoil positives as three- or four-layer photographic prints should also consider the greasy ink layer as an independent stratum.

It should be mentioned, therefore, that during instrumental analysis, it is important to precisely select the measurement points of the copy to maintain rigor in the interpretation of the spectra based on the presence of the ink layer in each analysed area (Dmin, Dmid, and Dmax). During this study, it has been verified that conducting the analyses with a spectrometer equipped with an attached microscope significantly increases the accuracy and reliability of the results, especially in the case of complex photographic samples like those analysed, where the distribution of materials can vary significantly across the surface.

In this article, it has been demonstrated that the external reflection mode is highly suitable for characterising bromoil prints using FTIR spectroscopy because, on one hand, from a methodological perspective, it is less invasive than the ATR mode as it avoids contact with the instrumentation, and on the other hand, this mode has proven useful for achieving the identification of the compounds in these works. The main drawback of using FTIR spectroscopy in external reflection mode is related to the interpretation of spectral data, which can exhibit significant distortions in bands shape, absorption frequency, and intensity compared to transmittance spectra.

These anomalies, produced by various factors such as absorption indices (*k*) and refraction indices (n) and surface roughness, which affect the extent of contributions from specular (or surface) and diffuse (or volume) reflection [28], can lead to Reststrahlen bands (or inverted bands) when the sample contains a compound with $k \gg 1$ [28,36]. This last distortion occurs frequently in bromoil samples in the analysis of the barium sulphate present in the barite layer and in iron oxide pigments.

The type of equipment used can also contribute significantly to the spectral response. Generally, the optical geometry of portable or mobile infrared fibre optic systems tends to maximise specular reflected light by working with an angle of incidence equal to the reflected angle [24]. However, in the infrared reflection measurements of works such as the photographic prints analysed, both surface reflection and volume reflection are generally active, affecting the spectral characteristics depending not only on the optical properties of the materials but also on the surface roughness. Regarding morphological properties, surfaces with particle sizes larger than the wavelength of infrared radiation provide a greater amount of specular light, while rougher surfaces, with particle dimensions similar to the wavelength of infrared radiation, mainly generate diffuse reflected radiation [24]. Optical properties, in terms of absorption and refraction indices, will influence the degree of light penetration and, therefore, the contribution of surface and volume reflection.

Consequently, in the analysis of complex samples, such as bromoil prints, where surface properties are very heterogeneous, different types of distortions may coexist in the same spectrum due to the derived shape, Reststrahlen effect, and intensity enhancement, making it impossible to apply Kramers–Kronig and Kubelka–Munk corrections [45], significantly complicating their interpretation.

In this study, this was particularly recurrent in the identification of barium sulphate from the barite and iron-oxide-based pigments. In the bromoils analysed, the cellulose and gelatine bands of the spectra obtained by reflection were comparable without much difficulty to the known bands of these materials in ATR mode.

Finally, it is particularly relevant to note the advantages and limitations that have been detected during the use of different types of instrumentation, as they have directly influenced the results. In the case of desktop spectrometers, first, the geometry or design characteristics of the equipment must be considered, not only to properly position the work in the spectrometer but also for the selection of the areas and measurement points of the samples. These devices are designed for analysing small samples, which can present difficulties for the light source to access more internal areas of the work, clearly restricting the selection of analysis points. Partly, this was one of the reasons why access to a portable device was sought to overcome this limitation inherent to the desktop equipment.

Moreover, and very importantly, the components (different sensitivities in varied spectral ranges) and the configuration of the spectrometer that may influence the extent of the spectral range must be considered. That is, depending on the limit set for the spectral range, IR bands of certain compounds that may be present may not be visible because their absorptions fall outside the established range. This is especially relevant when using the external reflection mode for certain compounds, such as with barium sulphate or iron oxides, because spectral distortions and band shifts occur in "unusual" ranges of the spectrum, which can lead to erroneous conclusions. In the present study, this circumstance occurred with the desktop equipment, with which the inverted bands associated with the mentioned compounds, appearing in wavenumber ranges of the lower range, could not be observed because the limit of the spectral range of this equipment was established around 700 cm^{-1}. The limit of the spectral range, however, is not necessarily directly related to the model (desktop or mobile) but rather to its features and configuration.

To conclude, although all major compounds have been detected with both devices, the desktop equipment has shown superior resolution, allowing greater precision in identifying minor bands associated with secondary compounds or physicochemical phenomena related to the state of conservation and aging of the samples, such as the appearance of absorptions that can be related to the degree of the oxy-polymerisation of the drying oils in the greasy ink.

3. Methodological Part

3.1. Photographic Artifacts

This study is based on 16 photographic bromoil samples, of which 13 are original historic prints can be dated between the late 1920s and 1960s, and 3 are contemporary prints produced in 2009–2010 (see Tables 1 and 2). The selection of works studied constitutes a sufficiently varied sample from both technical and material perspectives, regardless of the production era of the copy, which inherently introduces an implicit technical and material variable. The study includes bromoil prints composed of different tonalities and types of inks, some combined in the same work, and a significant variety of photographic supports of different formats, finishes, and surface textures.

Table 1. List of selected bromoil works as objects of study.

Historic Bromoil Photographic Samples			
Sample ID	**Author**	**Approximate Date**	**Collection**
Cortina_001	Josep Cortina Molist	Late 1940s to approx. 1950s	Josep Maria Cortina private collection
Cortina_002	"	"	"
Cortina_003	"	"	"
Cortina_004	"	"	"
Cortina_005	"	"	"
Janini_018	Joaquim Pla Janini	Undated (probably 1930s or earlier)	Vives Pla Family private collection
Janini_020	"	"	"
Janini_021	"	Undated (between 1920 and 1960)	"
Janini_026	"	Undated (between 1920 and 1960)	"
Janini_028	"	Undated (probably 1936 or earlier)	"
Janini_030	"	Undated (probably 1931 or earlier)	"
Janini_039	"	Undated (probably early 1940s or later)	"
Janini_050	"	Undated (probably 1935 or earlier)	"
Janini_051	"	Undated	"
Balanya_001	Jaume Balanyà	2009–2010	Jaume Balanyà private collection
Balanya_002	"	"	"
Balanya_003	"	"	"

Table 2. Images of the selected bromoil works as objects of study.

The historic bromoil samples come from two different collections and correspond to two authors of Spanish nationality. These copies were produced during the period of greatest expansion of the bromoil technique in Spain, covering the different phases that characterise the pictorialist movement in this country [1,3,46–48]. On one hand, the sample includes eight bromoil prints produced by the bromoil master Dr. Joaquim Pla Janini (1879–1970), considered the foremost reference for the bromoil technique in Spain and internationally recognised [3,47]. On the other hand, the historic bromoil sample also includes a total of five works by Josep Cortina Molist (1918–1995), an amateur photographer who was associated with the Agrupació Fotogràfica de Catalunya (A.F.C.), of which Dr. Pla Janini was the director during the years from 1927 to 1930, a period of peak activity for the A.F.C. [49]. In both cases, the selected works come from the private collection of the heirs of the aforementioned photographers.

The historic bromoil prints are an authentic reference: analyses from original works allow for a reliable understanding of the techniques and materials used during the height of pictorialism, as well as their evolution throughout the 20th century. This not only contributes to a more robust and reliable research methodology but also provides data on how these prints may have changed or deteriorated over time.

All the selected works of Dr. Joaquim Pla Janini have been corroborated as bromoil prints by the information written on the backs of the works, primarily annotations made by the artist himself, as well as stamps and labels from exhibitions and salons where the photographer presented the works (see Figure 28). The selected works from the collection of Josep Cortina Molist are recognised as bromoils because of the direct testimony of the artist's son, the person in charge of the collection, who has loaned the works for the study.

Figure 28. Participation labels from salons and exhibitions located on the back of the bromoil work *Janini_030*, including the I Salon International d'Art Photographique de Poitiers (November 1931), the XIII Salon International d'Art Photographique de Bruxeles (1931), and the exhibition organised by the Club Deportivo de Bilbao (December 1931).

The technical excellence of Joaquim Pla Janini's works is internationally recognised as a benchmark of high artistic quality in the bromoil process. This is supported by many of his contributions to specialised magazines of the time such as *Sombras*, *Luz*, *La Fotografía*, *El Progreso Fotográfico*, *Revista Fotográfica*, and *Arte Fotográfico*, the monograph dedicated to his person in the magazine *Art de la Llum* from October 1934 [50], or his participation in the first exhibition at the Parés Gallery in Barcelona in April 1928 or the I International Salon organised by the Agrupació Fotogràfica de Catalunya in 1929 [51], demonstrating his reputation and recognition as a leading figure of the pictorialist movement. His continued practice of bromoil reflects an artistic trajectory that spans various evolutionary stages of pictorialism and is representative of the materials and general procedures used in bromoil production in other European countries, as evidenced by the documentation of his private collection and the consultative sources preserved there, including English, German, and Italian manuals [7,52,53]. In fact, it was common for Catalan and Spanish bromoilists to draw information from manuals from other countries, especially England, Germany, and Italy. Moreover, there are very few examples of manuals and guides in Spanish and even fewer in Catalan. To give another example, among some of the manuals preserved by Josep Maria Casals i Ariet, one of the leading exponents of bromoil transfer and pictorialism in Catalonia, are Dr. Mayer's manual [7] and the publication by Symes [53]. In addition, one of the most important magazines of the time in the photographic field, *El Progreso Fotográfico*, was the translated edition of the famous magazine directed by Professor Rodolfo Namias, *Il Progresso Fotografico*, published in Turin, Italy.

On the other hand, it is very interesting to include part of the production of an amateur photographer like Josep Cortina Molist, who serves as the counterpart to Joaquim Pla Janini and represents the entire collective of enthusiasts for artistic photography that so characterised the associativism of the era, closely linked to the pictorialist movement. Like so many other amateur photographers of the time, Cortina also participated in the activities and salons of photography so common in that era, and his bromoil photographs connect him with the pictorialist aesthetic. Like many other enthusiasts, Mr. Cortina Molist did not achieve the recognition within the artistic circles that he probably deserved, while his works demonstrate a great mastery of photographic technique. In fact, the associativism of the time allowed amateur authors to access the knowledge of bromoil process that was disseminated by the most important authors in Europe, and therefore, it is plausible to think that they frequently used materials and methods similar to those used by the Spanish and European masters of the bromoil process.

The three contemporary bromoil prints selected for this study were created by the renowned photographer Jaume Balanyà, born in Barcelona in 1949. He is a professional photographer with 50 years of experience who has participated in exhibitions, won awards, carried out reportages, and published several books [54–56] specialising in old processes, such as salt paper, cyanotype, and bromoil. The bromoil prints loaned for this study are part of a series of 31 bromoil photographs that were exhibited for the first time in 2010 at the Altair bookstore and, subsequently, at the Saleta of Casa Golferichs in Barcelona during January and February 2011.

Incorporating contemporary professional-quality bromoil samples into the study provides the opportunity to understand the evolution of the methods and materials used in a photographic process that has nearly disappeared. On the other hand, they serve as a comparative reference to the historic copies to study the changes derived from the aging processes that affect bromoil prints and how these are reflected in the results of the FTIR analyses.

3.2. Experimental Part

Analyses using Fourier transform infrared spectroscopy (FTIR) were carried out in external reflection mode with two different devices, a desktop spectrometer and a mobile spectrometer.

Measurements with the desktop equipment were conducted at the Scientific and Technological Centers of the University of Barcelona (CCiTUB) (Barcelona, Spain) using a ThermoFisher Scientific® (Waltham, MA, USA) Nicolet iN10 MX spectrometer with an imaging microscope, equipped with a liquid nitrogen cooled mercury-cadmium-telluride (MCT) detector, KBr beam splitter, and globar light source. Measurements were carried out in external reflection mode, with an aperture of 100 µm × 100 µm, in the range of 4000 to 675 cm^{-1}, with a resolution of 4 cm^{-1} and 64 scans, and spectra were plotted as log(1/R).

Analyses conducted with the mobile equipment were performed with instrumentation provided by the MOLAB program and were made with an Alpha mobile spectrometer (Bruker Corporation, Billerica, MA, USA), which contains a high-sensitivity pyroelectric DLaTGS (deuterated L-alanine-doped triglycine sulphate) detector at ambient temperature. External reflection of an area approximately 5 mm in diameter was measured in a spectral range of 7500 to 375 cm^{-1} and a spectral resolution of 4 cm^{-1} over 300 scans, and spectra were plotted as log(1/R).

In the resulting spectra from the analyses of the bromoil prints, the Kramers–Kronig (KK) correction was not viable due to the combined response of surface and/or volume reflection attributable to the material, morphological, and surface complexity of the samples [24]. The heterogeneous nature of the bromoil prints, which includes variations in texture, material composition, and overlapping layers, significantly complicates the application of the Kramers–Kronig relations, a correction that requires a more homogeneous, shiny/polished (highly specular sample response) medium, which limited the use of the KK algorithm in this case [14,25].

For processing FTIR spectra obtained with the Nicolet iN10 MX equipment, Omnic 9.2.86 software was used, and for the spectra collected with the Bruker® Alpha mobile equipment, MOVIDA (MObile laboratory VIsualization Data) software was used [57].

Seven of the eight works by Janini were analysed using both spectrometers (desktop and mobile), except for *Janini_051*, which was analysed only with the mobile equipment. The works by Cortina and Balanyà were studied only with the desktop spectrometer. The locations of the points analysed with each device were recorded on measurement maps for each work.

In all the works, measurements were taken in areas of maximum image density (Dmax), equivalent to the blacks of the image, areas of minimum image density (Dmin), corresponding to the whites or highlights, and areas of medium image density or mid-greys (Dmid) (see Figure 29), following the methodology proposed by Reilly [41] for the morphological study of historic photographs. In the works where it was possible, the masked peripheral margins made equivalent to a pure Dmin area, that is, without any trace of greasy ink, were also analysed.

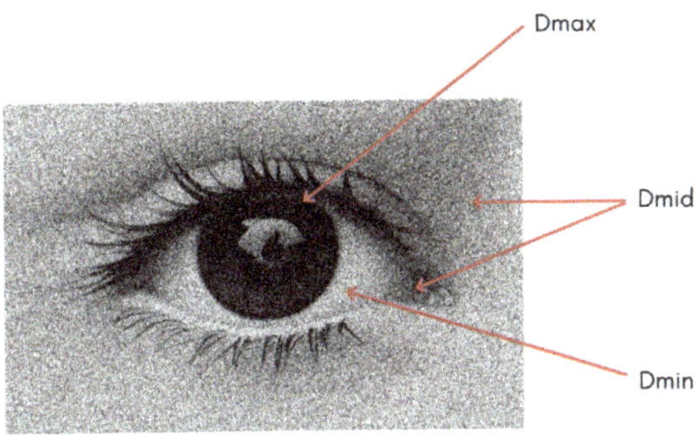

Figure 29. Graphic representation of the image areas of a bromoil copy corresponding to Dmax (maximum image density), which includes blacks and deep shadows with the maximum concentration of pigment from the grease ink: Dmid (medium image density), or midtones/grays, which have varying amounts of pigment (grease ink) proportional to the image density; and Dmin (minimum image density), or white areas, with little or no pigment presence. Top view.

Using the desktop equipment, a minimum of nine measurement points were set for each work: three measurements in Dmax areas, three in Dmid areas, and three in Dmin areas.

Additionally, whenever possible, three measurements were taken on the reverse side of the primary support, as well as other possible areas of interest (retouches, deteriorated areas, etc.). In the case of the mobile equipment, the number of points per area was reduced, but in all cases, a minimum of one point was measured for each representative area of the image (Dmin, Dmid, and Dmax), along with the reverse of the primary support.

The same methodology was used for the optical microscopy observation of the bromoils. A USB digital surface microscope, Hot® HT—60L, with Hotviewer® v2.7.9 software was used to carry out inspections at 80× and 180×. A revision of the works at 100× was also carried out with an Olympus® BX51 optical microscope coupled to a camera, Optika® 4083.B0.5, to take images using the Optika® Vision Lite programme.

The system for securing the works during measurement had to be adapted to the geometric specifics of each device used. For example, for the desktop equipment, originally designed for analysing small samples, a precision lift platform system with conservation cardboard support adapted to the device was developed to allow horizontal securing of the works, ensuring stability and avoiding vibrations, as the original stage of the desktop spectrometer was insufficient to optimally support the works. Glass weights were added around the analysis area to prevent any interference from vibrations of the sample during measurement. In the case of the mobile equipment, the securing was done vertically using an easel that held a Plexiglas® sheet, where the work was fixed with Melinex® mounting corners and strips.

4. Conclusions

In this study, 16 bromoil prints were analysed using FTIR spectroscopy in reflection mode. The selected bromoil samples cover a broad temporal range (~1920–2010) and feature a diverse profile of authors. This includes a heterogeneous spectrum of possible bromoil typologies with eventual variations in the materials used and differences in the fabrication processes, making it plausible to represent diverse compositions and structures through characterisation. This study was conducted using two FTIR instruments, both in external reflection mode: a desktop device and a mobile one.

The results demonstrate that FTIR spectroscopy in external reflection mode is a valuable technique for the study and characterisation of bromoil prints, both in terms of the results it provides and its status as a non-destructive and non-invasive technique. We can confirm that the methodology used for the study of bromoil samples has allowed us to clearly detect the compounds present and make attributions about the strata to which they belong. Cellulose has been described as the major component of the paper support; we have been able to identify the barite layer when this stratum is present in the work immediately above the paper and recognise gelatine and discern its function from the usual photographic colloids where the image-forming material is suspended in it, and the oils and some pigments have been detected in the most superficial layer as image-forming materials (Prussian blue and iron oxides). Regarding the advantages and limitations of the equipment used for this study, we can assert that desktop instrumentation usually provides better spectrum resolution, especially for the identification of oils and their degree of polymerisation, but the mobile instrumentation, which generally covers a broader spectral range, has been decisive in detecting barium sulphate, the main component of the barite layer.

This study shows that the characteristic compounds of bromoil prints that can be identified using external reflection FTIR spectroscopy are the drying oil and some iron-based pigments of the greasy ink of the image and the gelatine layer, onto which the ink is placed. It is demonstrated that, for these to be reliably attributed to the main layers of a bromoil positive and, therefore, be key identifiers of this photographic process, the image-forming materials (oil and pigments) must be detected with a response intensity directly proportional to the image density. Additionally, the gelatine layer should show a response inversely proportional to the image density in terms of band definition and, in some cases, relative intensity, as it is located beneath the ink stratum.

This article resolves some knowledge gaps and contradictions present in the specialised literature: it reveals that it is possible to find bromoils without a barite layer and refutes the idea that this stratum is a key identifier of bromoils, a notion widely accepted in the previous literature. Furthermore, following our study, we propose reconsidering the use of the term colloid or emulsion in bromoil copies to refer to the gelatine layer and suggest understanding it more as a matrix or intermediary stratum intended to retain the image ink. After a deep exploration of the phenomena that occur during the process and production of bromoil prints, the first model for representing the structure of bromoil copies based on evidence from instrumental FTIR spectroscopy analysis is proposed. From these results, it is suggested that bromoils can be described as processes of three or four layers, depending on the presence or absence of the barite layer in each specific case.

No significant variations in the main compounds have been found according to author, context, and/or era. Variations have only been found attributable to the degree of oxidation and polymerisation of the oils, in accordance with the age of the copies, but are not decisive for the identification of the process.

Author Contributions: Conceptualisation, R.F.-J., C.R.-R., A.N.-V. and M.O.-F.; methodology, R.F.-J.; validation, C.R.-R. and A.N.-V.; formal analysis, C.R.-R.; investigation, R.F.-J.; resources, M.O.-F.; data curation, R.F.-J.; writing—original draft preparation, R.F.-J.; writing—review and editing, R.F.-J., C.R.-R. and A.N.-V.; visualisation, R.F.-J.; supervision, M.O.-F.; project administration, M.O.-F.; funding acquisition, M.O.-F. and R.F.-J.: contributed to the global approach, looked for works to be analysed, managed access to the artworks. R.F.-J., carried out all microscopic analysis, carried out all FTIR analysis at CCiTUB, developed analysis and interpretation of all FTIR spectra, wrote the initial manuscript, and was responsible for visualisation of the manuscript. C.R.-R.: contributed to the global approach and to the conception or design of the work, managed access to the artworks, participated in FTIR analysis and interpretation of data, and contributed significantly to the manuscript. A.N.-V.: contributed to the global approach and to the conception or design of the work, participated in FTIR analysis and interpretation of data, and contributed significantly to the manuscript. M.O.-F.: managed access to the works, conducted the project administration, provided and managed access to infrastructure, was responsible for funding acquisition, participated in the writing process, and supervised the research development. All authors have read and agreed to the published version of the manuscript.

Funding: Financial support by the Access to Research Infrastructures activity in the H2020 Programme of the EU (IPERION HS Grant Agreement n. 871034) is gratefully acknowledged.

Institutional Review Board Statement: Not applicable.

Informed Consent Statement: Not applicable.

Data Availability Statement: Raw data were generated at Scientific and facilities of the Research Group of Conservation of Cultural Heritage, University of Barcelona. Derived data supporting the findings of this study are available from the corresponding author on request.

Acknowledgments: We would like to express our sincerest gratitude to the owners of the collections under study, who were so gracious as to grant us access, thereby making this study possible due to their considerable generosity and willingness to collaborate. We thank the Mobile LABoratory (MOLAB) SCITEC (Istituto di Scienze e Tecnologie Chimiche "Giulio Natta", from the Consiglio Nazionale delle Ricerche (CNR) for providing us access to the mobile spectrometer. In particular, we would like to express our gratitude to Letizia Monico for her help in taking spectra with the mobile equipment and for her comments on these results. We would also like to thank the Conservation of Heritage research group (UB) and specially its IP, Gema Campo-Francés, for all the facilities provided to carry out this research.

Conflicts of Interest: The authors declare no conflicts of interest.

References

1. Martínez, C. *Tendencias Fotográficas en España Entre 1900–1940;* L'Arc de Bera: Barcelona, Spain, 2008; p. 239.
2. Zelich, C. La fotografia pictorialista a Espanya, 1900–1936. In *La Fotografia Pictorialista a Espanya, 1900–1936;* [Exhibition Catalogue]; Fundació "la Caixa": Barcelona, Spain, 1998; pp. 13–25.
3. Sougez, M.L. *Historia de la Fotografía,* 8th ed.; Ediciones Cátedra: Madrid, Spain, 2001; p. 580.
4. Insenser, E. Fotografía y arte. In *La fotografía en España en el Periodo de Entreguerras. 1914–1939;* Curbet Comunicacio Gràfica S.L.; Centre de Recerca i Difusió de la Imatge (CRDI), Ajuntament de Girona, Eds.; Ajuntament de Girona: Girona, Spain, 2000; pp. 117–145.
5. García-Felguera, M.S. La fotografia vol ser art. El pictorialisme. In *Bromolis. Fotografia Pictorialista de Joaquim Pla Janini;* [Exhibition Catalogue]; Museu Marítim de Barcelona, Ed.; Museu Marítim de Barcelona: Barcelona, Spain, 2022; pp. 7–30.
6. Namias, R. *Il Processo Bromolio Ovvero la Bromoleotipia;* Il Progresso Fotográfico: Firenze, Italy, 1916; p. 109.
7. Mayer, E. *Bromoil Printing and Bromoil Transfer;* American Photographic Publishing: Boston, MA, USA, 1923; p. 191.
8. Image Permanence Institute. Graphics Atlas. 2024. Available online: http://www.graphicsatlas.org/ (accessed on 25 October 2024).
9. Pavão, L. *Conservación de Colecciones de Fotografía;* Cuadernos técnicos; Junta de Andalucía, Instituto Andaluz del Patrimonio Histórico, Centro Andaluz de la Fotografía: Granada, Spain, 2001; p. 271.

10. Stulik, D.; Kaplan, A. Silver Gelatin. In *The Atlas of Analytical Signatures of Photographic Processes*; Getty Conservation Institute: Los Angeles, CA, USA, 2013. Available online: https://www.getty.edu/conservation/publications_resources/pdf_publications/pdf/atlas_silver_gelatin.pdf (accessed on 30 October 2024).
11. Walker, J.M.; Berrie, B.H. Influence of image density and interfacial surfaces on ER-FTIR spectra of 19th century photographic print processes. *J. Cult. Herit.* **2023**, *63*, 1–10. [CrossRef]
12. Estupiñán Méndez, D.; Allscher, T. Advantages of external reflection and transflection over ATR in the rapid material characterization of negatives and films via FTIR spectroscopy. *Polymers* **2022**, *14*, 808. [CrossRef] [PubMed]
13. Liu, G.-L.; Chen, T.-C.; Kazarian, S.G. Identification of black and white antique photographs via non-destructive microscopic and spectroscopic techniques coupled with chemometric approaches. *Microchem. J.* **2024**, *200*, 110297. [CrossRef]
14. McClelland, A.; Bulat, E.; Bernier, B.; Murphy, E.L. Specular reflection FTIR: A non-contact method for analyzing coatings on photographs and other cultural materials. *J. Am. Inst. Conserv.* **2020**, *59*, 123–136. [CrossRef]
15. McCabe, C.; Glinsman, L.D. Understanding Alfred Stieglitz'platinum and palladium prints: Examination by X-Ray fluorescence spectrometry. In *Research Techniques in Photographic Conservation: Proceedings of the Conference in Copenhagen, 14–19 May 1995*; Koch, M.S., Padfield, T., Johnsen, J.S., Kejser, U.B., Eds.; Royal Danish Academy of Fine Arts: Copenhagen, Denmark, 1995; pp. 31–40.
16. Applegarth, J. Spectroscopic Analysis of the Traditional Cyanotype Process and its Impact on Art. Undergraduate Honors Thesis Collection, Butler University Libraries, Indianapolis, IN, USA, 2017; p. 31. Available online: https://digitalcommons.butler.edu/ugtheses/390 (accessed on 29 October 2024).
17. Kaplan, A.; Stulik, D. Pushing the Limits of the Identification of Photographs: Variants of the Gum Bichromate Process. *Top. Photogr. Preserv.* **2013**, *15*, 190–206.
18. Khanjian, H.P.; Stulik, D.C. Infrared spectroscopic studies of photographic material. In *Conservation Science 2002: Papers from the Conference Held in Edinburgh, Scotland, 22–24 May 2002*; Townsend, J., Eremin, K., Adriaens, A., Eds.; Archetype Publications: London, UK, 2003; pp. 195–200.
19. Stulik, D.C.; Kaplan, A. A new scientific methodology for provenancing and authentication of 20th century photographs: Nondestructive approach. In Proceedings of the 9th International Conference on NDT of Art, Jerusalem, Israel, 25–30 May 2008.
20. Scatigno, C.; Teodonio, L.; Di Rocco, E.; Festa, G. Spectroscopic Benchmarks by Machine Learning as Discriminant Analysis for Unconventional Italian Pictorialism Photography. *Polymers* **2024**, *16*, 1850. [CrossRef]
21. Buti, D.; Domenici, D.; Miliani, C.; Sáiz, C.G.; Espinoza, T.G.; Villalba, F.J.; Verde Casanova, A.; Sabía de la Mata, A.; Romani, A.; Presciutti, F.; et al. Non-invasive investigation of a pre-Hispanic Maya screenfold book: The Madrid Codex. *J. Archaeol. Sci.* **2014**, *42*, 166–178. [CrossRef]
22. Miliani, C.; Rosi, F.; Brunetti, B.G.; Sgamellotti, A. In situ noninvasive study of artworks: The MOLAB multitechnique approach. *Acc. Chem. Res.* **2010**, *43*, 728–738. [CrossRef]
23. Horn, K.; Walker, G.; Winton, V.; Ramanaidou, E.; Hamlett, C.; Hamlett, B. Field characterisation of rock art paintings using non-invasive reflectance spectroscopy in the search for organic paint binders at Genealogy and Stickman Rockshelters in the Weld Range (Western Australia). *J. Archaeol. Sci. Rep.* **2020**, *34*, 102617. [CrossRef]
24. Miliani, C.; Rosi, F.; Daveri, A.; Brunetti, B.G. Reflection infrared spectroscopy for the noninvasive in situ study of artists' pigments. *Appl. Phys. A Mater. Sci. Process.* **2012**, *106*, 295–307. [CrossRef]
25. Monico, L.; Rosi, F.; Miliani, C.; Daveri, A.; Brunetti, B.G. Non-invasive identification of metal-oxalate complexes on polychrome artwork surfaces by reflection mid-infrared spectroscopy. *Spectrochim. Acta Part A Mol. Biomol. Spectrosc.* **2013**, *116*, 270–280. [CrossRef] [PubMed]
26. Walker, J.; Hodgkins, R.; Berrie, B. On the surface: Reflectance FTIR spectroscopy in cultural heritage research. *Microsc. Microanal.* **2021**, *27* (Suppl. 1), 2800–2804. [CrossRef]
27. Brunetti, B.; Miliani, C.; Rosi, F.; Doherty, B.; Monico, L.; Romani, A.; i Sgamellotti, A. Non-invasive investigations of paintings by portable instrumentation: The MOLAB experience. *Toppics Curr. Chem.* **2016**, *374*, 10. [CrossRef] [PubMed]
28. Invernizzi, C.; Rovetta, T.; Licchelli, M.; Malagodi, M. Mid and near-infrared reflection spectral database of natural organic materials in the cultural heritage field. *Int. J. Anal. Chem.* **2018**, *2018*, 7823248. [CrossRef]
29. Poletto, M.; Ornaghi Junior, H.L.; Zattera, A.J. Native cellulose: Structure, characterization and thermal properties. *Materials* **2014**, *7*, 6105–6119. [CrossRef]
30. Liu, C.; Sun, R. Chapter 5—Cellulose. In *Cereal Straw as a Resource for Sustainable Biomaterials and Bifuels. Chemistry, Extractives, Lignins, Hemicelluloses and Cellulose*; Elsevier: Amsterdam, The Netherlands, 2010; pp. 131–167.
31. Hospodarova, V.; Singovszka, E.; Stevulova, N. Characterization of cellulosic fibers by FTIR spectroscopy for their further implementation to building materials. *Am. J. Anal. Chem.* **2018**, *9*, 303–310. [CrossRef]
32. Derrick, M.R.; Stulik, D.; Landry, J.M. *Infrared Spectroscopy in Conservation Science*; Getty Publications: Los Angeles, CA, USA, 1999; p. 248.
33. Barraza-Garza, G.; Barraza-Garza, G.; De La Rosa, L.A.; Martínez-Martínez, A.; Castillo-Michel, H.; Cotte, M.; Alvarez-Parrilla, E. La microespectroscopía de infrarrojo con transformada de fourier (FTIRM) en el estudio de sistemas biológicos. *Rev. Latinoam. Quim.* **2013**, *41*, 125–148.
34. Stulik, D.; Kaplan, A. Carbon. In *The Atlas of Analytical Signatures of Photographic Processes*; Getty Conservation Institute: Los Angeles, CA, USA, 2013. Available online: https://www.getty.edu/conservation/publications_resources/pdf_publications/pdf/atlas_carbon.pdf (accessed on 22 October 2024).

35. Meilunas, R.J.; Bentsen, J.G.; Steinberg, A. Analysis of aged paint binders by FTIR spectroscopy. *Stud. Conserv.* **1990**, *35*, 33–51. [CrossRef]
36. Arrizabalaga, I.; Gómez-Laserna, O.; Bustamante, J.; Arana, G.; Carrero, J.A.; Rodríguez, A.; Madariaga, J.M. Diffuse reflectance FTIR database for the interpretation of the spectra obtained with a handheld device on built heritage materials. *Anal. Methods* **2014**, *7*, 1061–1070. [CrossRef]
37. Banou, P.; Boyatzis, S.; Choulis, K.; Theodorakopoulos, C.; Alexopoulou, A. Oil Media on Paper: Investigating the Interaction of cold-pressed linseed oil with paper supports with FTIR analysis. *Polymers* **2023**, *15*, 2567. [CrossRef]
38. Lazzari, M.; Chiantore, O. Drying and oxidative degradation of linseed oil. *Polym. Degrad. Stab.* **1999**, *65*, 303–313. [CrossRef]
39. Romero, M.T.; Ferrer, N. Fourier transform infrared spectroscopy applied to the characterisation of 17th–20th century calcographic and xylographic inks. *Microchim. Acta* **1999**, *131*, 237–245. [CrossRef]
40. Vahur, S.; Knuutinen, U.; Leito, I. ATR-FT-IR spectroscopy in the region of 500–230 cm^{-1} for identification of inorganic red pigments. *Spectrochim. Acta Part A Mol. Biomol. Spectrosc.* **2009**, *73*, 764–771. [CrossRef] [PubMed]
41. Reilly, J.M. *Care and Identification of 19th-Century Photographic Prints*; Eastman Kodak Company: Rochester, NY, USA, 1986; p. 116.
42. Mortimer, F.J.; Coulthurst, S.L. *The Oil and Bromoil Processes*, 2nd ed.; Hazell, Watson & Viney: London, UK, 1912; p. 96.
43. Namias, R. *Positivos a las Tintas Grasas. Óleo y Bromóleo*; Garriga, R., Translator; Bailly-Bailliere: Madrid, Spain, 1934; p. 54.
44. Whalley, G.E. *Bromoil and Transfer*; Fountain Press: London, UK, 1961; p. 87.
45. Griffiths, P.; de Haseth, J.A. *Fourier Transform Infrared Spectrometry*, 2nd ed.; Wiley-Interscience: Hoboken, NJ, USA, 2007; 560p.
46. Bayé, C.R. Una aproximación al pictorialismo en Catalunña. *Cuerpo, alegoría y paisaje en la fotografía de Vilatobà, Pla Janini, Masana y Casals i Ariet. Commun. Pap. Media Literay Gend. Stud.* **2017**, *6*, 53–78.
47. Fontcuberta, J. *Historias de la Fotografía Española. Escritos 1977–2004*; Gustavo Gili: Barcelona, Spain, 2008; p. 396.
48. Sánchez Vigil, J.M. Las revistas de fotografía y su aportación al estudio de la documentación fotográfica. Un modelo: La Fotografía. *Bol. Anabad* **2007**, *57*, 71–84.
49. Formiguera, P. Joaquim Pla Janini. La fotografia apassionada. *Kesse Butlletí Cercle D'estudis Històrics I Soc. Guillem Oliver* **1996**, *3*, 7–11.
50. Carbonell, C. El Dr. Joaquim Pla, figura representativa de la fotografia catalana contemporània. *Art Llum* **1934**, *17*, 152–153.
51. Covarsí, L. Joaquim Pla Janini. In *Bromolis. Fotografia Pictorialista de Joaquim Pla Janini*; [Exhibition Catalogue]; Museu Marítim de Barcelona, Ed.; Museu Marítim de Barcelona: Barcelona, Spain, 2022; pp. 65–81.
52. Whalley, G.E. *An Introduction to Bromoil and Transfer*; James A. Sinclair & Co.: London, UK; p. 15.
53. Symes. Bromoil and Transfer. In *A Revised and Enlarged Edition of Perfection in the Pigment Processes*; Fountain Press: London, UK, 1948; p. 54.
54. Balanyà, J. *La Nostra Barcelona (1960–2014)*; Balanyà, J., Ed.; Self-published: Barcelona, Spain, 2014; p. 125.
55. Balanyà, J. *Nepal 1982. Per les Valls del Khumbu i de Katmandú*; Balanyà, J., Ed.; Self-Published: Barcelona, Spain, 2023; p. 127.
56. Font i Rodon, M.; Balanyà, J. *Paisatges de la Serra de Tramuntana. Mallorca*; Triangle Postals, S.L.: Barcelona, Spain, 2021; p. 272.
57. Amat, A.; Miliani, C.; Brunetti, B.G. Non-invasive multi-technique investigation of artworks: A new tool for on-the-spot data documentation and analysis. *J. Cult. Heritage* **2013**, *14*, 23–30. [CrossRef]

molecules

MDPI

Article

Hats Off to Modeling! Profiling Early Synthetic Dyes on Historic Woolen Samples with ATR-FTIR Spectroscopy and Multivariate Curve Resolution–Alternating Least Square Algorithm

Tiziana Forleo [1,2], Lorena Carla Giannossa [1,3,*], Anna De Juan Capdevila [4], Giovanni Lagioia [5] and Annarosa Mangone [1,3]

1 Department of Chemistry, University of Bari Aldo Moro, 70126 Bari, Italy; tiziana.forleo@ispa.cnr.it (T.F.); annarosa.mangone@uniba.it (A.M.)
2 Institute of Sciences of Food Production, National Research Council (CNR), 70126 Bari, Italy
3 Centro Interdipartimentale Laboratorio di Ricerca per la Diagnostica dei Beni Culturali, via Orabona 4, 70126 Bari, Italy
4 Department of Chemical Engineering and Analytical Chemistry, Universitat de Barcelona, 08028 Barcelona, Spain; anna.dejuan@ub.edu
5 Department of Economics, Management and Business, University of Bari Aldo Moro, 70126 Bari, Italy; giovanni.lagioia@uniba.it
* Correspondence: lorenacarla.giannossa@uniba.it

Abstract: This research focuses on analyzing wool samples dyed with synthetic dyes from the early 20th century. A methodology to identify and distinguish wool fibers dyed with azo, triphenylmethane, and xanthene dyes, which are no longer in use, using the ATR-FTIR spectra, is presented. Firstly, the dataset was subjected to PCA, which revealed the similarities and differences among the samples, illustrating a distribution pattern based on dye classes. MCR-ALS was employed to extract the spectral profiles of the dyed fibers, thereby enhancing the efficacy of the analytical techniques and extracting the comprehensive information from a single instrument. The combination of ATR-FTIR spectroscopy with chemometric methods, such as PCA and MCR-ALS, has proven to be an effective strategy for identifying and differentiating wool fibers dyed with early azo, triphenylmethane, and xanthene dyes. This approach has demonstrated particular effectiveness in enabling rapid analysis without requiring sampling or pretreatment. Moreover, the analysis is supported by thorough bibliographic research on these no longer used colorants. In order to maximize the potential of non-destructive spectroscopic techniques, such as ATR-FTIR, the approach used has proven to be crucial. This study underscores how chemometric techniques expand the capabilities of spectroscopy, extracting extensive information from a single instrument and aligning with the goals of cultural heritage analysis.

Keywords: ESDs; early synthetic dyes; historic materials; ATR-FTIR spectroscopy; MCR-ALS modeling

1. Introduction

In this study, samples of wool dyed with early dyes from the pattern book "*Färbungen auf Damen-Wool-Hüten / Teintures sur Chapeaux de laine pour dames / Dyeings on Ladies' Wool Hats*" published by the company Leopold Cassella & Co. (Frankfurt am Main, Germany) were analyzed. The investigated pattern book (Figure 1) is part of a massive and diversified compendium located at the Commodity Science Museum of the Department of Economics, Management, and Business Law at the University Aldo Moro of Bari (Italy) [1–3]. It is a sample book of synthetic dyes utilized for coloring woolen fabrics, particularly for the creation of women's hats. It is divided into two sections. The first part provides a list of samples, the dye used, the percentage of dye used in each sample, and the etching process. The second part contains the samples presented as colored wool flaps.

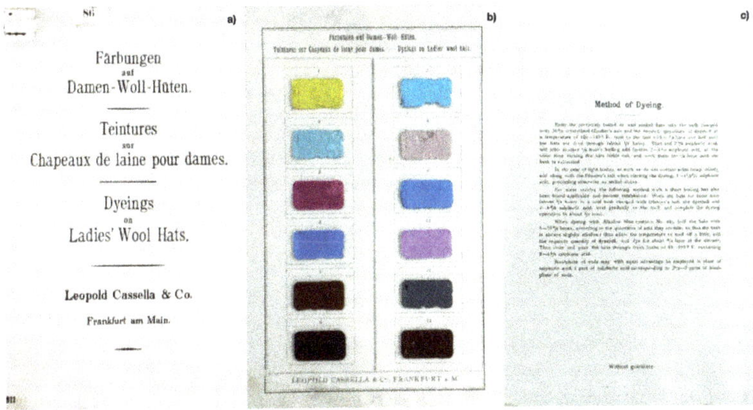

Figure 1. (**a**) The front cover of the pattern book; (**b**) an example page with dyed woolen pieces; (**c**) the page explaining the process of dyeing.

The early synthetic dyes (ESDs) used to color the samples belong to three groups: the triphenylmethane dyes, the azo dyes, and the xanthene dyes.

Triphenylmethane dyes have a basic structure consisting of three aryl groups linked to a methyl group. They were among the first dyes to be developed, with fuchsine introduced in 1859 and crystal violet in 1883 [4,5].

Azo dyes are characterized by the presence of an azo group (−N=N) and linking aromatic rings. They generally contain one or more SO_3^- Na^+ groups, making them soluble in water and enabling them to bind to the surfaces of polar materials like wool, cotton, and nylon. They can also have two azo groups (diazo) or more [6,7]. The xanthene dyes, as the name suggests, are characterized by a xanthene core [6,7]. Most of the early synthetic dyes are no longer commercially available due to impractical production processes and unsatisfactory properties for textile dyeing. Consequently, they have been progressively replaced over time [7].

This study is not only informative, but it also contributes to completing the scenario for future research into the diagnosis of this sample type. As mentioned earlier, analyses in cultural heritage, especially with samples like these, necessitate methods with several characteristics, such as the following: it must be non-destructive, respecting the physical integrity of the samples; it must be fast, allowing for the analysis of similar objects or several parts of the same one; it must be universal, capable of analyzing different parts of the same sample with a single instrument, with no or minimal pre-processing; it must be versatile, offering both local and average composition of heterogeneous samples; and it must be sensitive and multi-elemental. Finding a method that satisfies all these requirements is complicated, and approaches based on the principle of "primum non nocere" remain a priority, offering the least destructive techniques that achieve the trade-off between the absence of damage and the maximum representativeness of the sample analyzed [8–12]. While chromatography-based techniques are the typically preferred methods of analysis, their inherently destructive nature limits their applicability in cultural heritage diagnostics [13].

Other techniques may prove useful in the study of dye tissue samples; however, they are also destructive or micro-destructive [14–16], and in some cases, the extraction of the dye is also required [14,16,17].

Consequently, research efforts are increasingly turning towards spectroscopic techniques [1,11,18,19].

Among spectroscopic techniques, Attenuated Total Reflectance (ATR)-FTIR spectroscopy is commonly used to determine the composition of fibers rather than colorants and their general condition [18,20]. On the other hand, UV-VIS and Raman spectroscopies are employed for colored fibers. In particular, Raman spectroscopy allows for the identification of both the type of fiber and the color of the fiber [17,18].

In addition, the use of chemometrics, which gives a considerable boost to the discrimination of colored fibers, has grown considerably in recent years [1,14,19].

Colored textiles are important samples for understanding critical issues related to scientific and technological advances, historical progress, and cultural exchanges that have taken place over centuries [21,22].

Microscopic and molecular knowledge provides information on degradation and thus guidelines for the conservation and restoration of artifacts [21,22]. In addition, it is essential to reconstruct the technological processes that led to the creation of artifacts in order to understand their capabilities, economic aspects, social significance, and cultural context [21,22]. The study of colored textiles also plays an important role in forensic science. In this field, it is not always necessary to absolutely identify the dye molecules but rather to differentiate between colored fibers in order to establish similarities or differences between trace evidence and compatible items [16,23].

In this study, ATR-FTIR was employed for sample analysis. Subsequently, the dataset underwent two chemometric approaches: Principal Component Analysis (PCA), to explore and observe similarities and differences among the samples, and Multivariate Curve Resolution–Alternating Least Square (MCR-ALS), to identify the spectral profile of the sample.

The primary objective of this study is to combine ATR-FTIR spectroscopy with PCA and MCR-ALS to identify the spectral profile of wool fibers dyed with synthetic dyes from the early 1900s. By employing ATR spectroscopy to identify colored fibers, this study aims to maximize the amount of information that can be obtained from a specific experimental technique, thereby enhancing the understanding of a sample using a single type of instrument.

2. Results and Discussion

This section reports the results obtained from the application of chemometric methods on the ATR-FTIR spectra to identify the spectral profile of wool fabrics dyed with different dyes (triphenylmethane, azo, xanthene). First, a PCA has been applied to explore the dataset and highlight the similarities and differences among the dyes. Afterwards, an MCR-ALS analysis allows us to understand the specific composition of the dyes.

The first part is dedicated to the study of wools colored with pure dyes with a concentration greater than 1%, as reported in the pattern book. The aims are to discriminate fabrics colored with different dyes and to identify the spectral profile of the wool fibers dyed.

The second part is dedicated to the study of the most complex dyed fabrics colored with unknown dyes and mixtures of two dyes, known and unknown (each with a concentration higher than 1%) to test the method with more complex samples.

2.1. Fabrics Colored with One Dye

The samples analyzed are presented in Table 1.

Figure 2 shows, for each sample, the average of the three acquisitions made as the most representative spectrum. We can notice the differences between them. Each color seems to have a distinct spectral shape; just look at the difference, for example, among the GG and the CS in the range between 1000 and 1100 cm^{-1}, although they have a similar structure.

Table 1. Dataset A: fabrics with pure dyes with known composition.

	Code in the Pattern Book	Dye and Concentration	Identifying Name of the Sample	Family Structure	Structure	References
	10	2% Acid Violet 4RS	AV	Tryphenylmethane	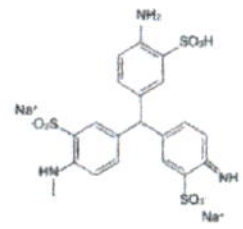	[24–26]
	22	3% Cyanole Extra Pat.	CY	Triphenylmethane	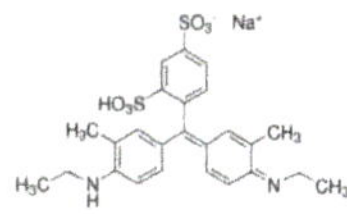	[27,28]
	31	2% Orange GG	GG	Azo	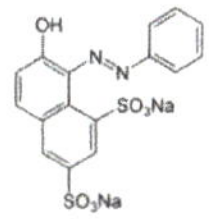	[29]
	23	3% Crystal Scarlet 6R	CS	Azo	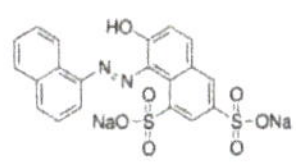	[25,30]
	26	2% Irisamine G pat.	IG	Xanthene	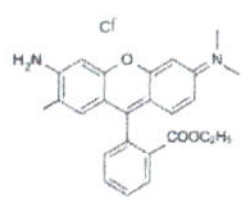	[28]
	3	1.5% Rose Bengale Extra N.	RB	Xanthene	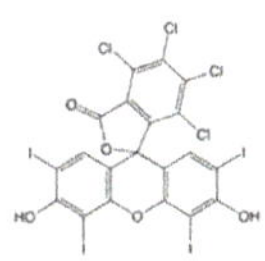	[25,28]

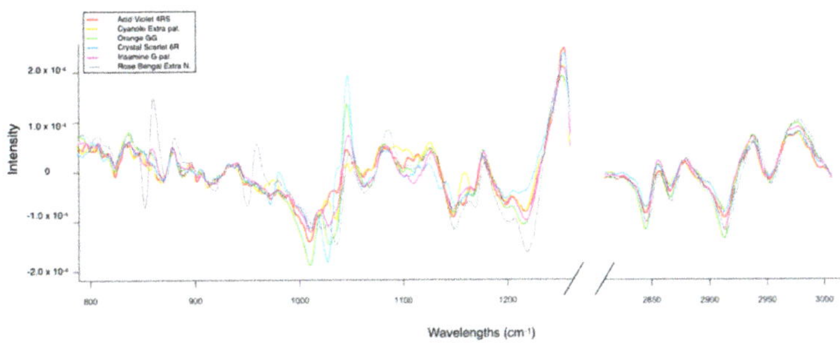

Figure 2. The spectra of all fabrics colored with pure dyes. The spectra are pretreated with Savitzky–Golay 1st derivative filter and are normalized.

2.1.1. The Exploratory Analysis

A PCA was performed as an exploratory analysis to visualize the similarities and differences between samples. The first four components describe 90.4% of the total variance (PC1 46.8%, PC2 23.7%, PC3 14.9, PC4 5.2%). From score plots in Figure 3, it can be seen that the dyed wool samples are grouped based on the type of dye and the distance between replicates, which could give us information about inhomogeneity. For example, only one replicate of GG tends to get away from the others in plots a–c. The plots also suggest that the samples can be separated by functional group according to PC2: the triphenylmethane dyes are approximately located at PC2-negative and the azo and xanthene dyes at PC2-positive.

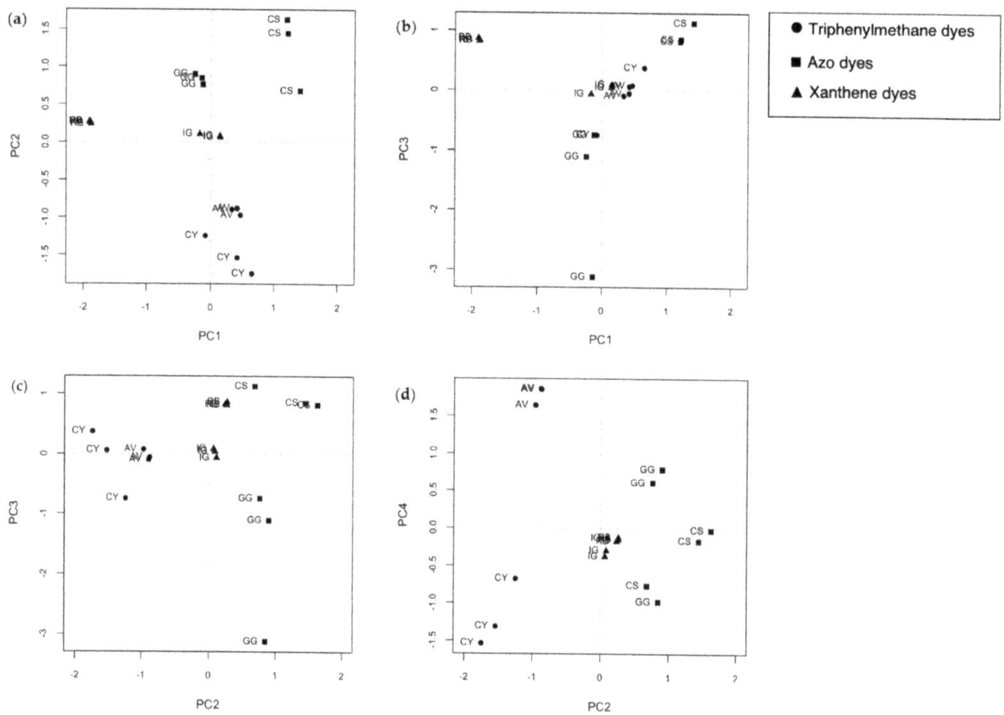

Figure 3. Score plots (**a**) PC1 vs. PC2; (**b**) PC1 vs. PC3; (**c**) PC2 vs. PC3; (**d**) PC2 vs. PC4. Each sample is labeled by the code presented in Table 1; the symbols indicate the a priori belonging class.

2.1.2. MCR-ALS

After the exploratory analysis, the application of classical MCR-ALS was carried out. The number of components estimated by a singular value decomposition (SVD) was four. The initial estimates used were those obtained from a method based on SIMPLISMA, which was applied to the squared first-derivative corrected spectra, and MCR-ALS was performed by applying the constraint of non-negativity in concentration profiles. Moreover, with four components, a better data fit (LOF% = 11.0, r^2 = 98.8) was achieved. Figure 4 shows the resolved pure spectra (ST) in the bottom plot, and the information related to the concentration profiles (C) in the top plot. In the latter plot, every component is represented using bars to provide an easier visualization of the composition in every sample measured.

The first and second components (in red and green, respectively) are related to the azo dyes. Both components present the pure spectra with several peaks in the range between 1030 and 1200 cm^{-1}, attributed to the S-O vibration characteristic of azo dyes. The component shown in red is very dominant in the fabric dyed CS, whereas the second component is more present in the GG sample. The clear distinction between the first two components is consistent with the results of the PCA, showing the different characteristics of CS and GG (samples were far away from each other). Furthermore, the replicates of the different samples generally have a similar distribution of components (see bar distribution). In cases where the differences in intensity are greater, this could be an indication of sample inhomogeneity, as is the case with GG, which is in agreement with the PCA results. The third component (in violet) is very dominant in the xanthene dyes. The sample RB is totally described by this component. Otherwise, in the case of IG, the contributions of other components are required to define the structure.

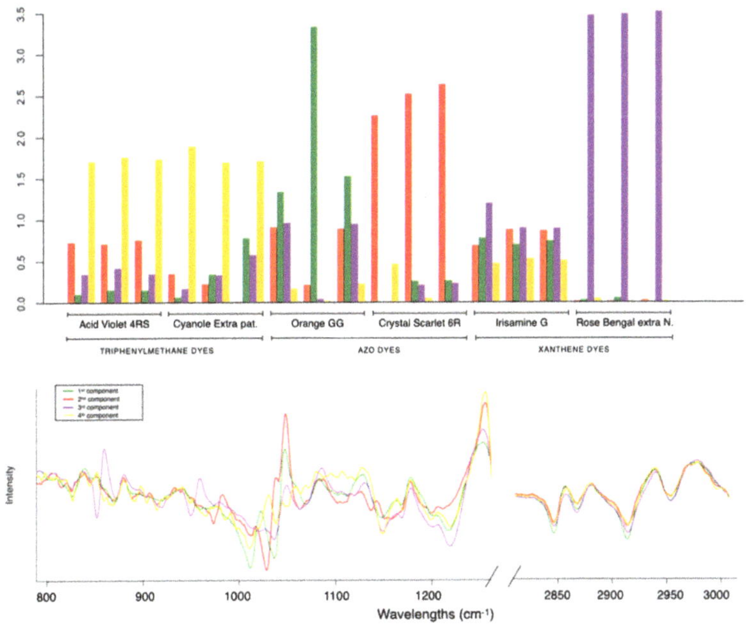

Figure 4. The resolved pure spectra (ST) (**bottom** plot) and the information related to the concentration profiles (C) (**top** plot).

The identified spectral features for each component can be found in the Supplementary Materials (Table S1, Refs. [31–38]). In particular, it can be noticed at 927 cm^{-1} of the in-plane bending of the fused xanthene rings, and 1234 cm^{-1} of the stretching C-O-C presented in the structure [36]. The resolved signature shows a clear presence of peaks at 955 cm^{-1}, characteristic of the bond C-halogen of the RB. The fourth component (in yellow) is predominant in the triphenylmethane dyes. In this case, the high similarity of the contributions of the concentration profiles to AV and CE corresponds to the high similarity of their structures. In addition, the presence of signals attributable to the aromatic part in the region before 1000 cm^{-1} and the S-O stretching at 1028 and 1054 cm^{-1} can be observed.

2.2. *Investigation on Known and Unknown, Pure and Mixture Dyes in Fabrics*

In this section, the data matrix of fabrics colored with pure dyes (with known and unknown structures) and the binary mixture of dyes are analyzed (Table 2 added to Table 1). All the dyes considered show a concentration higher than 1%.

2.2.1. The Exploratory Analysis

A PCA was performed on the dataset. The score plots in Figure 5 show the distribution of all samples in the space of PC1-PC2 (a) and, for a better visualization, the tridimensional distribution in spaces PC1-PC2-PC3 (b) and PC2-PC5-PC7 (c).

Table 2. Fabrics with a mixture of dyes or unknown samples that constitute dataset B with the samples in Table 1.

	Code in the Pattern Book	Dye and Concentration	Identifying Name of the Sample	Family Structure (Dye 1)	Structure (Dye 1)	Ref.	Family Structure (Dye 2)	Structure (Dye 2)	Ref.
Mixtures									
	36	3.5% Solid Blue R.2% Cyanole Green 6G	SC	Unknown	Unknown	[27,30]	Tryphen	Unknown	[25,28]
	57	3% Cyanole Green 6G1.5% Fast Yellow S	CF	Tryphen	Unknown	[27,30]	Azo	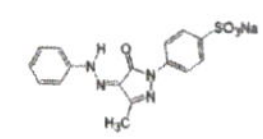	[25,28]
	35	2.5% Orange GG1.25% Lanafuchsine SG	OL	Azo	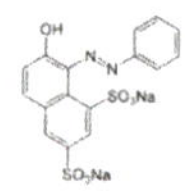	[31]	Azo	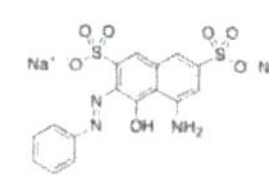	[39]
Unknown Pures									
	24	10% Naphthol Black SG	NB	Unknown	Unknown	[30]	-	-	
	58	8% Hat Black BBV	HB	Unknown	Unknown	[30]	-	-	

The samples with pure dyes analyzed previously (AV, CY, GG, CS, IG, and RB) tend to group again according to functional groups. The pure dyes of unknown composition (NB and HB) are localized farther than the other samples in all the plots, but always close to the azo dyes. This behavior suggests a different structure. Moreover, they are placed near the azo and the azo–azo dyes, and this could be consistent with the hypothesis that they could be diazo dyes. When looking at the mixtures, the sample stained with Cyanole Green 6G and Solid Blue R (SC, empty circle) is far away from all the others in plots (a) and (b) but close to the sample stained with only triphenylmethanes in plot (c). This sample contains Cyanole Green 6G, which is a triphenylmethane, but it also contains Solid Blue R, which seems to belong to another class of dye, as previously stated. The mixture of azo–azo dyes, OL, is located in the same area as the azo dyes. The mixture of triphenylmethane and azo dyes, CF, is close to the azo dyes in plot (c).

2.2.2. MCR-ALS

The MCR-ALS algorithm was performed on the data matrix. The number of components estimated with SVD is seven. The initial estimates used are those obtained by the SIMPLISMA method applied to the corrected squared spectra, as previously stated. The MCR-ALS algorithm is applied with the following constraint: non-negativity on the concentration profiles. Optimized MCR-ALS was performed. The results provided an LOF of 9.2% and a r^2 of 99.2%.

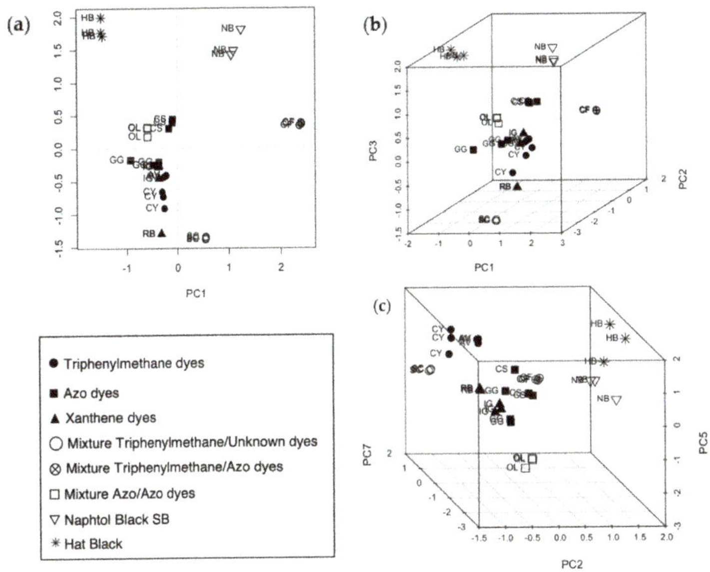

Figure 5. (**a**) Bidimensional score plot PC1-PC2. (**b**) Tridimensional score plot PC1-PC2-PC3. (**c**) Tridimensional score plot PC2-PC5-PC7.

Figure 6 shows the resolved concentration profiles (C) and the resolved pure spectra (ST). The interpretation of all the spectra profiles obtained is presented in Table S2 in the Supplementary Materials.

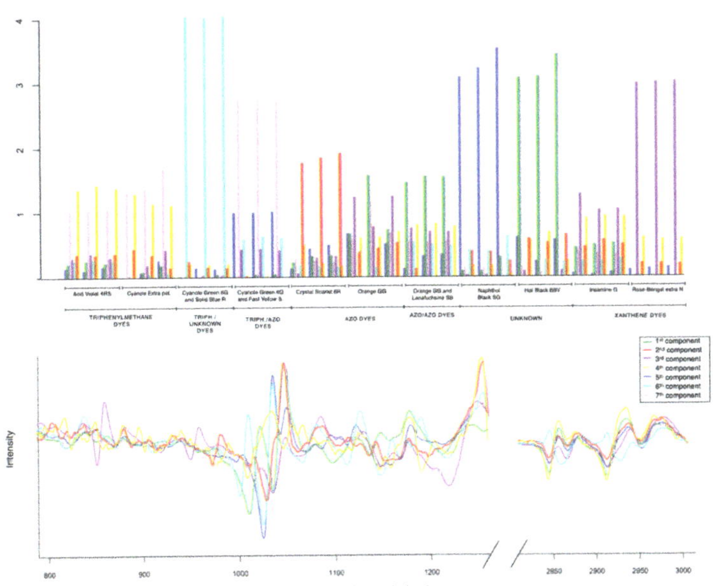

Figure 6. Resolved concentration profiles (C) and resolved pure spectra (ST).

As stated before, the same components are found plus additional ones. The first component (in green) is more related to a replicate of GG and to OL and HB samples. This result, according to the PCA, suggests that the spectral profile of HB is very similar to these types of azo dyes. Once again, GG is shown as an inhomogeneous sample. As before, the second (in red) and third (in purple) components are dominant in the CS and in the sample with xanthene dyes. The third component is also present in the description of two replicates of the GG sample. The fourth component (in yellow) is related to triphenylmethane dyes. This also contributes to the description of the samples with triphenylmethane dyes and xanthene dyes. This result is in complete agreement with the PCA. The two families share the core structure of xanthene, as shown by the peak at 926 cm^{-1} (f) of the in-plane motion of the fused xanthene ring and at 1171 cm^{-1} [39], attributed to C-C-H bending.

The triphenylmethane dyes are also described by the seventh component, which is dominant in sample CF, with the mixture of triphenylmethane and azo dyes. The fifth component (in blue) is dominant in the NB and the sixth component (in cyan) is dominant in the SC. The seventh component is related to the CF, which contains a mixture of triphenylmethane and azo dyes. This component is also present in samples containing only triphenylmethane dyes.

3. Materials and Methods

3.1. The Samples

For the purposes of this study, we have focused on the wool samples with a dye concentration higher than 1%. In the first part of the evaluation, we focused on the samples shown in Table 1. Among the triphenylmethane dyes there is Acid Violet 4RS, which is described in the literature as the sodium salt of dimethylrosaniline trisulfonic acid and is also known as Red Violet 4RS [24–26].

Another triphenylmethane dye is Cyanole Extra Pat., which, as the name suggests, was patented by Cassella Leopoldo & Co. It is the sodium salt of m-oxydiethyldiaminophenyl-ditolyl-carbinol disulphonic acid [27,28]. Orange GG, an azo dye also known as Orange G, is formed by azo coupling between aniline and 2-(6,8-disulphonic acid)-naphthol [29]. The Crystal Scarlet 6R, also known as Crystal Ponceau 6R or New Coccine R, is another azo dye with azo groups conjugating 2-(6,8-disulphonic acid)-naphthol with naphthalene [25,30].

Among the xanthene dyes, there is Rose Bengal Extra N, which is tetra-iso-dichlorofluoresceine, also used in biological and industrial applications [25,27].

Finally, Irisamine G is a dye patented by Cassella Leopoldo & Co. and an analog of Rhodine 3G [28].

In the second part, the samples in Table 2 are added to those in Table 1.

There are three samples stained with a mixture of two dyes. The first is dyed with Solid Blue R, an unknown dye that seems to belong to the family of Induline [25] in the literature, and with Cyanole Green 6G, a triphenylmethane dye whose structure is unknown in the literature [28]. The second is dyed with Cyanole Green 6G and an azo dye, Fast Yellow S [28]. The third is dyed with two azo dyes, Orange GG and Lanafuchsine SG [28,29].

Then, there are two pure black dyes whose structure and family are unknown in the literature. Usually, they belong to the diazo family [28].

Unfortunately, in the pattern book, there is no sample of undyed wool.

3.2. ATR-FTIR Spectroscopy

The ATR-FTIR spectra were obtained using a Spectrum Two FTIR spectrometer (Perkin Elmer, Milan, Italy) with 32 scans, a resolution of 2 cm^{-1}, and scanning from 4000 to 400 cm^{-1}.

The background spectra were recorded against air. For each sample, three measurements were taken at three different points on the surface of the fabric sample.

The averaged raw FTIR-ATR spectrum of all dyed yarns, with identification of the signals attributable to wool, is presented in SM (Table S3, Figure S1, Refs. [40–44]).

3.3. Data Pre-Processing and Chemometrics

The spectra obtained by the ATR-FTIR spectroscopy were pre-processed with the Savitzky–Golay first derivative filter and normalized. The chemometrics methods applied were the PCA and the MCR.

Principal Component Analysis (PCA) is the most well-known multivariate exploratory data analysis tool. It is an unsupervised decomposition method that performs a data projection from the original high-dimensional space to a lower-dimensional one defined by a set of new variables known as principal components (PCs).

It reduces the dimensionality of the dataset and minimizes information redundancy and noise. Moreover, it enables visualization and simplifies the understanding of relationships between complex variables, thus facilitating the interpretation of the data [45–47].

The term Multivariate Curve Resolution (MCR) refers to a family of chemometric methods that, starting from measurements containing signals from various components (the raw data matrix D), provide a bilinear model of the individual, chemically significant contributions [48,49]:

$$D = C\,S^{T} + E$$

The equation reflects the bilinear nature of spectroscopic measurements as expressed by Beer–Lambert's law, where the variation in a dataset containing the spectra of a multicomponent system, D, can be expressed as the product of the pure spectra, S^T, of the individual components and their respective concentration profiles, C (Figure 7).

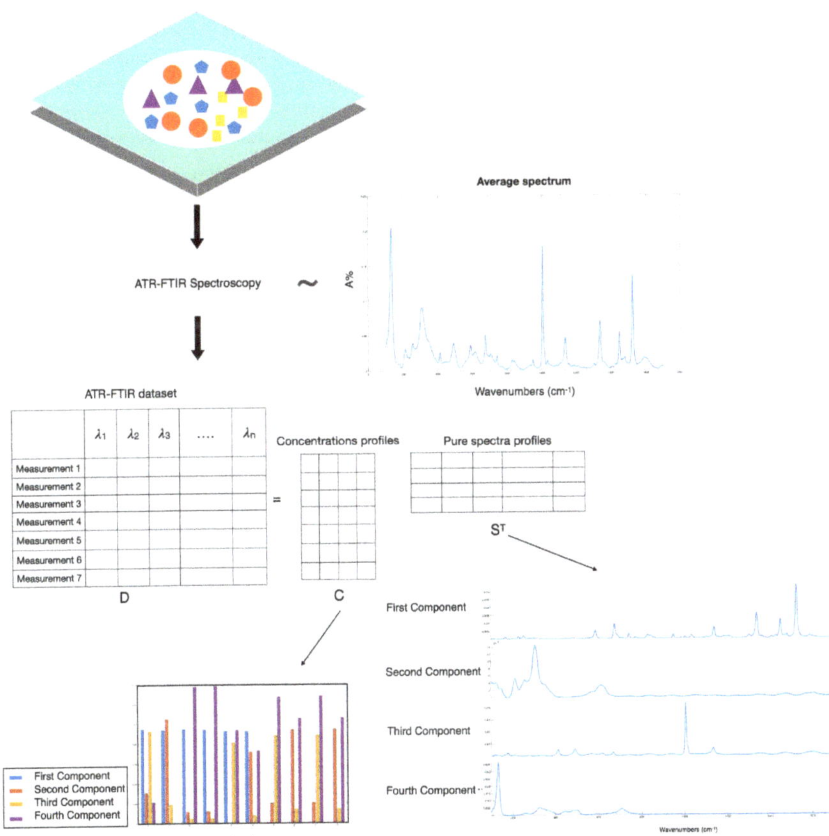

Figure 7. Example of the Multivariate Curve Resolution applied to a dataset of ATR-FTIR spectra.

The matrix decomposition in a bilinear model can be achieved using different algorithms [48,49]. In this work, we employ the alternating least squares algorithm (ALS). The complete method is called MCR-ALS.

One limitation of this method is that, mathematically, the original dataset can be described by different combinations of concentration and spectral profiles, a phenomenon known as ambiguity, that manifests in several ways (permutation ambiguity, intensity ambiguity, rotational ambiguity). The introduction of constraints in the MCR-ALS analysis is crucial because they incorporate both chemical and mathematical information to yield a meaningful concentration and spectral profiles while reducing or eliminating ambiguity in the MCR solutions [48,49]. These constraints encompass chemical and physical properties that are systematically presented in the component profiles within the dataset. In each iterative cycle, the calculated profiles are adjusted to comply with predetermined conditions imposed by the selected constraints. Their choice and the way they are applied are active steps of the analyst who knows the samples, the dataset, their properties, and the profiles to resolve. In this work, we apply only the non-negativity constraint, which enforces that the profiles consist of positive values. It can be applied individually to concentration profiles, pure spectra, or both.

The algorithm proceeds through several steps:

Determining the number of components in a system, which may be known or determined using methods such as PCA. Both MCR and PCA aim to describe the maximum variance present in the dataset. This is typically assessed using the Single Value Decomposition (SVD) algorithm. The analysis is further complemented by studying the noise pattern in the scores and loadings profiles. The final MCR model should give an optimal fit and chemically resolved profiles [48,49].

Generating initial estimates of C and S^T, which can be pure spectra or concentration profiles derived from prior knowledge (e.g., known spectra of one or more components) or from systematic behavior in the concentration direction. In this work, to generate initial estimates, a purest variable selection method called SIMPLISMA is used. This method aims to select the dataset row (sample) that is most different from the others and provides initial estimates of spectra and concentration profiles.

Optimizing the spectra and concentration profiles through alterative and iterative least squares, seeking convergence while adhering to constraints. Convergence is assessed using the following parameters of lack of fit and explained variance [48,49]:

$$\%LOF = 100 \cdot \sqrt{\frac{\sum_{i,j} e_{ij}^2}{\sum_{i,j} d_{ij}^2}}$$

$$\%EV = 100 \cdot \sqrt{1 - \frac{\sum_{i,j} e_{ij}^2}{\sum_{i,j} d_{ij}^2}}$$

Here, e_{ij} represents the residuals associated with the matrix decomposition using the MCR of the original dataset (d_{ij}).

Optimization is deemed successful when the iteratively obtained model no longer significantly improves over consecutive iterations [48,49].

All evaluations were carried out under MATLAB (R2023a, Mathworks, Natick, MA, USA) [50] environment, using MCR-ALS 2.0 toolbox [51] and PLS toolbox (Eigenvector Research, Wenatchee, WA, USA) [52].

4. Conclusions

The combined use of ATR spectroscopy with chemometric methods, such as PCA and MCR-ALS, proves valuable in swiftly identifying and distinguishing wool fibers dyed with early azo, triphenylmethane, and xanthene dyes. This approach has been shown to be particularly effective for quick analysis without the need for sampling or pretreatment.

The PCA was performed as an exploratory analysis to visualize similarities and differences between samples and shows a grouping based on the types of dyes. The MCR-ALS approach allowed us to extract the spectral profile of colored fibers, maximizing the ability of the analytical techniques used and extracting the maximum information from a single instrument.

It can be confidently stated that the applied method is highly effective for analyzing simple systems, such as fabrics dyed with a single dye. However, further improvements are required to enhance its performance when applied to more complex systems, particularly those involving more dyes.

To the best of our knowledge, this is the first instance where fabric samples have been analyzed using ATR-FTIR and chemometric methodologies with the aim of extracting information beyond the fibers themselves, including the types of dyes used.

This methodology, requiring no sampling or pretreatment, streamlines the analysis process. Thorough bibliographic research supports the identification of these outdated colorants. This study highlights how chemometric techniques amplify the capabilities of spectroscopy, extracting abundant information from a single instrument and aligning with the ideal cultural heritage analysis.

Supplementary Materials: The following supporting information can be downloaded at https://www.mdpi.com/article/10.3390/molecules29194651/s1, Table S1: Interpretation of all the pure spectra obtained with MCR-ALS; Table S2: Interpretation [a–h] of all the pure spectra obtained with MCR-ALS for the samples presented in Table S2; Figure S1: The average raw ATR-FTIR spectrum of all dyed yarns; Table S3: Identification of the signals attributable to wool.

Author Contributions: Conceptualization, T.F. and L.C.G.; methodology, T.F., L.C.G., A.D.J.C. and A.M.; software, A.D.J.C. and T.F.; validation, L.C.G., A.D.J.C. and A.M.; formal analysis, A.D.J.C. and T.F.; investigation, T.F. and L.C.G.; data curation, T.F., L.C.G., A.D.J.C. and A.M.; writing—original draft preparation—T.F. and L.C.G.; writing—review and editing—L.C.G., A.D.J.C., T.F., G.L. and A.M.; visualization, T.F.; supervision, L.C.G., A.D.J.C. and A.M. All authors have read and agreed to the published version of the manuscript.

Funding: A.D.J.C. acknowledges funding from the Spanish government through the project PID2023-146465NB-I00.

Institutional Review Board Statement: Not applicable.

Informed Consent Statement: Not applicable.

Data Availability Statement: The data presented in this study are available on request from the corresponding author.

Conflicts of Interest: The authors declare no conflicts of interest.

References

1. Caggiani, M.C.; Forleo, T.; Pojana, G.; Lagioia, G.; Mangone, A.; Giannossa, L.C. Characterization of silk-cotton and wool-cotton blends pattern books by Fiber Optics Reflectance Spectroscopy. The booming market of first synthetic textiles dyes in early 20th Century. *Microchem. J.* **2022**, *175*, 107178. [CrossRef]
2. Gentile, L.; Giannossa, L.C.; Mangone, A.; Lagioia, G.; Palazzo, G. From Byssus Threads to Pinna nobilis Sea-Silk: A Fiber Chemical and Structural Characterization. *Acad. Mater. Sci.* **2023**, *1*, 6123. [CrossRef]
3. Montenegro, V. *Il Patrimonio Storico-Scientifico Dell'università Degli Studi di Bari Aldo Moro*; Giuseppe Laterza Ed.: Bari, Italy, 2017; pp. 135–142.
4. Mikheev, Y.A.; Guseva, L.N.; Ershov, Y.A. The nature of chromaticity of triphenylmethane, xanthene, phtalocyanine, and thiazine dyes, Russ. *J. Phys. Chem. A* **2010**, *84*, 1778–1791.
5. Available online: https://www.britannica.com/technology/dye/Development-of-synthetic-dyes (accessed on 23 July 2024).
6. Benkhaya, S.; M'rabet, S.; Harfi, A.E. Classifications, properties, recent synthesis and application of azo dyes. *Heliyon* **2020**, *6*, e03271. [CrossRef] [PubMed]
7. Longoni, M.; Gavazzi, M.; Monti, D.; Bruni, S. Early synthetic textile dyes of the late 19th century from the "Primo Levi" Chemistry Museum (Rome): A multitechnique analytical investigation. *J. Cult. Herit.* **2023**, *59*, 131–139. [CrossRef]
8. Tamburini, D.; Sabatini, F.; Berbers, S.; van Bommel, M.R.; Degano, I. An Introduction and Recent Advances in the Analytical Study of Early Synthetic Dyes and Organic Pigments in Cultural Heritage. *Heritage* **2024**, *7*, 1969–2010. [CrossRef]

9. Mazzeo, R.; Roda, A.; Prati, S. Analytical chemistry for cultural heritage: A key discipline in conservation research. *Anal. Bioanal. Chem.* **2011**, *399*, 2885–2887. [CrossRef] [PubMed]
10. Trojanowicz, M. Modern chemical analysis in archaeometry. *Anal. Bioanal. Chem.* **2008**, *391*, 915–918. [CrossRef]
11. Janssens, K.; Van Grieken, R. *Non-Destructive Microanalysis of Cultural Heritage Materials, Wilson & Wilson's Comprehensive Analytical Chemistry*; Elsevier: Amsterdam, The Netherlands, 2004; Volume XLII, pp. 1–10.
12. Lahanier, C.; Preusser, F.D.; Van Zelst, L. Study and conservation of museum objects: Use of classical analytical techniques, Section I. Introduction: Analytical problems in art and archeology. *Nucl. Instrum. Methods Phys. Res. B* **1986**, *14*, 1–9. [CrossRef]
13. Ding, L.; Gong, T.; Wang, B.; Yang, Q.; Liu, W.; Pemo, R.; Metok, T. Non-invasive study of natural dyes in textiles of the Qing Dynasty using fiber optic reflectance spectroscopy. *J. Cult. Herit.* **2021**, *47*, 69–78. [CrossRef]
14. Kokot, S.; Gilbert, C. Application of Drift Spectroscopy and Chemometrics to the Discrimination of dye mixtures extracted from fibres from worn clothing. *Analyst* **1994**, *119*, 671–676. [CrossRef]
15. Di Tullio, V.; Doherty, B.; Capitani, D.; Miliani, C.; Greco, E.; Ciliberto, E.; Rossi, L.; Proietti, N. NMR spectroscopy and micro-analytical techniques for studying the constitutive materials and the state of conservation of an ancient Tapa barkcloth from Polynesia, is. Wallis. *J. Cult. Herit.* **2020**, *45*, 379–388. [CrossRef]
16. Smith, G.D.; Esson, J.M.; Chen, V.J.; Hanson, R.M. Forensic dye analysis in cultural heritage: Unraveling the authenticity of the earliest Persian knotted-pile silk carpet, Forensic Science International. *Synergy* **2021**, *3*, 100130. [CrossRef] [PubMed]
17. Petroviciu, I.; Teodorescu, I.C.; Vasilca, S.; Albu, F. Transition from Natural to Early Synthetic Dyes in the Romanian Traditional Shirts Decoration. *Heritage* **2023**, *6*, 505–523. [CrossRef]
18. Goodpater, J.V.; Liszewski, E.A. Forensic Analysis of dyed textile fibers. *Anal. Bioanal. Chem.* **2009**, *394*, 2009–2018. [CrossRef] [PubMed]
19. Li, M.; Han, G.; Jiang, W.; Zhou, C.; Zhang, Y.; Wang, S.; Su, J.; Li, X. Rapid identification of plant- and chemical-dyed cotton fabrics using the near-infrared technique. *Text. Res. J.* **2020**, *90*, 2275–2283. [CrossRef]
20. Geminiani, L.; Campione, F.P.; Corti, C.; Giussani, G.; Gorla, G.; Luraschi, M.; Recchia, S.; Rampazzi, L. Historical silks: A novel method to evaluate their condition with ATR-FTIR spectroscopy and Principal Component Analysis. *J. Cult. Herit.* **2024**, *67*, 9–22. [CrossRef]
21. Chen, K.; Leona, M.; Vo-Dinh, T. Surface-enhanced Raman Scattering for identification of organic pigments and dyes in works of art and cultural heritage material. *Sens. Rev.* **2007**, *27*, 109–120. [CrossRef]
22. Degano, I.; Ribechini, E.; Modugno, F.; Colombini, M.P. Analytical Methods for the Characterization of Organic Dyes in Artworks and in Historical Textiles. *Appl. Spectrosc. Rev.* **2009**, *44*, 363–410. [CrossRef]
23. Lewis, S.W. 11—Analysis of dyes using chromatography. In *Woodhead Publishing Series in Textiles Identification of Textile Fibers*; Houck, M.M., Ed.; Woodhead: Cambridge, UK, 2009.
24. Norton, T.H. *Artificial Dyestuffs Used in the United States*; Department of Commerce of the United States: Washington, DC, USA, 1916; pp. 139–140.
25. Cooke, E.I.; Cooke, R.W.I. *Handbook of Chemical Synonyms and Trade Names*; Gardner, W., Ed.; CRC Press: Boca Raton, FL, USA, 2019
26. Derz, F.W. *Chemproductindex A-G*, 2020th ed.; de Gruytier: Berlin, NY, USA, 1976.
27. Adam, E. *Chemisch-Technische Untersuchungsmethoden*; Berl, E., Lunge, G., Eds.; Springer: Berlin/Heidelberg, Germany, 2013.
28. Von Georgievics, G.; Grandmougin, E. *A Text-Book of Dye Chemistry. The Chemistry of Dyestuffs*; D Van Nostrand Company: New York, NY, USA, 1920.
29. Veitch, F.P.; Donk, M.G. *Wood Turpentine: Its Production, Refining Properties, and Uses*; U.S. Department of Agriculture, Burean of Chemistry: Washington, DC, USA, 1911.
30. Hurst, G.H. *A Dictionary of Coal Tar Colours*; Heywood and Company: London, UK, 1892.
31. Coates, J. *Interpretation of Infrared Spectra, A Practical Approach, Encyclopedia of Analytical Chemistry*; Meyers, R.A., Ed.; John Wiley & Sons Ltd.: Chichester, UK, 2000; pp. 10815–10837.
32. Doherty, B.; Vagnini, M.; Dufourmantelle, K.; Sgamellotti, A.; Brunetti, B.; Miliani, C. A vibrational spectroscopic and principal component analysis of triarylmethane dyes by comparative laboratory and portable instrumentation. *Spectrochim. Acta Part A Mol. Biomol. Spectrosc.* **2014**, *121*, 292–305. [CrossRef] [PubMed]
33. Ahmed, F.; Dewani, R.; Pervez, M.K.; Mahboob, J.; Aoomro, S.A. Non-destructive FT-IR analysis of mono azo dyes. *Bulg. Chem. Commun.* **2016**, *48*, 71–77.
34. Zubiri, I.B.; Carré, A. Giving a new status to a dyes collection: A contribution to the chromotope project. *Heritage* **2023**, *6*, 2202–2219. [CrossRef]
35. Prati, S.; Milosevic, M.; Sciutto, G.; Bonacini, I.; Kazarin, S.G.; Mazzeo, R. Analysis of trace amounts of dyes with a new enhanced sensitivity FTIR spectroscopic technique: MU-ATR (metal underlayer ATR spectroscopy). *Anal. Chim. Acta* **2016**, *941*, 67–79. [CrossRef]
36. Xiong, T.; Zhang, Y.; Donà, L.; Gutiérrez, M.; Möslein, A.F.; Babal, A.S.; Amin, N.; Civalleri, B.; Tan, J. Tunable fluorescein-encapsulated zeolitic imidazolate framework-8 nanoparticles for solid-state lighting. *ACS Appl. Nano Mater* **2021**, *4*, 10321–10333. [CrossRef]
37. Al-Rubaie, L.A.R.; Mhessn, R.J. Synthesis and Characterization of Azo Dye Para Red and New Derivatives. Available online: https://onlinelibrary.wiley.com/doi/10.1155/2012/206076 (accessed on 20 September 2024).

38. Muthukumar, M.; Thalamadai, K.M.; Bhaskar Raju, G. Electrochemical removal of CI Acid orange 10 from aqueous solutions. *Sep. Purif. Technol.* **2007**, *55*, 198–205. [CrossRef]
39. Wang, L.; Roitberg, A.; Meuse, C.; Gaigalas, A.K. Raman and FTIR spectroscopies of fluorescein in solutions. *Spectrochim. Acta Part A Mol. Biomol. Spectrosc.* **2001**, *57*, 1781–1791. [CrossRef] [PubMed]
40. Jiang, Z.; Li, W.; Wang, Y.; Wang, Q. Second-Order Derivation Fourier Transform Infrared Spectral Analysis of Regenerated Wool Keratin Structural Changes. *AATCC J. Res.* **2022**, *9*, 43–48. [CrossRef]
41. ATR-FT-IR Spectrum of Wool. Available online: https://spectra.chem.ut.ee/textile-fibres/wool/ (accessed on 17 September 2024).
42. Geba, M.; Lisa, G.; Ursescu, C.M.; Olaru, A.; Spiridon, I.; Leon, A.L.; Stanculescu, I. Gamma irradiation of protein-based textiles for historical collections decontamination. *J. Therm. Anal. Calorim.* **2014**, *118*, 977–985. [CrossRef]
43. Kan, C.W.; Chan, K.; Yuen, C.W.M. Surface characterization of Low Temperature Plasma treated wool fiber—The effect of the nature of gas. *Fibers Polym.* **2004**, *5*, 52–58. [CrossRef]
44. Kissi, N.; Curran, K.; Vlachou-Mogire, C.; Fearn, T.; McCullough, L. Developing a non-invasive tool to assess the impact of oxidation on the structural integrity of historic wool in Tudor tapestries. *Herit. Sci.* **2017**, *5*, 49. [CrossRef]
45. Massart, D.L.; Vandeginste, B.G.M.; Deming, S.M.; Michotte, Y.; Kaufman, I. *Chemo-Metrics: A Textbook. Data Handling in Science and Technology*; Elsevier: Amsterdam, The Netherland, 2001.
46. Todeschini, R. *Introduzione Alla Chemiometria*; Edises: Naples, Italy, 2003.
47. Ballabio, D.; Consonni, V. Classification tools in chemistry. Part 1: Linear models. PLS-DA. *Anal. Methods* **2013**, *16*, 3790–3798. [CrossRef]
48. De Juan, A.; Jaumot, J.; Tauler, R. Multivariate Curve Resolution (MCR). Solving the mixture analysis problem. *Anal. Methods.* **2014**, *14*, 4964–4976. [CrossRef]
49. De Juan, A.; Tauler, R. Multivariate Curve Resolution—Alternating Least Squares for Spectroscopic Data. *Data Handl. Sci. Technol.* **2016**, *30*, 5–51.
50. Available online: https://it.mathworks.com/products/matlab.html (accessed on 20 September 2024).
51. Available online: https://mcrals.wordpress.com/ (accessed on 20 September 2024).
52. Available online: https://it.mathworks.com/products/connections/product_detail/pls-toolbox.html (accessed on 20 September 2024).

molecules

MDPI

Article

Functionalization of Artwork Packaging Materials Utilizing Ag-Doped TiO_2 and ZnO Nanoparticles

Tilde de Caro [1,*,†], Roberta Grazia Toro [1,†], Luminita Cassone [2], Francesca Irene Barbaccia [2,3], Camilla Zaratti [4], Irene Angela Colasanti [2], Mauro Francesco La Russa [5] and Andrea Macchia [2,4]

1 CNR-ISMN, Istituto per lo Studio dei Materiali Nanostrutturati, Strada Provinciale 35 d n. 9, 00010 Rome, Italy; robertagrazia.toro@cnr.it
2 Youth in Conservation of Cultural Heritage (YOCOCU APS), Via T. Tasso 108, 00185 Rome, Italy; luminitacassone.lc@gmail.com (L.C.); colasanti.1748988@studenti.uniroma1.it (I.A.C.); andrea.macchia@unical.it (A.M.)
3 Department of Technological Innovation Engineering, Digital Technologies for Industry 4.0, International Telematic University Uninettuno, Corso Vittorio Emanuele II 39, 00186 Rome, Italy
4 Lab4Green, Via T. Tasso 108, 00185 Rome, Italy; zaratti.1920859@studenti.uniroma1.it
5 Department of Biology, Ecology and Earth Sciences (DIBEST), University of Calabria, Via Pietro Bucci, Arcavacata, 87036 Rende, Italy; mlarussa@unical.it
* Correspondence: tilde.decaro@cnr.it
† These authors contributed equally to this work.

Abstract: Most of the artworks stored in museums are often kept in inappropriate climatic and environmental conditions that facilitate the formation and growth of microorganisms, such as fungi, which are responsible for many types of biodegradation phenomena. To mitigate and prevent these deteriorative processes, functionalized packaging materials can be used for the storage and handling of artworks. The aim of this study was to develop a potential anti-biodeterioration coating suitable for packaging purposes. TiO_2 and ZnO doped with different amounts of Ag (0.5 wt%, 1 wt%, and 3 wt%) were synthesized and dispersed in polyvinyl alcohol (PVA) and acrylic resin (Paraloid B72), then applied on different types of packaging materials (cellulose and the high-density spunbound polyethylene fiber Tyvek®, materials that are frequently used as packaging in museums). Analytical investigations (SEM/EDS, Raman, FTIR, and XRD) were employed to assess dispersion on the packaging material. Furthermore, resistance against biodeteriogens was assessed using *Cladosporium* sp., a bioluminometer, to define the biocidal efficacy.

Keywords: storage; packaging; smart; museum; nanomaterials; TiO_2; ZnO; Ag doping

1. Introduction

The preservation of cultural heritage artworks is of utmost importance, as these represent a testimony to the past and, more importantly, a legacy for future generations, a legacy that is up to us to maintain and preserve.

One of the problems that artworks face is fungal colonization and development. This is a typical issue for museums, particularly those housed in historic buildings with an unstable environment or extreme interior temperatures [1–4]. Air conditioning systems can control climatic factors, but this method is susceptible to fungi contamination and is difficult to operate in large areas due to high energy consumption [5]. As a result, some microorganisms and fungal species thrive in hot and humid conditions, within both historic buildings and modern climate-controlled structures [6,7]. These pathogens infiltrate museum spaces and storage through air currents or hitch a ride on visitors' clothing, hair, bodies, or shoes [8]. Even a single contaminated item entering the museum environment can act as a carrier of contamination [9], leading to the surface degradation of artworks and impacting both coloration and structural integrity [10].

Additionally, artworks may be biologically attacked by microorganisms during their transportation from exhibition halls to storage or when they are loaned to other museum entities for temporary exhibitions. The most common microorganisms, specifically fungi, that attack and destroy artworks are *Aspergillus* spp., *Penicillium* spp., *Cladosporium* spp., *Mucor*, *Alternaria*, *Rhizopus* spp., and *Fusarium* [11–13].

Microbial and fungal metabolic processes oxidize, hydrolyze, and decompose organic artworks. Cellulose-based materials such as paper, photographs, fabrics, canvas paintings, and wood sculptures are susceptible to "foxing" and selective fungal infections on hemicellulose portions of the material [14–17]. However, microbes are not selective; they can attack inorganic materials, leading to structural problems and chromatic alterations [18–20]. Fungal activity, for example, alters porosity and surface compactness and can cause weight loss in stone structures [21,22].

Several scientific investigations have examined possible strategies to avoid the spread of microorganisms on artworks and in museum rooms and storage. Methods include chemical biocide attacks and physical mechanisms such as laser or microwave treatments to create unfavorable environmental conditions for their growth and proliferation [23–28]. Despite the different strategies employed to fight microorganisms, spore transmission by external vectors continues to be an important driver of contamination [29].

Therefore, safeguarding museum artworks against fungal attacks by using appropriate packaging could be essential to the long-term preservation of these art pieces [30–32]. Packaging materials should operate as a barrier to biocontamination, limiting the spread of spores within museum rooms and storage. The type of support and protection required for the item will be determined by the storage environment and the artwork itself; therefore, both should be considered.

Traditional packaging, while useful for protecting the material from stresses during transport and variations in irradiance and hygrometric conditions, may still allow microorganism spores to pass through, allowing them to proliferate on the artwork if the conditions are favorable [33,34].

To avoid contamination of works of art, the packaging utilized should, in addition to being appropriate for the type of art, be functionalized against microbes. Antimicrobial packaging technologies to prevent microbial growth have been widely studied and employed in food packaging. Several antimicrobial packaging technologies, involving different materials, have been investigated [35–41].

In food packaging, the formulation of an active protection system can provide additional protection compared to simple classical packing methods. Depending on their different applications, packaging can provide different functions, making it highly customizable [42,43]. For example, some innovative solutions to extend the shelf life of food by inhibiting or retarding microbial growth use the action of antimicrobial compounds incorporated within the packaging system [42].

Polymeric materials are commonly utilized in the production of this type of packaging. Creating multilayered packaging is the best way to protect food. However, this approach can be costly; an alternative is to incorporate compounds, such as nanoparticles, into the polymeric material to achieve a similar level of protection [44]. A tailored solution for the cultural heritage field still needs to be investigated [45].

Non-acid-free materials such as cellulose paper or Tyvek® (a material produced by Du Pont (Wilmington, DE, USA) made of polyethylene fibers (PE) or rigid plastics such as polycarbonates (Makrolon™) and acrylics (Perspex™, Plexiglas™)), as well as foams such as latex, are all examples of materials that should be avoided [31,46–48].

The aim of this study is to functionalize commonly used packaging materials for cultural heritage by using nanoparticles (NPs) that exhibit antimicrobial properties. Through the incorporation of NPs in two different mediums, the goal is to develop antifungal coatings that can be applied on packaging material to inhibit the growth of microorganisms.

Based on antibacterial, antifouling, and photocatalytic properties, TiO_2 and ZnO NPs were chosen to increase the protective biocide efficacy of packaging materials [23,30,49,50].

Previous research on ZnO NPs has highlighted a correlation between reduced NP dimensions and increased concentration in inhibiting fungal growth [51]. TiO_2 and ZnO were further investigated by incorporating Ag as a doping agent to enhance their effectiveness by increasing their adhesion on the chosen substrate [52,53].

Ag-doped TiO_2 and ZnO NPs were synthesized using simple and environmentally friendly sol-gel methods, devoid of toxic and harmful solvents. To enhance adhesion between the substrates (a paper sheet and Tyvek®) and the coating, Ag-doped NPs were dispersed in two polymeric media consisting of polyvinyl alcohol (PVA) and acrylic resin (Paraloid B72).

Poly(vinyl alcohol) (PVA) and poly(methyl methacrylate) (PMMA) are the most commonly studied polymers for incorporating inorganic nanofillers. Nanomaterials are increasingly seen as superior alternatives to traditional additives for enhancing polymer properties.

The use of nanoparticles as polymer fillers aligns with the current strong interest in developing and applying new composite materials. Polymer nanocomposites effectively combine the unique size-dependent capabilities of nanoparticles with the desirable qualities of host polymers, such as versatility in design, long-term stability, and reprocessability. While the properties of nanocomposites generally reflect the combined attributes of their inorganic and organic components, synergistic interactions can sometimes lead to entirely new characteristics [54,55].

PVA is a non-toxic, water-soluble polymer widely utilized as a matrix material for synthesizing highly transparent nanocomposites due to its film-forming capability, biocompatibility, and cost efficiency. Its chemical formula is $[CH_2\text{-}CH\ (OH)]n$, where n denotes the number of polymerizations. As reported by Huang et al. [56], the concentration of PVA is a critical factor because low concentrations generally lead to coatings with poor adhesion, while high concentrations cause PVA to become viscous and unsuitable for spraying.

Paraloid B72 was developed as a surface coating and offers various benefits, including strength and hardness without brittleness. Paraloid B72 is also characterized by a high glass transition temperature (Tg) and is less likely to experience cold flow [57]. It has been reported that by adding nano-TiO_2 particles to high-density polyethylene and other materials, it is possible to obtain useful novel materials for packing applications [58].

There are several studies about the application of Paraloid B72 functionalized with NPs for the consolidation and protection of stone surfaces; however, there are very few studies focused on the packaging of artworks [59–63].

The biocidal efficacy depends both on the physicochemical properties of the particles and their interaction with the medium used for the coating formulation. The physicochemical properties of TiO_2 and ZnO can vary depending on the material's primary particle size, crystallinity, and formulation. Different properties have been found to cause different toxicological responses. The biocidal efficacy of both titanium dioxide and zinc oxide nanoparticles, as well as doped formulations, is closely related to their structure, which must be characterized for their application [64–66].

This type of multifunctional coating is not new to the cultural heritage sector. However, in this study, we have developed and applied coatings for the first time to enhance the packaging used in museums.

The structure and composition of the synthesized materials were investigated by using different analytical investigations, and biocide efficacy against fungi was defined using a bioluminometer. The experimental results show that the synthesized composite materials are able to improve the packaging used in museums to preserve cultural heritage.

2. Results and Discussion

2.1. Characterization of TiO_2-Ag NPs

TiO_2-Ag NPs were synthesized using no harmful solvents and non-toxic reagents. The success of the adopted procedure was first approximately assessed by observing the morphological change in color of the obtained powders, ranging from white to brown, finally changing to deep grey upon increasing the amount of Ag doping (Figure 1).

Figure 1. Stereomicroscope images of TiO_2, TiO_2-Ag (0.5%), TiO_2-Ag (1%), and TiO_2-Ag (3%).

Structural characterization of the obtained TiO_2-Ag NPs was performed by X-ray diffraction measurements (Figure 2). Diffraction peaks positioned at 2θ of 25.31°, 36.92°, 37.82°, 38.55°, 48.05°, 53.98°, 55.09°, 62.79°, 68.89°, 70.56°, and 75.41° correspond to the pure TiO_2 in the anatase phase (A), and were assigned to (101), (103), (004), (112), (200), (105), (211), (204), and (116) crystallographic planes (JCPDS card no. 21-1272) [67]. Moreover, the appearance of typical diffraction peaks (111), (200), and (220) positioned at 2θ of 38.10°, 44.39°, and 64.70 indicated the face-centered cubic metallic Ag crystal structure (JCPDS Card no. 04-0836) [68]. The characteristic XRD peak positions of anatase were scarcely affected by Ag doping, thus suggesting that Ag NPs did not enter the TiO_2 crystal structure, causing distortion in the TiO_2 crystal lattice, but were mainly on the surface of TiO_2 NPs. Furthermore, XRD measurements also indicated the presence of rutile (R) as a secondary phase in all Ag-doped samples, characterized by diffraction peaks at 2θ of 23.43°, 36.05°, 41.25°, and 56.61°, which are related to (110), (101), (111), and (220) crystallographic planes (JCPDS card no. 21-1276) [69]. To determine the effect of doping with silver on the crystallite size of the nanoparticles, the mean crystallite size was calculated from the highest diffraction peak and full width at half maximum (FWHM) using the Debye–Scherrer equation (Equation (1)):

$$D\ (\text{nm}) = 0.9\lambda/\beta \text{Cos}\theta \quad (1)$$

where β is the FWHM (Full width at half maximum), λ is the X-ray wavelength of Cu-Ka (1.5406 A°), and θ is the diffraction angle of the particles. The average crystallite sizes of pure and Ag-doped TiO_2 NPs were calculated by the Scherrer equation using the full width at half maximum (FWHM) of the (101) diffraction peak. The corresponding values are reported in Table 1. Calculations suggested that TiO_2 crystallite size varied in the range of 12–13 nm, with a slight reduction in size upon increasing the amount of Ag doping (Table 1).

Figure 3a shows the Raman spectra for different TiO_2-Ag NPs. In the same picture, the Raman spectrum of undoped TiO_2 NPs is shown for comparison. According to factor group analysis, TiO_2 in the anatase phase has six Raman active modes (A1g + 2B1g + 3Eg) that are visible at 145 cm^{-1} (Eg), 197 cm^{-1} (Eg), 399 cm^{-1} (B1g), 513 cm^{-1} (A1g), 519 cm^{-1} (B1g), and 639 cm^{-1} (Eg) [70]. A smaller particle size affects the force constants and vibrational amplitudes of the nearest neighboring bonds, resulting in observed Raman shifts from 144 to 145 cm^{-1} [71].

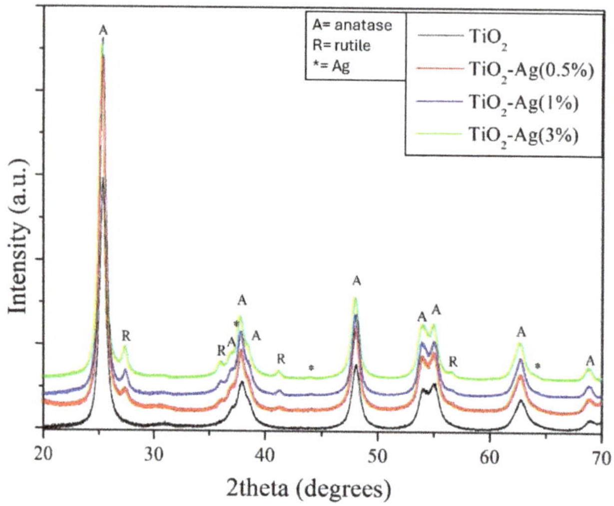

Figure 2. X-ray diffraction measurements of TiO_2, TiO_2-Ag (0.5%), TiO_2-Ag (1%), and TiO_2-Ag (3%). (A: anatase, R: rutile, * Ag).

Table 1. TiO_2 crystallite size (D), calculated by the Debye–Scherrer equation, and optical bandgap energies (Eg), calculated by the K–M equation.

Nanoparticles	TiO_2			
	Undoped	**0.5% Ag**	**1% Ag**	**3% Ag**
D (nm)	12.33	13.09	12.89	12.15
Eg (eV)	3.00	2.99	2.92	2.88

Figure 3a highlights that the Raman spectra of all the synthesized TiO_2-Ag and undoped TiO_2 NPs have the characteristic Raman active modes of the anatase phase. In addition, the Raman spectra of doped samples report the presence of a peak at 122 cm^{-1}, characteristic of Ag^0, which is likely to be formed during thermal annealing in air [72]. Therefore, the Raman spectra of TiO_2 and TiO_2-Ag did not exhibit any discernible differences in crystalline structure.

The optical absorption of Ag-doped and undoped TiO_2 NPs was measured with diffuse reflectance UV/Vis absorption spectroscopy and compared in Figure 3b. As can be seen, for the undoped TiO_2 NPs (the black curve in Figure 3b), the absorption spectrum showed typical semiconductor absorption properties, with strong absorption in the ultraviolet region and weaker absorption in the visible light region. Compared to pure TiO_2, the absorption intensity in the visible region was much higher after Ag doping, which indicates that the Ag doping process enhances the absorption efficiency of the TiO_2 photocatalyst in visible light. Furthermore, the absorption spectra showed significant absorption in the visible region proportional to Ag doping, and absorption potential became stronger upon increasing the Ag doping concentration. In addition, the absorption edges of the Ag-doped TiO_2 NPs were significantly red-shifted compared to the undoped TiO_2. This behavior, already reported in previous studies [73], was attributed either to the surface plasmon absorption of Ag nanoparticles or enhanced photon path length in the surface region of TiO_2 NPs. The optical bandgap energies (Eg) of the synthesized samples were calculated according to the K–M equation and the values are reported in Table 1. Eg value was affected by Ag doping and its value started to decrease upon increasing doping concentration (Figure 3c), reaching the value of 2.88 eV for the sample with 3 wt% of Ag doping.

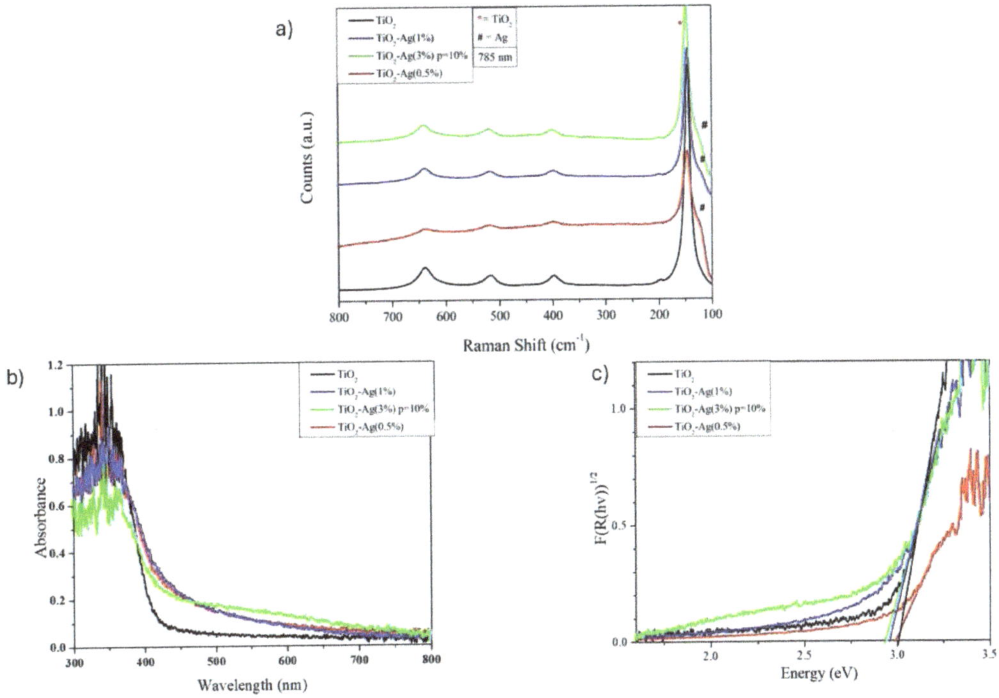

Figure 3. Spectra of TiO_2 and TiO_2-Ag (0.5%, 1%, 3%) acquired from different analyses, respectively. (a) Micro-Raman spectra; (b) UV–Vis spectra; (c) Kubelka–Munch.

2.2. Characterization of ZnO-Ag NPs

In the case of Ag-doped ZnO NPs, the color of the synthesized powders changed with the dopant, varying from white for the undoped ZnO to gold-ochre for Ag-doped samples (Figure 4).

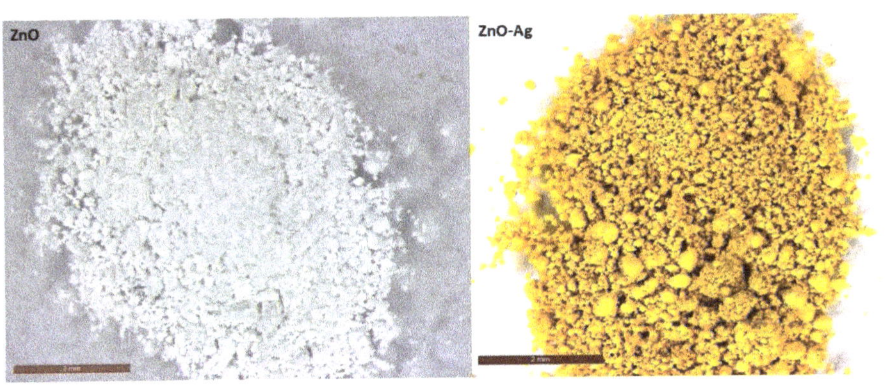

Figure 4. Stereomicroscope images of ZnO (left) and ZnO-Ag NPs (right).

Structural characterization carried out by X-ray Diffraction experiments confirmed the formation of ZnO in all the analyzed samples. XRD patterns (Figure 5d) showed the presence of peaks at 31.75°, 34.41°, 36.25°, 47.54°, 56.58°, 62.85°, 66.39°, 67.92°, and 69.06°, which can be related to the (100), (002), (101), (102), (110), (103), (200), (112), and (201) reflections of the hexagonal-type ZnO structure (JCPDS Card no. 043-0002) [74]. Furthermore, in the XRD pattern of Ag-doped ZnO with a nominal content of Ag equal to 1% and 3 wt%, two other peaks were clearly visible in addition to the ZnO peaks. These peaks at 2θ values of 38.10° and 44.38° can be attributed to the (111) and (200) reflections of the cubic Ag structure. The reflections characteristic of Ag were not detected in the 0.5% ZnO-Ag, probably due to the low concentration of Ag dopant. As a dopant, Ag can exist as a substituent for Zn^{2+} in the ZnO lattice or as an intermediate space atom [75]. As a substitute for Zn^{2+}, the Ag dopant should cause a peak shift in the corresponding XRD. Furthermore, none of the doped ZnO NP XRD patterns showed any significant shifts in peak positions. As suggested by Kareem et al. [75], Ag particles are likely to segregate at the grain boundaries of ZnO crystallites rather than going to the lattice of ZnO.

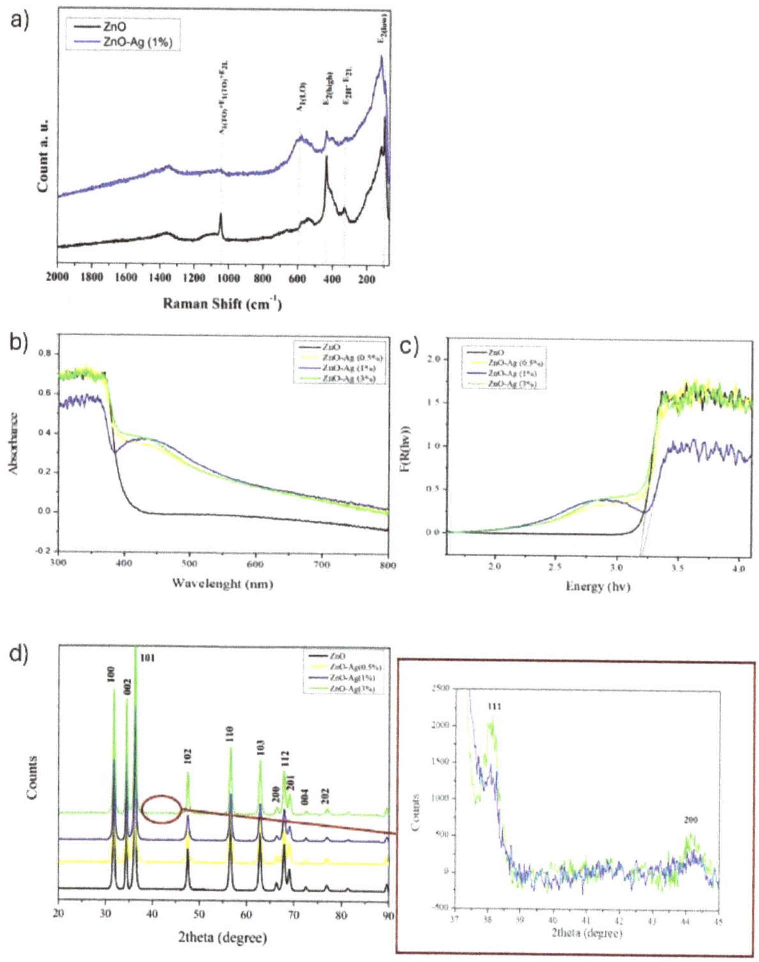

Figure 5. Spectra of ZnO and doped ZnO with Ag, respectively. (**a**) Micro-Raman spectra; (**b**) UV–VIS spectra recorded by diffuse reflectance spectroscopy; (**c**) Kubelka–Munch spectra; (**d**) XRD spectra.

The effect of doping on the crystallite size of the nanoparticles was evaluated by using the Debye–Scherrer equation. The average sizes of undoped ZnO, 0.5% ZnO-Ag, 1% ZnO-Ag, and 3% ZnO-Ag are reported in Table 2. As can be seen, Ag doping causes a sensible decrease in crystallite size [76].

Table 2. ZnO crystallite size (D), calculated by the Debye–Scherrer equation, and optical bandgap energies (Eg) calculated by the K–M equation.

Nanoparticles	ZnO			
	Undoped	**0.5% Ag**	**1.0% Ag**	**3% Ag**
D (nm)	22.96	20.54	18.96	21.37
Eg (eV)	3.21	3.19	3.21	3.14

The structure of ZnO-Ag powders was further investigated by μ-Raman spectroscopy. Raman spectra are reported in Figure 5a: the analysis defined shifts in position and intensity between pure and doped ZnO [77–79].

According to factor group analysis, the active Raman modes of the ZnO phase are visible at 100 cm^{-1} (E_{2L}), 331 cm^{-1} (E_{2h}-E_{2L}), 437 cm^{-1} (E_{2H}), and 580 cm^{-1} ($A1_{LO}$) [80].

The intensity of the peak at 437 cm^{-1} decreases with the addition of Ag because the ZnO structure becomes more defective; nonetheless, the peak at 578 cm^{-1} is marked by an increase in intensity, shift, and the broadening of peaks due to the reduction in crystallite size.

The reduction in crystallite size exerts an influence on the force constants and vibrational amplitudes of adjacent bonds. This, in turn, induces the red-shift and broadening of peaks in the resulting spectra of doped ZnO-Ag.

Diffuse reflectance spectroscopy (UV–Vis DRS) was employed to examine the light absorption properties of undoped and Ag-doped ZnO NPs. For all the samples, absorption increased sharply in the ultraviolet region due to the occurrence of band gap absorption. The UV–Vis spectra exhibited a shift in band edge, depending on the variation in Ag content. In the case of Ag-doped samples, the absorption of visible light is dominated by surface plasmon resonance (SPR) resulting from Ag nanoparticles.

Using the KM equation, the optical band gap values (Eg) were calculated, and they are presented in Table 2. Band gap energy is reduced with increased Ag doping for all nanoparticles. The observed variation in band gap values of ZnO (Figure 5c) with Ag doping may be attributed to the creation of oxygen vacancies, which introduce new intermediate energy levels between the valence (VB) and the conduction (CB) bands, responsible for reducing the final bandgap [81].

2.3. *Application of Nanoparticles on Packaging*

First, 1 wt% Ag-doped TiO_2 and 1 wt% Ag-doped ZnO NPs were dispersed in PVA and in a commercial acrylic resin, namely Paraloid B72. Then, they were applied onto paper and polyethylene foils using an airbrush from Aerograph. For comparison, undoped TiO_2 and ZnO were also used.

Figure 6 shows the dispersion of nanoparticles (NPs) on polyethylene (PE) and paper substrates under UV light observation. Under UV light, all formulations are characterized by a dark orange fluorescence, except for TiO_2, where fluorescence is absent, probably due to its high UV absorption ability.

Figures 7 and 8 report the observations collected by optical microscopy on treated paper foils. A whitish appearance was evident in the films containing both TiO_2 and TiO_2-Ag NPs. Additionally, a different distribution of particles was observed depending on the type of resin. In both cases, Paraloid B72-based formulations resulted in a better distribution of nanoparticles in comparison to PVA, which, in turn, showed a higher tendency for particle aggregation.

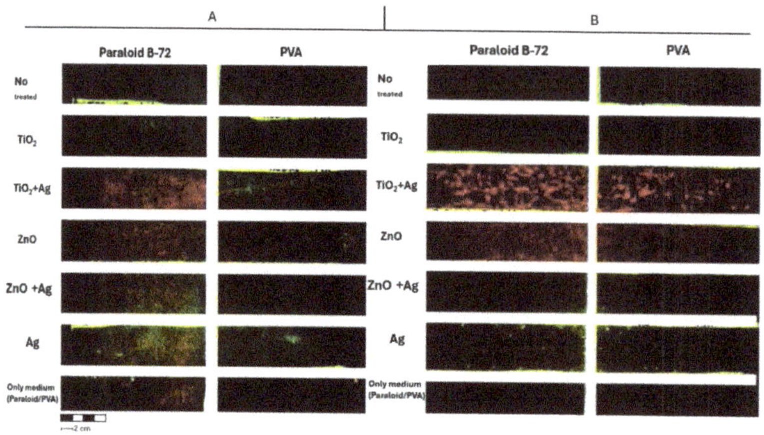

Figure 6. NP UV analysis images of the paper sample (**A**) and the Tyvek® sample (**B**).

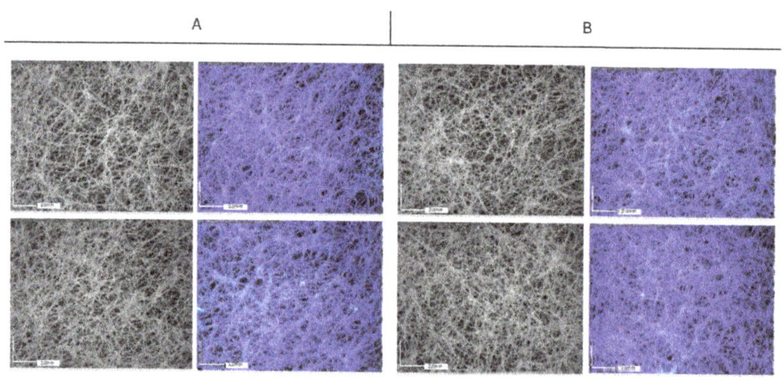

Figure 7. Paper packaging. Optical microscopy images (50×) acquired using visible light (**left**) and UV light (**right**). (**A**) Paraloid B72 medium (**top**) and PVA (**bottom**) + TiO_2 NPs; (**B**) Paraloid B72 medium (**top**) and PVA (**bottom**) + TiO_2-Ag NPs.

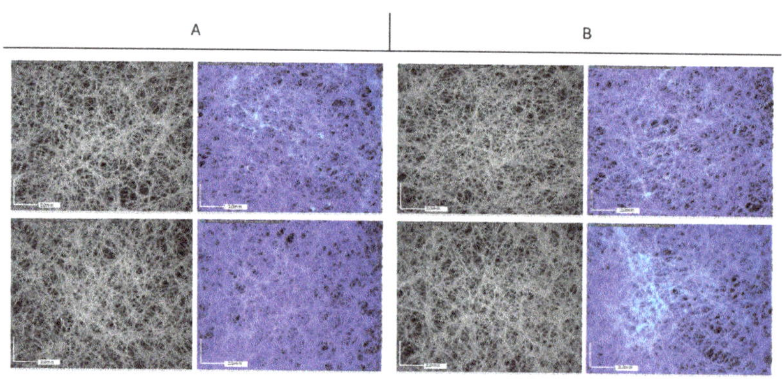

Figure 8. Paper Packaging. Optical microscopy images (50×) acquired using visible light (**left**) and UV light (**right**). (**A**) Paraloid B72 medium (**top**) and PVA (**bottom**) + ZnO NPs; (**B**) Paraloid B72 medium (**top**) and PVA (**bottom**) + ZnO-Ag NPs.

Optical microscope observations collected on Tyvek® (Figures 9 and 10) have highlighted the non-uniform distribution of the NPs, which tend to aggregate mainly around the wrinkles of the substrate, adhering to the surface irregularities of the substrate and consequently aggregating in limited areas.

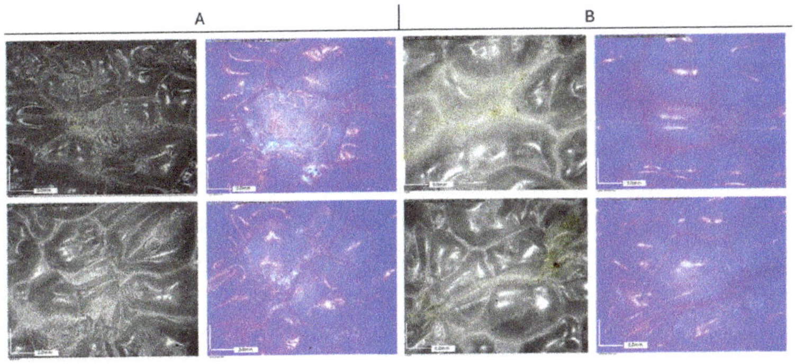

Figure 9. Tyvek® packaging. Optical microscopy images (50×) acquired under visible light (**left**) and UV light (**right**). (**A**) Paraloid B72 medium (**top**) and PVA (**bottom**) + TIO_2 NPs; (**B**) Paraloid B72 medium (**top**) and PVA (**bottom**) + TIO_2-Ag NPs.

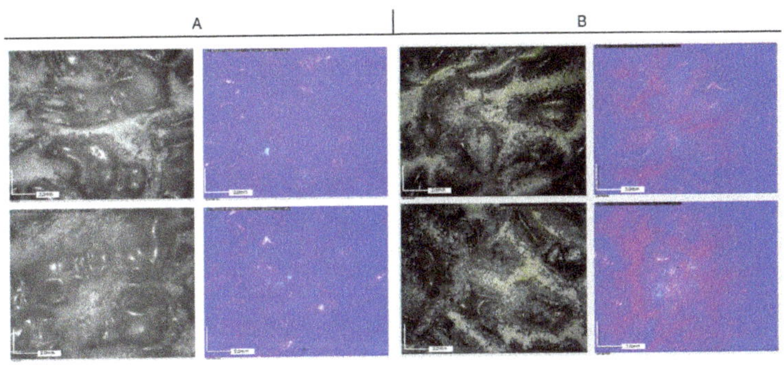

Figure 10. Tyvek® packaging. Optical microscopy images (50×) acquired under visible light (**left**) and UV light (**right**). (**A**) Paraloid B72 medium (**top**) and PVA (**bottom**) + ZnO NPs; (**B**) Paraloid B72 medium (**top**) and PVA (**bottom**) + ZnO-Ag NPs.

In both cases, a better distribution of nanoparticles is achieved when Paraloid B72 is used as the dispersant medium.

Table 3 shows the total chromatic variation calculated as the color difference between the substrates with coatings compared to the coating treatment with nanoparticles.

Table 3. ΔE between the substrates with coatings compared to the coating treatment with nanoparticles. SD < 0.5.

Coating	Particles	Paper	Tyvek®
Paraloid B72	TiO_2	2.58	3.91
	TiO_2-Ag	3.44	5.60
	ZnO	4.73	4.01
	ZnO-Ag	3.75	6.61
PVA	TiO_2	3.23	5.39
	TiO_2-Ag	3.55	5.94
	ZnO	2.81	4.72
	ZnO-Ag	3.37	5.64

Comparing the Tyvek® substrate to paper, only minor and noticeable chromatic variations (fewe than five) were found.

The coatings of ZnO and TiO_2 induce a chromatic variation that is imperceptible to the naked eye compared to the coatings with doped Ag particles. The range of values between 2.58 and 5.64 indicates that all nanoparticles are present on all treated surfaces.

The surface morphology of the treated samples was analyzed by scanning electron microscopy (SEM). Figure 11 displays images of the paper sample and Figure 12 shows images of Tyvek® surfaces coated with doped ZnO-Ag and TiO_2-Ag NPs. SEM investigation confirmed that Zn-Ag nanoparticles dispersed in both PVA and Paraloid were more evenly distributed than TiO_2-Ag-based formulations, which instead showed a sensible tendency to aggregate.

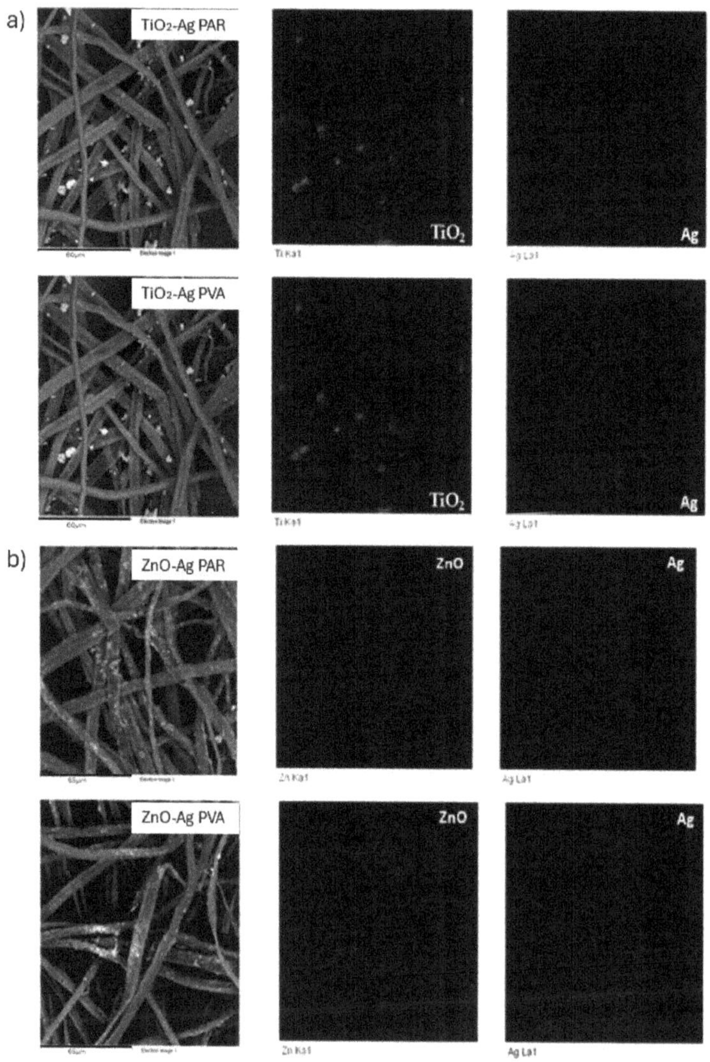

Figure 11. SEM analysis of NP distribution on packaging paper; (**a**) spatial distribution of TiO_2-Ag NPs in Paraloid B72 (TiO_2-Ag PAR) (**top**) and PVA (TiO_2-Ag PVA) (**bottom**); (**b**) spatial distribution of ZnO-Ag NPs in Paraloid B72 (ZnO-Ag PAR) (**top**) and PVA (ZnO-Ag PVA) (**bottom**).

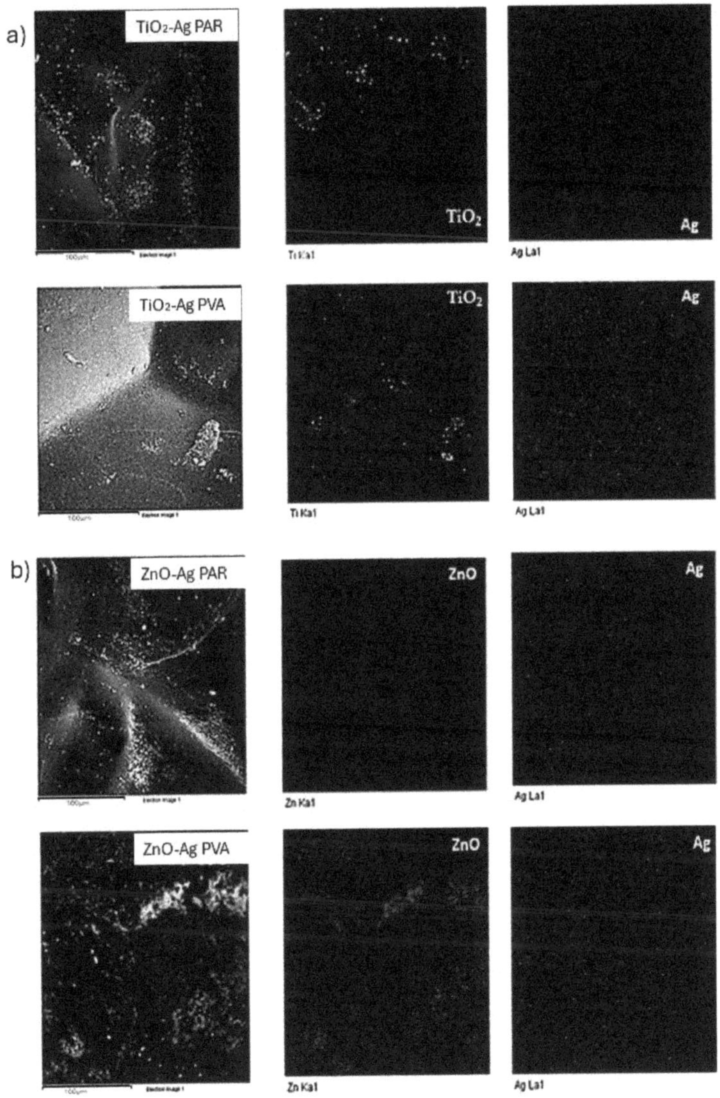

Figure 12. SEM analysis of NP distribution on PE; (**a**) spatial distribution of TiO_2-Ag NPs in Paraloid B72 (TiO_2-Ag PAR) (**top**) and PVA (TiO_2-Ag PVA) (**bottom**); (**b**) spatial distribution of ZnO-Ag NPs in Paraloid B72 (ZnO-PAR) (**top**) and PVA (ZnO-Ag PVA) (**bottom**).

μ-Raman spectroscopy results confirmed the successful application of TiO_2-Ag and ZnO-Ag NPs on both Tyvek® and paper samples: in all Raman spectra, it is possible to identify the characteristic peak at 144 cm^{-1}, 437 cm^{-1}, 2886 cm^{-1}, and 2892 cm^{-1}, typical of TiO_2, ZnO, PVA, and Paraloid B72, respectively (Figure S1, Table S1).

3. Microbiological Analysis

The results from the fungal attack (Figure 13) define a significant proliferation of microbial species on untreated surfaces. The ATP (adenosine triphosphate, a marker for surface contamination of biological origin) data obtained indicate substantial differences in microbial vitality reduction between the two substrates considered in the experiment, with greater efficacy observed for coatings applied on paper compared to Tyvek®. This outcome is evident when employing PVA as a medium. The most notable variances were observed in the use of the two media; samples treated with Paraloid B72, as opposed to those treated with PVA, demonstrate a greater inhibition of biological growth for an equivalent amount of added nanoparticles. Regarding nanoparticles, the film characterized by the presence of TiO_2 does not inhibit growth, although a slight decrease in ATP is still observed. The most favorable outcomes are achieved by functionalizing the media with TiO_2 and ZnO doped with silver. Even the use of simple zinc oxide allows for a reduction in biological growth, albeit with less efficacy compared to the same particle doped with Ag. In all highlighted cases, complete inhibition of microbiological growth is not achieved. Silver-doped zinc oxide nanoparticles exhibit superior antibacterial activity compared to pure zinc oxide nanoparticles, as reported in the literature [82].

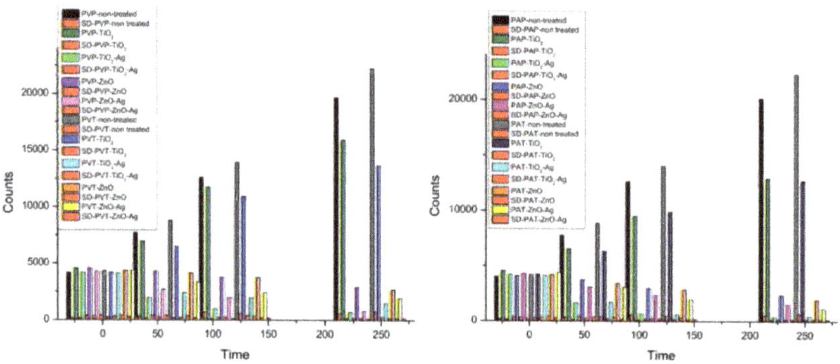

Figure 13. ATP values for several packing materials treated with the different experimental coatings.

The incorporation of the metal dopant enhances the photocatalyst's efficiency by reducing band gap energy and decreasing the rate of electron–hole recombination. The inefficacy of pure TiO_2 as a biocide is due to its absorption in the ultraviolet region, where its photocatalytic activity is very high. By doping with metals, the absorption region of TiO_2 is extended to the visible part of radiation. Doping facilitates the formation of hydroxyl and superoxide radicals, which are strong oxidizers; these radicals are the basis of the biocidal properties observed in doped TiO_2 compared to undoped TiO_2 [83]. Incubation of the samples was carried out under visible light illumination. The ineffectiveness of undoped TiO_2 compared to undoped ZnO is attributed to the absence of UV radiation in the conducted tests.

4. Materials and Methods

4.1. Reagents

Titanium (IV) Butoxide (reagent grade, 97% Sigma-Aldrich, Milan, Italy), Zinc Acetate dihydrate (ACS reagent, ≥98% Sigma Aldrich), $AgNO_3$ (ACS reagent, ≥99.0% Sigma-Aldrich), Citric Acid (ACS reagent, ≥99.5% Sigma-Aldrich), and Ethanol (99.8% Sigma-Aldrich) were used without any further purification. Double distilled water was used in all experiments. Cellulose paper packaging was provided by IMAR Italia and Tyvek® (polyethylene) was provided by DuPont (Wilmington, DE, USA).

4.2. Synthesis of Nanoparticles

4.2.1. Synthesis of TiO_2-Ag Nanoparticles

TiO_2-Ag NPs with different amounts of Ag doping (0.5, 1, and 3 wt%) were synthesized using a simple and green sol-gel approach in accordance with the procedure reported by Souza et al. [84], which was opportunely modified. First, 6.8 mL of titanium butoxide

$[Ti(OBu)_4]$ was mixed with 2.9 mL of ethanol in a beaker and stirred vigorously for 30 min (solution A). In another beaker, 4.80 g of citric acid and 39.6 mL of deionized water (solution B) were mixed and a pH of ~2.0 was measured. Solution A was added dropwise to solution B under stirring, and the resulting solution was kept at 80 °C under reflux and constant stirring for 2 h to ensure the complete hydrolysis of the titanium precursor. The resulting clear and colorless hydrosol solution was allowed to cool down to room temperature. A certain volume of NH_4OH 1 M was slowly added to TiO_2 hydrosol until a pH of 5 was measured. After that, different amounts of $AgNO_3$ were added, and the obtained hydrosols were sonicated at 80 W for 1 h at a temperature of 60 °C. Brownish Ag-doped TiO_2 powders were obtained by solvent evaporation at 100 °C for 24 h. Crystalline Ag-doped TiO_2 NPs were finally synthesized after two-step thermal annealing at 400 °C for 1 h (heating ramp: 1 °C/min) and then at 500 °C for 2 h (heating ramp 1 °C/min) (Figure 1). TiO_2 NPs, except for $AgNO_3$, were obtained following the same procedure.

4.2.2. Synthesis of ZnO-Ag Nanoparticles

Ag-doped ZnO NPs were obtained by the following procedure: in a typical reaction process, 0.02 mol of $Zn(NO_3)_2 \cdot 6H_2O$ was dissolved in 30 mL of distilled water (DW) at room temperature and under continuous stirring. Different amounts of 0.1 M $AgNO_3$ solution were added to the clear solution to obtain a nominal dopant concentration of 0.5, 1, and 3 wt%, followed by the addition of a few drops of 1 M NaOH solution until a pH of 11 was reached. The obtained solution was first kept under agitation for 30 min at room temperature and then heated at 60 °C under reflux for 6h. The white precipitate was cooled down to room temperature and then was collected by centrifugation at 8000 rpm for 10 min and washed three times with water and ethanol. Finally, the powders were dried at 70 °C in a furnace for 12 h. No further thermal treatment was performed (Figure 4). Undoped ZnO NPs were obtained by the same procedure without the addition of $AgNO_3$.

4.3. *Preparation of TiO_2-Ag/PVA and ZnO-Ag/PVA Formulations*

The application of coating formulations is very simple (Figure 14). After NP synthesis, they were added to a PVA water solution to form an isotropic % by ultrasonication for 1 h. Then, the dispersion was sprayed onto the paper or PE surface and dried at room temperature.

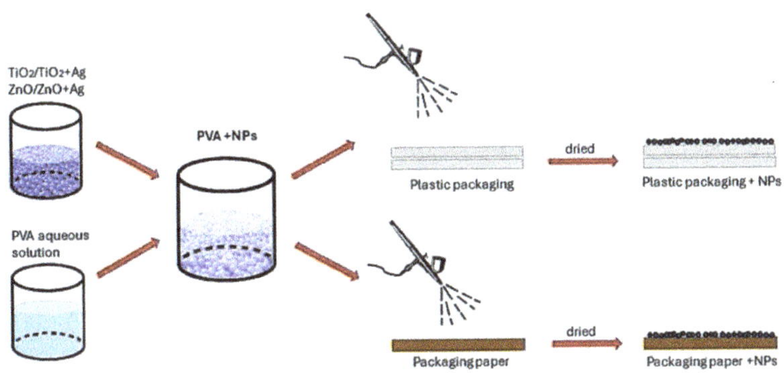

Figure 14. Procedure of TiO_2/TiO_2+Ag and ZnO/ZnO+Ag NP application on packaging materials.

For the preparation of the antimicrobial formulations, only TiO_2-Ag and ZnO-Ag NPs with a nominal content of Ag equal to 1 wt% were used. In addition, undoped TiO_2 and ZnO NPs were used as references. A fixed amount of nanoparticles (0.1 g) was dispersed in distilled water (10 mL) and ultrasonicated for 30 min at a power of 80 W in order to obtain a better dispersion of the nanoparticles in water. At this stage, a water solution of PVA (1 wt%) was added to the dispersion of nanoparticles to reach a final volume of 50 mL. The final solution was further stirred for 1 h at room temperature. Finally, after it was homogeneously stirred, the solution was sprayed on the surface of the paper and Tyvek® by an aerographer.

To gain insight into the effects of PVA, two other formulations were set up using different concentrations of the PVA solution, namely 5 wt% and 10 wt%. Finally, the sprayed surfaces were allowed to dry at room temperature for 24 h.

The same procedure was followed for the preparation of other similar formulations in which the commercial acrylic resin Paraloid B72 was used as a substitution for PVA.

4.4. Methods

The optical properties of the NPs were defined using stereomicroscope analysis, performed with a Leica M125C stereomicroscope equipped with a DMC4500 USB digital camera (Leica, Wetzlar, Germany) and using a Jasco V630 UV–VIS spectrophotometer (Jasco, Tokyo, Japan) with a double ray, characterized by a silicon photodiode detector equipped with two sources of radiation. One of these is a deuterium lamp (190–350 nm) and the other is a halogen lamp (330–110 nm). To analyze the structure and crystallinity phases of the NPs, XRD analysis was carried out by means of a BrukerD8 advance theta-theta diffractometer (Bruker, Ettlingen, Germany) with a one millimeter point collimator with x, y, z adjustment, by laser and 3D detector.

The film formed by the coating was analyzed by UV fluorescence using a full spectrum NX500 28.2 MP BSI CMOS camera (Samsung, Suwon-si, Republic of Korea) with a Madetec UV lamp (Tec Lighting, Brea, CA, USA), at a wavelength of 365 nm.

To observe the distribution of NPs on the surface, digital microscope analysis was performed with Dino-lite AM4113t-FVW (Dino-Lite Digital Microscope, Bangkok, Thailand) in VIS and UV light (395 nm).

Colorimetric analysis was performed with a 3Nh TS7700 portable spectrophotometric colorimeter (3NH, Guangzhou, China). Spectrocolometry applied to the coatings was used to determine whether the treatment induced any color changes on the surface, highlighting the presence or absence of the treatment itself.

For the analysis of coatings and nanomaterials, micro-Raman spectroscopy was chosen because it is an accurate and noninvasive technique. The Raman spectra were obtained using a Renishaw RM 2000 Raman microscope (Renishaw, Hongkong, China) conjugated with a Leica optical microscope and equipped with a silicone CCD detector. The laser radiation was focused with the 50× and 100× lenses. The excitation source was a 785 nm diode laser. The actual power at the sample was 1.58 mW. The diffraction grating was 1800 L/mm. The accumulation time was 10 s with 10 repetitions. Two Edge filters blocked the Rayleigh-scattered light below 100 cm^{-1}. For this reason, the study of ultra-low wavenumber Raman spectra in the region of <100 cm^{-1} is overlooked. To investigate the superficial morphology of the samples, an SEM TESCAN instrument was used, equipped with an X INCA 300 microprobe and a four-sensor retro diffuse electron detector (BSE) (TESCAN, Brno, Czechia). The beam acceleration voltage was set between 20 KeV and 30 Kev, which is sufficient to exceed the critical electron energy of the X-ray emission (quantitative resolution limit of 0.2% by weight).

To assess the stability to biological exposure of the tested formulations, the different coatings applied on paper and Tyvek® were contaminated with pathogenic agents from *Cladosporium* sp., a widely distributed species in indoor environments [28,29] that belongs to the family of Dematiaceae. The samples were incubated 3 times at 60 °C for 240 min. Incubation of the samples was carried out under visible light illumination to simulate museum lighting conditions. The stability to biological attack of several coatings was defined by ATP measurements using a Lumitester PD 30 (Kikkoman, Chiba, Japan).

5. Conclusions

Within museums, not only are there exhibition sections for public viewing but there are also storage areas where artworks that do not find space in the exhibition areas are housed.

The biological degradation of materials is a critical concern in preserving cultural heritage, and the development of innovative nanoparticle-based packaging represents a promising solution to mitigate antifungal attacks on stored artworks.

This study has demonstrated the potential use of nanoparticulate biocides for functionalizing materials utilized in the packaging of artworks. For this purpose, Ag-doped ZnO and Ag-doped TiO_2 NPs were synthesized and dispersed in two different polymeric mediums, namely PVA and Paraloid. This doping ensures enhanced biocidal efficacy of the particles, albeit with a chromatic variation nearly imperceptible to the human eye. SEM/EDS analysis highlighted the presence of nanoparticles on the surface of paper and Tyvek®, with a better homogeneous distribution of ZnO-Ag NPs than TiO_2-Ag NPs. Ag-doped TiO_2 exhibited the highest biocidal efficacy, attributed to its visible light-induced photocatalytic activity resulting from the doping process. Conversely, undoped TiO_2 demonstrated no biocidal efficacy due to the absence of UV light-induced photocatalytic activation. Both doped and undoped ZnO nanoparticles exhibited good biocidal efficacy, although complete inhibition of microbial vitality (ATP) was not achieved in all cases.

This study contributes to the ongoing efforts to develop effective strategies for preserving cultural heritage materials during transport and storage, addressing the pressing issue of biological degradation.

Supplementary Materials: The following supporting information can be downloaded at: https://www.mdpi.com/article/10.3390/molecules29153712/s1. Figure S1. Micro-Raman spectra: (a) paper after TiO_2 NPs and adhesives application; (b) paper after ZnO NPs and adhesives application; (c) Tyvek® after TiO_2 NPs and adhesives application; (d) Tyvek® after ZnO NPs and adhesives application; were PAP is refer to Paraloid on paper, PAT is Paraloid on Tyvek®, PVP is PVA on paper and PVT is PVA on Tyvek®. Table S1. Raman peaks of packaging material studied. Refs. [84–94] are cited in the Supplementary Materials.

Author Contributions: Conceptualization, T.d.C., R.G.T. and A.M.; methodology, T.d.C., R.G.T., A.M. and L.C.; validation L.C., F.I.B. and C.Z.; formal analysis, I.A.C.; investigation, T.d.C., R.G.T. and L.C.; resources, A.M.; data curation, T.d.C., R.G.T. and A.M.; writing—original draft preparation, T.d.C., R.G.T. and A.M.; writing—review and editing, F.I.B., I.A.C., A.M. and M.F.L.R.; visualization, C.Z.; supervision, T.d.C. All authors have read and agreed to the published version of the manuscript.

Funding: This research received no external funding.

Institutional Review Board Statement: Not applicable.

Informed Consent Statement: Not applicable.

Data Availability Statement: The original contributions presented in the study are included in the article/Supplementary Material.

Acknowledgments: The authors acknowledge Daniela Ferro for helpful SEM discussions and the YOCOCU APS organization for the support and for the materials used in the experimentation.

Conflicts of Interest: The authors declare no conflicts of interest.

References

1. Ferdyn-Grygierek, J.; Grygierek, K. HVAC control methods for drastically improved hygrothermal museum microclimates in warm season. *Build. Environ.* **2019**, *149*, 90–99. [CrossRef]
2. Christensen, J.E.; Kollias, C.G. Hygrothermal Evaluation of a Museum Storage Building based on Actual Measurements and Simulations. *Energy Procedia* **2015**, *78*, 651–656. [CrossRef]
3. Sterflinger, K. Fungi: Their role in deterioration of cultural heritage. *Fungal Biol. Rev.* **2010**, *24*, 47–55. [CrossRef]
4. Grabek-Lejko, D.; Tekiela, A.; Kasprzyk, I. Risk of biodeterioration of cultural heritage objects, stored in the historical and modern repositories in the Regional Museum in Rzeszow (Poland). A case study. *Int. Biodeterior. Biodegrad.* **2017**, *123*, 46–55. [CrossRef]
5. Bakker, A.; Siegel, J.A.; Mendell, M.J.; Prussin, A.; Marr, L.; Peccia, J. Bacterial and fungal ecology on air conditioning cooling coils is influenced by climate and building factors. *Indoor Air* **2019**, *30*, 326–334. [CrossRef] [PubMed]
6. Kavkler, K.; Gunde-Cimerman, N.; Zalar, P.; Demšar, A. Fungal contamination of textile objects preserved in Slovene museums and religious institutions. *Int. Biodeterior. Biodegrad.* **2015**, *97*, 51–59. [CrossRef]

7. de Carvalho, H.P.; Mesquita, N.; Trovão, J.; Rodríguez, S.F.; Pinheiro, A.C.; Gomes, V.; Alcoforado, A.; Gil, F.; Portugal, A. Fungal contamination of paintings and wooden sculptures inside the storage room of a museum: Are current norms and reference values adequate? *J. Cult. Herit.* **2018**, *34*, 268–276. [CrossRef]
8. Lazaridis, M.; Katsivela, E.; Kopanakis, I.; Raisi, L.; Mihalopoulos, N.; Panagiaris, G. Characterization of airborne particulate matter and microbes inside cultural heritage collections. *J. Cult. Herit.* **2018**, *30*, 136–146. [CrossRef]
9. Romero, S.M.; Giudicessi, S.L.; Vitale, R.G. Is the fungus Aspergillus a threat to cultural heritage? *J. Cult. Herit.* **2021**, *51*, 107–124. [CrossRef]
10. Branysova, T.; Demnerova, K.; Durovic, M.; Stiborova, H. Microbial biodeterioration of cultural heritage and identification of the active agents over the last two decades. *J. Cult. Herit.* **2022**, *55*, 245–260. [CrossRef]
11. Scarlat, I.P.; Lazăr, V.; Stepa, R.; Chifiriuc, M.C.; Dițu, L.M.; Haiducu, M. Seasonal monitoring of microbiological air contamination from a museum—A case study. *MATEC Web Conf.* **2017**, *121*, 11–17. [CrossRef]
12. Skóra, J.; Gutarowska, B.; Pielech-Przybylska, K.; Stępień, Ł.; Pietrzak, K.; Piotrowska, M.; Pietrowski, P. Assessment of microbiological contamination in the work environments of museums, archives and libraries. *Aerobiologia* **2015**, *31*, 389–401. [CrossRef] [PubMed]
13. Scarlat, I.; Haiducu, M.; and Stepa, R. Air Contamination With Fungals In Museum. *ACTA Univ. Cibiniensis* **2015**, *66*, 165–168. [CrossRef]
14. Szulc, J.; Otlewska, A.; Ruman, T.; Kubiak, K.; Karbowska-Berent, J.; Kozielec, T.; Gutarowska, B. Analysis of paper foxing by newly available omics techniques. *Int. Biodeterior. Biodegrad.* **2018**, *132*, 157–165. [CrossRef]
15. Goodell, B.; Winandy, J.E.; Morrell, J.J. Fungal Degradation of Wood: Emerging Data, New Insights and Changing Perceptions. *Coatings* **2020**, *10*, 1210. [CrossRef]
16. Trovão, J.; Portugal, A. Current Knowledge on the Fungal Degradation Abilities Profiled through Biodeteriorative Plate Essays. *Appl. Sci.* **2021**, *11*, 4196. [CrossRef]
17. Atwa, D.M.; Ibrahim, S.; Stani, C.; Birarda, G.; Ali, N.; Abdullah, E.; Vaccari, L.; Grenni, P.; Visca, A.; Badran, Y.; et al. Biodeterioration Assessment of a Unique Old Pharaonic Kingdom Wooden Statue Using Advanced Diagnostic Techniques. *Appl. Sci.* **2022**, *12*, 7020. [CrossRef]
18. Macchia, A.; Cerafogli, E.; Rivaroli, L.; Colasanti, I.A.; Aureli, H.; Biribicchi, C.; Brunori, V. Marble Chromatic Alteration Study Using Non-Invasive Analytical Techniques and Evaluation of the Most Suitable Cleaning Treatment: The Case of a Bust Representing Queen Margherita di Savoia at the U.S. Embassy in Rome. *Analytica* **2022**, *3*, 406–429. [CrossRef]
19. De Leo, F.; Antonelli, F.; Pietrini, A.M.; Ricci, S.; Urzì, C. Study of the euendolithic activity of black meristematic fungi isolated from a marble statue in the Quirinale Palace's Gardens in Rome, Italy. *Facies* **2019**, *65*, 18. [CrossRef]
20. Macchia, A.; Aureli, H.; Prestileo, F.; Ortenzi, F.; Sellathurai, S.; Docci, A.; Cerafogli, E.; Colasanti, I.A.; Ricca, M.; La Russa, M.F. In-Situ Comparative Study of Eucalyptus, Basil, Cloves, Thyme, Pine Tree, and Tea Tree Essential Oil Biocide Efficacy. *Methods Protoc.* **2022**, *5*, 37. [CrossRef]
21. McNamara, C.J.; Mitchell, R. Microbial deterioration of historic stone. *Front. Ecol. Environ.* **2005**, *3*, 445–451. [CrossRef]
22. Gaylarde, C.C.; Baptista-Neto, J.A. Microbiologically induced aesthetic and structural changes to dimension stone. *NPJ Mater. Degrad.* **2021**, *5*, 33. [CrossRef]
23. Cappitelli, F.; Cattò, C.; Villa, F. The Control of Cultural Heritage Microbial Deterioration. *Microorganisms* **2020**, *8*, 1542. [CrossRef] [PubMed]
24. Menegaldo, M.; Livieri, A.; Isigonis, P.; Pizzol, L.; Tyrolt, A.; Zabeo, A.; Semenzin, E.; Marcomini, A. Environmental and economic sustainability in cultural heritage preventive conservation: LCA and LCC of innovative nanotechnology-based products. *Clean. Environ. Syst.* **2023**, *9*, 100124. [CrossRef]
25. Kakakhel, M.A.; Wu, F.; Gu, J.D.; Feng, H.; Shah, K.; Wang, W. Controlling biodeterioration of cultural heritage objects with biocides: A review. *Int. Biodeterior. Biodegrad.* **2019**, *143*, 104721. [CrossRef]
26. Macchia, A.; Aureli, H.; Biribicchi, C.; Docci, A.; Alisi, C.; Prestileo, F.; Galiano, F.; Figoli, A.; Mancuso, R.; Gabriele, B.; et al. In Situ Application of Anti-Fouling Solutions on a Mosaic of the Archaeological Park of Ostia Antica. *Materials* **2022**, *15*, 5671. [CrossRef]
27. Mekonnen, H.; Bires, Z.; Berhanu, K. Practices and challenges of cultural heritage conservation in historical and religious heritage sites: Evidence from North Shoa Zone, Amhara Region, Ethiopia. *Herit. Sci.* **2022**, *10*, 172. [CrossRef]
28. Castrillón Rivera, L.E.; Palma Ramos, A.; Castañeda Sánchez, J.I.; Drago Serrano, M.E. Origin and Control Strategies of Biofilms in the Cultural Heritage. In *Antimicrobials, Antibiotic Resistance, Antibiofilm Strategies and Activity Methods*, 1st ed.; Kirmusaoglu, S., Ed.; IntechOpen: London, UK, 2019; Volume 1, pp. 1–24. [CrossRef]
29. Ruga, L.; Orlandi, F.; Fornaciari, M. Preventive Conservation of Cultural Heritage: Biodeteriogens Control by Aerobiological Monitoring. *Sensors* **2019**, *19*, 3647. [CrossRef]
30. Franco-Castillo, I.; Hierro, L.; de la Fuente, J.M.; Seral-Ascaso, A.; Mitchell, S.G. Perspectives for antimicrobial nanomaterials in cultural heritage conservation. *Chem* **2021**, *7*, 629–669. [CrossRef]
31. Gorgolis, G.; Ziemann, S.; Kotsidi, M.; Paterakis, G.; Koutroumanis, N.; Tsakonas, C.; Anders, M.; Galiotis, C. Novel Graphene-Based Materials as a Tool for Improving Long-Term Storage of Cultural Heritage. *Materials* **2023**, *16*, 3528. [CrossRef]
32. Abdel-Hamied, M.; Abdelhafez, A.A.M.; Abdel-Maksoud, G. Consolidation materials used with illuminated and non-illuminated paper manuscripts and historical leather bindings: A review. *Pigment. Resin Technol.* 2024, *ahead-of-print*. [CrossRef]

33. Carboni, E. *Cultura in Viaggio. Quando la Movimentazione Diventa Un'arte*, 12th ed.; Museologia Scientifica: Calenzano, Italy, 2018; pp. 58–65.
34. Novak, M.; Grau-Bove, J.; De Stefani, C.; Checkley-Scott, C.; Kraševec, I.; Kralj Cigić, I.; Elnaggar, A. Evaluation and modelling of the environmental performance of archival boxes, part 1: Material and environmental assessment. *Herit. Sci.* **2024**, *12*, 24. [CrossRef]
35. Alkassfarity, A.N.; Yassin, M.A.; Abdel Rehim, M.H.; Liu, L.; Jiao, Z.; Wang, B.; Wei, Z. Modified cellulose nanocrystals enhanced polycaprolactone multifunctional films with barrier, UV-blocking and antimicrobial properties for food packaging. *Int. J. Biol. Macromol.* **2024**, *261*, 129871. [CrossRef]
36. Sarfraz, M.H.; Hayat, S.; Siddique, M.H.; Aslam, B.; Ashraf, A.; Saqalein, M.; Khurshid, M.; Sarfraz, M.F.; Afzal, M.; Muzammil, S. Chitosan based coatings and films: A perspective on antimicrobial, antioxidant, and intelligent food packaging. *Prog. Org. Coat.* **2024**, *188*, 108235. [CrossRef]
37. Zagloul, H.; Dhahri, M.; Bashal, A.H.; Khaleil, M.M.; Habeeb, T.H.; Khalil, K.D. Multifunctional Ag_2O/chitosan nanocomposites synthesized via sol-gel with enhanced antimicrobial, and antioxidant properties: A novel food packaging material. *Int. J. Biol. Macromol.* **2024**, *264*, 129990. [CrossRef]
38. Alqarni, L.S.; Alghamdi, A.M.; Elamin, N.Y.; Rajeh, A. Enhancing the optical, electrical, dielectric properties and antimicrobial activity of chitosan/gelatin incorporated with Co-doped ZnO nanoparticles: Nanocomposites for use in energy storage and food packaging. *J. Mol. Struct.* **2024**, *1297*, 137011. [CrossRef]
39. Emir, A.A.; Yildiz, E.; Kirtil, E. Peppermint oil-infused polylactic acid films: A novel approach for antimicrobial and biodegradable food packaging. *Polym. Eng. Sci.* **2024**, *64*, 2943–2955. [CrossRef]
40. Bertuzzi, M.A.; Gamboni, J.E.; Slavutsky, A.M.; Ibarguren, C. Novel food packaging systems with antimicrobial agents from microbial source. In *Food Packaging and Preservation*, 1st ed.; Jaiswal, A.K., Shankar, S., Eds.; Elsevier: Amsterdam, The Netherlands, 2024; pp. 91–111. [CrossRef]
41. Zeng, J.; Ma, Y.; Li, P.; Zhang, X.; Gao, W.; Wang, B.; Xu, J.; Chen, K. Development of high-barrier composite films for sustainable reduction of non-biodegradable materials in food packaging application. *Carbohydr. Polym.* **2024**, *330*, 121824. [CrossRef]
42. Wyrwa, J.; Barska, A. Innovations in the food packaging market: Active packaging. *Eur. Food Res. Technol.* **2017**, *243*, 1681–1692. [CrossRef]
43. Yao, Q.; Huang, F.; Lu, Y.; Huang, J.; Ali, M.; Jia, X.Z.; Zeng, X.A.; Huang, Y.Y. Polysaccharide-based food packaging and intelligent packaging applications: A comprehensive review. *Trends Food Sci. Technol.* **2024**, *147*, 104390. [CrossRef]
44. Fatima, S.; Khan, M.R.; Ahmad, I.; Sadiq, M.B. Recent advances in modified starch based biodegradable food packaging: A review. *Heliyon* **2024**, *10*, e27453. [CrossRef]
45. Bandelli, D.; Mastrangelo, R.; Poggi, G.; Chelazzi, D.; Baglioni, P. New sustainable polymers and oligomers for Cultural Heritage conservation. *Chem. Sci.* **2024**, *15*, 2443–2455. [CrossRef]
46. Walker, S. Investigation of the properties of Tyvek®, pertaining to its use as a storage material for artifacts. In *The Conservation of Artifacts Made from Plant Materials*, 1st ed.; International Institute for Conservation of Historic and Artistic Works-Canadian Group (IIC-CG): Ottawa, ON, Canada, 1986; Volume 12, pp. 21–25.
47. Byrne, K. Handling, Packing, and Shipping. In *NPS Museum Handbook*, 1st ed.; part II; International Institute for Conservation of Historic and Artistic Works-Canadian Group (IIC-CG): Ottawa, ON, Canada, 2000; pp. 21–25.
48. Neda, K.E. A Study of Packaging the Museum Pieces for Transportation with an Emphasis on Textiles. *J. Packag. Sci. Tech.* **2023**, *13*, 1–10.
49. Adel, A.M.; Al-Shemy, M.T.; Diab, M.A.; El-Sakhawy, M.; Toro, R.G.; Montanari, R.; de Caro, T.; Caschera, D. Fabrication of packaging paper sheets decorated with alginate/oxidized nanocellulose-silver nanoparticles bio-nanocomposite. *Int. J. Biol. Macromol.* **2021**, *181*, 612–620. [CrossRef]
50. Adel, A.M.; Al-Shemy, M.T.; Diab, M.A.; El-Sakhawy, M.; Toro, R.G.; Cerri, L.; Caschera, D. Immobilization of TiO_2NP@ oxidized cellulose nanocrystals for paper-based active packaging materials. *Int. J. Biol. Macromol.* **2023**, *231*, 123270. [CrossRef]
51. Padmavathy, N.; Vijayaraghavan, R. Enhanced bioactivity of ZnO nanoparticles—An antimicrobial study. *Sci. Technol. Adv. Mater.* **2008**, *9*, 035004. [CrossRef]
52. Ruffolo, S.A.; Macchia, A.; La Russa, M.F.; Mazza, L.; Urzì, C.; De Leo, F.; Barberio, M.; Crisci, G.M. Marine Antifouling for Underwater Archaeological Sites: TiO_2 and Ag-Doped TiO_2. *Int. J. Photoenergy* **2013**, *251647*, 1–6. [CrossRef]
53. Pathak, T.K.; Kroon, R.E.; Craciun, V.; Popa, M.; Chifiriuc, M.C.; Swart, H.C. Influence of Ag, Au and Pd noble metals doping on structural, optical and antimicrobial properties of zinc oxide and titanium dioxide nanomaterials. *Heliyon* **2019**, *5*, e01333. [CrossRef]
54. Artesani, A.; Di Turo, F.; Zucchelli, M.; Traviglia, A. Recent advances in protective coatings for cultural heritage—An overview. *Coatings* **2020**, *10*, 217. [CrossRef]
55. Palanichamy, K.; Anandan, M.; Sridhar, J.; Natarajan, V.; Dhandapani, A. PVA and PMMA nano-composites: A review on strategies, applications and future prospects. *Mater. Res. Express* **2023**, *10*, 022002. [CrossRef]
56. Huang, J.; Lyu, S.; Fu, F.; Wu, Y.; Wang, S. Green preparation of a cellulose nanocrystals/polyvinyl alcohol composite superhydrophobic coating. *RSC Adv.* **2017**, *7*, 20152–20159. [CrossRef]
57. Brandrup, J.; Immergut, E.H.; Grulke, E.A. *Polymer Handbook*, 4th ed.; Wiley: Hoboken, NJ, USA, 2003; Volume 2.

58. Hussein Elsayed, N.; Najmi, A.H.; Mohamed, G.M.; Mohamed, A.A.; Al-Sayed, H.M.A. Preparation and characterization of Paraloid B-72/TiO_2 nanocomposite and their effect on the properties of polylactic acid as strawberry coating agents. *J. Food Saf.* **2020**, *40*, e12838. [CrossRef]
59. Gîfu, I.C.; Ianchiș, R.; Nistor, C.L.; Petcu, C.; Fierascu, I.; Fierascu, R.C. Polyelectrolyte Coatings—A Viable Approach for Cultural Heritage Protection. *Materials* **2023**, *16*, 2873. [CrossRef] [PubMed]
60. Kanth, A.P.; Soni, A.K. Application of nanocomposites for conservation of materials of cultural heritage. *J. Cult. Herit.* **2023**, *59*, 120–130. [CrossRef]
61. Aldosari, M.A.; Darwish, S.S.; Adam, M.A.; Elmarzugi, N.A.; Ahmed, S.M. Using ZnO nanoparticles in fungal inhibition and self-protection of exposed marble columns in historic sites. *Archaeol. Anthropol. Sci.* **2019**, *11*, 3407–3422. [CrossRef]
62. Abo Elgat, W.A.A.; Taha, A.S.; Böhm, M.; Vejmelková, E.; Mohamed, W.S.; Fares, Y.G.D.; Salem, M.Z.M. Evaluation of the Mechanical, Physical, and Anti-Fungal Properties of Flax Laboratory Papersheets with the Nanoparticles Treatment. *Materials* **2020**, *13*, 363. [CrossRef]
63. Karapanagiotis, I.; Hosseini, M. Superhydrophobic Coatings for the Protection of Natural Stone. In *Advanced Materials for the Conservation of Stone*, 1st ed.; Karapanagiotis, I., Hosseini, M., Eds.; Springer International: Cham, Switzerland; New York, NY, USA, 2018; pp. 1–25. [CrossRef]
64. De Matteis, V.; Cascione, M.; Rinaldi, R. Titanium dioxide: Antimicrobial surfaces and toxicity assessment. In *Titanium Dioxide (TiO_2) and Its Applications*; Elsevier: Amsterdam, The Netherlands, 2021; pp. 373–393. [CrossRef]
65. Ayorinde, T.; Sayes, C.M. An updated review of industrially relevant titanium dioxide and its environmental health effects. *JHM Lett.* **2023**, *4*, 100085. [CrossRef]
66. Sharma, P.; Kumari, R.; Yadav, M.; Lal, R. Evaluation of TiO_2 Nanoparticles Physicochemical Parameters Associated with their Antimicrobial Applications. *Indian J. Microbiol.* **2022**, *62*, 338–350. [CrossRef] [PubMed]
67. Li, W.; Liang, R.; Hu, A.; Huang, Z.; Zhou, Y.N. Generation of oxygen vacancies in visible light activated one-dimensional iodine TiO_2 photocatalysts. *RSC Adv.* **2014**, *4*, 36959–36966. [CrossRef]
68. Mardiansyah, D.; Badloe, T.; Triyana, K.; Mehmood, M.Q.; Raeis-Hosseini, N.; Lee, Y.; Sabarman, H.; Kim, K.; Rho, J. Effect of temperature on the oxidation of Cu nanowires and development of an easy to produce, oxidation-resistant transparent conducting electrode using a PEDOT:PSS coating. *Sci. Rep.* **2018**, *8*, 10639. [CrossRef]
69. Zhang, J.; Yu, X.; Guo, W.; Qiu, J.; Mou, X.; Li, A.; Liu, H. Construction of titanium dioxide nanorod/graphite microfiber hybrid electrodes for a high performance electrochemical glucose biosensor. *Nanoscale* **2016**, *8*, 9382–9389. [CrossRef] [PubMed]
70. Ohsaka, T. Temperature Dependence of the Raman Spectrum in Anatase TiO_2. *J. Physical. Soc. Jpn.* **1980**, *48*, 1661–1668. [CrossRef]
71. Choi, H.C.; Jung, Y.M.; Bin Kim, S. Size effects in the Raman spectra of TiO_2 nanoparticles. *Vib. Spectrosc.* **2005**, *37*, 33–38. [CrossRef]
72. de Caro, T.; Caschera, D.; Ingo, G.M.; Calandra, P. Micro-Raman innovative methodology to identify Ag–Cu mixed sulphides as tarnishing corrosion products. *J. Raman Spectrosc.* **2016**, *47*, 852–859. [CrossRef]
73. Tasić, N.; Ćirković, J.; Ribić, V.; Žunić, M.; Dapčević, A.; Branković, G.; Branković, Z. Effects of the silver nanodots on the photocatalytic activity of mixed-phase TiO_2. *J. Am. Ceram. Soc.* **2022**, *105*, 336–347. [CrossRef]
74. Shatnawi, M.; Alsmadi, A.M.; Bsoul, I.; Salameh, B.; Alna'washi, G.A.; Al-Dweri, F.; El Akkad, F. Magnetic and optical properties of Co-doped ZnO nanocrystalline particles. *J. Alloys Compd.* **2016**, *655*, 244–252. [CrossRef]
75. Kareem, M.A.; Bello, I.T.; Shittu, H.A.; Sivaprakash, P.; Adedokun, O.; Arumugam, S. Synthesis, characterization, and photocatalytic application of silver doped zinc oxide nanoparticles. *Clean. Mater.* **2022**, *3*, 100041. [CrossRef]
76. Chitradevi, T.; Jestin Lenus, A.; Jaya, N.V. Structure, morphology and luminescence properties of sol-gel method synthesized pure and Ag-doped ZnO nanoparticles. *Mater. Res. Express* **2020**, *7*, 015011. [CrossRef]
77. Mondal, A.; Pal, S.; Sarkar, A.; Bhattacharya, T.S.; Pal, S.; Singha, A.; Ray, S.K.; Kumar, P.; Kanjilal, D.; Jana, D. Raman investigation of N-implanted ZnO: Defects, disorder and recovery. *J. Raman Spectrosc.* **2019**, *50*, 1926–1937. [CrossRef]
78. Song, Y.; Zhang, S.; Zhang, C.; Yang, Y.; Lv, K. "Raman Spectra and Microstructure of Zinc Oxide irradiated with Swift Heavy Ion. *Crystals* **2019**, *9*, 395. [CrossRef]
79. Pal, B.; Giri, P.K. Defect Mediated Magnetic Interaction and High Tc Ferromagnetism in Co Doped ZnO Nanoparticles. *J. Nanosci. Nanotechnol.* **2011**, *11*, 9167–9174. [CrossRef] [PubMed]
80. Zeferino, R.S.; Flores, M.B.; Pal, U. Photoluminescence and Raman Scattering in Ag-doped ZnO Nanoparticles. *J. Appl. Phys.* **2011**, *109*, 014308. [CrossRef]
81. Chelouche, A.; Djouadi, D.; Merzouk, H.; Aksas, A. Influence of Ag doping on structural and optical properties of ZnO thin films synthesized by the sol–gel technique. *Appl. Phys. A* **2014**, *115*, 613–616. [CrossRef]
82. Akbarizadeh, M.R.; Sarani, M.; Darijani, S. Study of antibacterial performance of biosynthesized pure and Ag-doped ZnO nanoparticles. *Rend. Lincei Sci. Fis. Nat.* **2022**, *33*, 613–621. [CrossRef]
83. La Russa, M.F.; Macchia, A.; Ruffolo, S.A.; De Leo, F.; Barberio, M.; Barone, P.; Crisci, G.M.; Urzì, C. Testing the antibacterial activity of doped TiO_2 for preventing biodeterioration of cultural heritage building materials. *Int. Biodeterior. Biodegrad.* **2014**, *96*, 87–96. [CrossRef]
84. Souza, D.C.; Amorim, S.; Cadamuro, R.; Fongaro, G.; Peralta, R.A.; Peralta, R.M.; Li Puma, G.; Moreira, R.F.P.M. Hydrophobic cellulose-based and non-woven fabrics coated with mesoporous TiO_2 and their virucidal properties under indoor light. *Carbohydr Polym Technol. Appl.* **2022**, *3*, 100182. [CrossRef]

85. Schipporeit, S.; Mergel, D. Spectral decomposition of Raman spectra of mixed-phase TiO_2 thin films on Si and silicate substrates. *J. Raman Spectrosc.* **2018**, *49*, 1217–1229. [CrossRef]
86. Ibarrondo, I.; Balziskueta, U.; Martínez-Arkarazo, I.; García-Florentino, C.; Arana, G.; Azkarate, A.; Madariaga, J.M. Portable Raman can be the new hammer for architects restoring 20th-century built heritage elements made of reinforced concrete. *J. Raman Spectrosc.* **2021**, *52*, 109–122. [CrossRef]
87. Pigorsch, E. New insights into paper—Chemical paper analysis using Raman microscopy. *J. Raman Spectrosc.* **2021**, *52*, 78–84. [CrossRef]
88. Nie, B.; Zhou, Q.; He, J.; Yang, F. Chromatographic paper embedded with silver nanostructure as a disposable substrate for surface-enhanced Raman spectroscopy and catalytic reactor. *J. Raman Spectrosc.* **2015**, *46*, 211–216. [CrossRef]
89. Dangi, S.B.; Hashmi, S.Z.; Kumar, U.; Choudhary, B.L.; Kuznetsov, A.E.; Dalela, S.; Kumar, S.; Dolia, S.N.; Kumar, S.; Sofi, B.F.; et al. Exploration of spectroscopic, surface morphological, structural, electrical, optical and mechanical properties of biocompatible PVA-GO PNCs. *Diam. Relat. Mater.* **2022**, *127*, 109158. [CrossRef]
90. Yu, J.; Tian, H.; Huang, M.; Xu, X. Facile Synthesis of Ag NP Films via Evaporation-Induced Self-Assembly and the BA-Sensing Properties. *Foods* **2023**, *12*, 1285. [CrossRef] [PubMed]
91. Singh, W.I.; Sinha, S.; Devi, N.A.; Nongthombam, S.; Laha, S.; Swain, B.P. Investigation of chemical bonding and electronic network of rGO/PANI/PVA electrospun nanofiber. *Polym. Bull.* **2021**, *78*, 6613–6629. [CrossRef]
92. Bai, X.; Detalle, V. Time-Gated Pulsed Raman Spectroscopy with NS Laser for Cultural Heritage. *Heritage* **2023**, *6*, 1531–1540. [CrossRef]
93. Luo, C.-Q.; Ho, L.-P.; Ling, F.C.-C. The origin of additional modes in Raman spectra of ZnO:Sb films. *Phys. B Condens. Matter* **2020**, *593*, 412256. [CrossRef]
94. Phan, S.; Padilla-Gamiño, J.L.; Luscombe, C.K. The effect of weathering environments on microplastic chemical identification with Raman and IR spectroscopy: Part, I. polyethylene and polypropylene. *Polym. Test.* **2022**, *116*, 107752. [CrossRef]

molecules

MDPI

Article

Egg vs. Oil in the Cookbook of Plasters: Differentiation of Lipid Binders in Wall Paintings Using Gas Chromatography–Mass Spectrometry and Principal Component Analysis

Jana Nádvorníková [1,*], Václav Pitthard [2], Ondřej Kurka [1], Lukáš Kučera [1] and Petr Barták [1]

1 Department of Analytical Chemistry, Faculty of Science, Palacký University, 17. Listopadu 12, 779 00 Olomouc, Czech Republic; ondrej.kurka@upol.cz (O.K.); lukas.kucera@upol.cz (L.K.); petr.bartak@upol.cz (P.B.)

2 Conservation Science Department, Kunsthistorisches Museum, Burgring 5, 1010 Vienna, Austria; vaclav.pitthard@khm.at

* Correspondence: jana.nadvornikova02@upol.cz

Abstract: Wall paintings are integral to cultural heritage and offer rich insights into historical and religious beliefs. There exist various wall painting techniques that pose challenges in binder and pigment identification, especially in the case of egg/oil-based binders. GC-MS identification of lipidic binders relies routinely on parameters like the ratios of fatty acids within the plaster. However, the reliability of these ratios for binder identification is severely limited, as demonstrated in this manuscript. Therefore, a more reliable tool for effective differentiation between egg and oil binders based on a combination of diagnostic values, specific markers (cholesterol oxidation products), and PCA is presented in this study. Reference samples of wall paintings with egg and linseed oil binders with six different pigments were subjected to modern artificial ageing methods and subsequently analysed using two GC-MS instruments. A statistically significant difference (at a 95% confidence level) between the egg and oil binders and between the results from two GC-MS instruments was observed. These discrepancies between the results from the two GC-MS instruments are likely attributed to the heterogeneity of the samples with egg and oil binders. This study highlights the complexities in identifying wall painting binders and the need for innovative and revised analytical methods in conservation efforts.

Keywords: binding media; cholesta-3,5-dien-7-one; dicarboxylic acids; egg; gas chromatography–mass spectrometry; linseed oil; principal component analysis; P/S ratio

1. Introduction

Wall paintings are one of the oldest forms of artistic expression and are an important part of cultural heritage. Scientific research is being carried out to determine the material composition of paintings, as well as their ageing mechanisms and restoration impacts. It is important to ensure that conservation measures preserve the historical and artistic value of the paintings. Further, research has explored painting techniques to aid in the authentication and attribution of artworks [1,2]. In recent years, there has been a significant scientific effort to develop innovative and non-invasive analytical methods for studying wall paintings. Since these paintings are often located in remote or challenging locations, portable techniques are preferred [2–5]. Optical methods such as spectroscopy and microscopy (e.g., UV-induced fluorescence photography, XRF imaging, and Raman spectroscopy) are usually used, as they enable detailed examination without direct contact [6–8]. However, non-invasive methods may not always be sufficient. Micro-invasive and micro-destructive approaches can provide more comprehensive analysis while minimising the impact on the artwork during sample collection [2].

There are various techniques for wall painting, varying based on the plaster condition, binders, and pigments. A fresco method involves applying pigments to fresh, wet lime plaster. On the other hand, a secco is performed on dry plaster and requires binding media such as animal glues, casein, drying oils, or eggs [9–11]. Determining the binders is challenging due to their complex composition, minimal sample availability (<1 mg), and the low binder-to-pigment ratio (ca. 10% w/v) [7,12–14]. It is also affected by external factors such as climatic conditions and air pollutants, as well as internal interactions between the binders and pigments that contribute to degradation over time [15–17]. Accurate identification of the binding media is crucial for historical reconstruction and restoration efforts; this requires an understanding of the degradation products and sensitive analytical techniques [18–20].

Eggs served as a prevalent painting binder from ancient times to the 15th century, when they gradually gave way to drying oils [12,21–23]. The use of linseed, poppyseed, and walnut oils became prevalent in the 16th and 17th centuries, persisting as a tradition [12,24,25]. However, many paintings used a combination of both binders, either separately on distinct sections or together in the form of an oil and egg emulsion known as "tempera grassa" [12,26]. The decision to use both binders was primarily driven by visual considerations, as certain pigments were incompatible with oil, risking yellowing due to the oil ageing [25,27,28] or producing greenish hues when combined with egg yolk [19]. Indeed, analysis of egg and oil binders side by side can be a challenging task [29].

Gas chromatography combined with mass spectrometric detection (GC-MS) is a commonly used technique for identifying organic materials, especially lipidic binding media [13,21,30]. Fatty acid methyl ester (FAME) analysis forms the basis of lipid binder identification using GC-MS [31,32]. While the palmitic-to-stearic-acid ratio (P/S) was traditionally considered a reliable parameter, which remained constant during ageing [12,13,33,34], recent studies have highlighted its susceptibility to numerous factors. These include the sample age, composition, conservation efforts, cleaning processes [24,31,35], and the faster evaporation rate of selected saturated fatty acids [36]. Caution is advised when relying solely on the P/S ratio for identification due to sample dilution and instrument response variations [37]. Alternative parameters offer more precise identification. These include the azelaic-to-palmitic-acid ratio (A/P) [7,38,39]; the azelaic-to-other-dicarboxylic-acid ratio, such as suberic (Su) or sebacic (Seb) acid—A/Su, A/Seb [35,40]; the sum of the relative content of dicarboxylic acids (ΣD) [38]; and the oleic-to-palmitic-acid or -stearic-acid ratio (O/P, O/S) [16,40,41]. Differentiating between egg and oil can be aided by the A/P ratio and ΣD. Drying oils have a higher amount of unsaturated fatty acids compared to eggs. This leads to the production of more dicarboxylic acids, which are generated through degradation (oxidative) processes from these unsaturated acids and result in notably higher values in the A/P ratio and ΣD in drying oils [16]. However, these parameters may still be influenced by the film thickness, pigment type, oil source and processing, and the painting's storage and preservation history [13,21,42].

While GC-MS analysis of fatty acid methyl esters is a well-established practice in the literature, the use of fatty acid ratios for identifying lipidic binders still poses many challenges. Typically, the results are expressed as ratios of the chromatographic peak areas rather than in concentration units, leading to potential variations depending on the methodology employed. Consequently, the outcomes of different researchers may differ significantly, highlighting the limited reliability and accuracy of this evaluation method [43]. The purpose of this study is to demonstrate the extent of the potential discrepancies in the observed diagnostic values (A/P, A/Su, O/S, P/S, and ΣD) even when the data are obtained using identical methods and using two GC-MS instruments with nearly identical hardware configurations. Additionally, the study aims to propose a solution to this issue by employing a combination of markers based on principal component analysis. This approach is meant to highlight the issue of the substantial differences in the obtained values, as well as enable a more reliable differentiation between the egg and oil binders used in paintings. Finally, this study also stresses the importance of utilising reference samples of wall paintings for egg and oil binding media, which should undergo artificial aging to simulate historical samples and then can serve as a reliable benchmark reference for verifying the reliable diagnostic value ranges.

2. Results and Discussion

2.1. Derivatisation of the Samples

GC-MS analysis of lipidic binding media is based on the determination of fatty acids, which must first be released from the respective triacylglycerols [12]. The liberated fatty acids are converted into a suitable form for analysis, with the most common form being fatty acid methyl esters (FAMEs). Traditionally, FAMEs have been prepared via saponification followed by methylation, e.g., with diazomethane [21,44,45] or methanol under acidic catalysis [46]. The procedure of processing samples for analysis can be quite lengthy, as it involves several steps, such as hydrolysis, acidification, extraction, and derivatisation. This not only leads to a prolonged processing time but also increases the risk of sample loss or contamination during handling [47,48]. Quaternary ammonium hydroxides, especially *m*-(trifluoromethyl)phenyltrimethylammonium hydroxide (TFTMAH), have become increasingly popular, as they offer numerous advantages over other types of reagents [1,39,49]. TFTMAH is a one-step esterification/transesterification reagent that requires little to no sample preparation; thus, this procedure is faster than traditional methods. Additionally, it is less basic and does not require excessively high reaction temperatures, which reduces the likelihood of undesirable side reactions [50]. To facilitate the solubilisation of nonpolar analytes, TFTMAH was applied, together with toluene and methanol, and the mixture was heated to 60 °C, the optimal temperature for obtaining the maximum yields, as described previously [25]. Numerous fatty acids were identified in the model samples with egg and linseed oil binders (Figure S1 and Table S1 in the Supplementary Materials); however, particular attention was paid to azelaic, oleic, palmitic, sebacic, stearic, and suberic acids and their characteristic ratios (P/S, A/P, O/S, and A/Su).

2.2. Variations in the Diagnostic Values

The diagnostic ratios of fatty acids serve as widely recognised indicators for distinguishing between egg and oil binding media. However, the reports from the literature show significant variations in values. For instance, the reported P/S ratio ranges from 0.5 to 4.1 for egg binder [36,41] and from 1.2 to 2.0 for linseed oil [30,31]. Obviously, the variations are more pronounced in eggs due to the occurrence of specific egg components (yolk, white, or whole egg), their more complex composition, and the lower homogeneity of the samples [41]. Additionally, the parameters of the hardware instrumentation, not extensively addressed in previous studies, can influence the values. Preliminary experiments with various hardware setups indicated that the values could differ significantly depending on the used instrumentation and current conditions. Therefore, the model samples of the wall paintings were analysed using the same method in two laboratories, located in Olomouc (GC-MS 1) and in Vienna (GC-MS 2), using two nearly identical GC-MS instruments, differing only slightly in their column stationary phase and manufacturer. Both columns were nonpolar. A 5% diphenyl—95% dimethylpolysiloxane phase was used in GC-MS 1 (OPTIMA-5 MS) [51] and a phenyl arylene polymer, which is virtually equivalent to 5% diphenyl-95% dimethylpolysiloxane, was used in GC-MS 2 (DB-5ms, Low Bleed, Ultra Inert) [52].

2.3. Artificial Ageing

Reference samples of the wall paintings were subjected to artificial ageing to better understand the chemical changes in the composition of the lipid binders and the mechanisms that occur during ageing. These samples may experience certain mechanisms, such as photooxidative processes, cleavages, and degradation reactions, which normally occur over decades or even centuries. However, the laboratory can accelerate these processes by utilising UV light, humidity, and elevated temperatures [49,53]. Before and after each step of the artificial ageing process (NA = non-aged samples, UV1 = ageing for 504 h, UV2 = ageing for 1008 h, and UV3 = ageing for 1512 h), a group of samples was collected from a reference block of wall paintings (details are presented in the Materials and Methods section). It contained seven egg binder samples (one non-pigmented and six pigmented samples) and seven samples with a linseed oil binder (with the same pigments as in the egg samples). The diagnostic ratios of fatty acids and the sum of the relative representation of selected dicarboxylic acids—∑D (azelaic, sebacic, and suberic)—were monitored across the ageing steps, from unaged samples (NA) to the most aged (UV3). Significant differences in these parameters were observed not only between the ageing steps but also between the egg and oil binders. Table 1 shows the differences between the individual parameters in the egg and linseed oil binders. The average values and standard deviations of these parameters were calculated from the mentioned seven samples (egg/oil binder) with different pigments, with each sample injected three times in a total of twenty-one analyses.

Table 1. Changes in fatty acid ratios and the sum of the relative representation of dicarboxylic acids (∑D) during artificial ageing. Differences are observed between the reference samples with linseed oil and egg binding media and between the instrumentation GC-MS 1 and GC-MS 2. The average value of the selected ratio (or parameter ∑D) was calculated from seven samples of reference wall painting, each injected three times (n = 21). Standard deviations for each fatty acid ratio and parameter ∑D were calculated in the same way.

	GC-MS 1				GC-MS 2			
LO	NA	UV1	UV2	UV3	NA	UV1	UV2	UV3
P/S	1.3 ± 0.1	1.3 ± 0.2	1.3 ± 0.1	1.3 ± 0.1	1.4 ± 0.1	1.5 ± 0.1	1.5 ± 0.1	1.5 ± 0.1
A/P	0.4 ± 0.1	0.7 ± 0.3	0.9 ± 0.3	0.8 ± 0.3	0.2 ± 0.1	0.8 ± 0.1	0.5 ± 0.1	0.5 ± 0.2
O/S	1.8 ± 0.4	0.4 ± 0.1	0.3 ± 0.1	0.4 ± 0.3	2.2 ± 0.5	0.3 ± 0.1	0.19 ± 0.06	0.11 ± 0.07
A/Su	3.8 ± 1.1	1.8 ± 0.4	1.9 ± 0.4	1.9 ± 0.3	4.2 ± 1.3	2.1 ± 0.4	2.4 ± 0.4	2.1 ± 0.4
∑D	9.0 ± 3.5	13.4 ± 3.2	22.5 ± 3.3	25.7 ± 4.8	6.0 ± 2.6	29.5 ± 4.7	31.3 ± 7.3	35.4 ± 4.0
Egg	NA	UV1	UV2	UV3	NA	UV1	UV2	UV3
P/S	1.9 ± 0.2	1.8 ± 0.3	1.8 ± 0.3	1.8 ± 0.3	2.8 ± 0.4	2.9 ± 0.3	2.7 ± 0.3	2.9 ± 0.2
A/P	0.07 ± 0.01	0.2 ± 0.1	0.2 ± 0.05	0.2 ± 0.1	0.02 ± 0.01	0.12 ± 0.03	0.1 ± 0.04	0.13 ± 0.03
O/S	2.7 ± 0.3	0.2 ± 0.1	0.2 ± 0.05	0.2 ± 0.1	4.0 ± 0.8	0.05 ± 0.03	0.04 ± 0.02	0.05 ± 0.02
A/Su	3.1 ± 0.6	1.4 ± 0.2	1.3 ± 0.2	1.2 ± 0.1	4.4 ± 0.6	1.3 ± 0.2	1.4 ± 0.2	1.4 ± 0.3
∑D	2.0 ± 0.8	7.4 ± 3.1	7.9 ± 2.7	9.3 ± 2.1	0.9 ± 0.7	12.2 ± 1.9	13.5 ± 1.7	13.5 ± 3.1

∑D—sum of the relative representation of azelaic, sebacic, and suberic acid (%), LO—linseed oil, GC-MS 1—instrument located in Olomouc, GC-MS 2—instrument located in Vienna, NA—non-aged samples, UV1, UV2, UV3—steps of artificial ageing process.

2.4. Fatty Acids and Diagnostic Values

As can be seen in Table 1, the P/S ratio in the egg and linseed oil binding media samples remained remarkably consistent throughout and even prior to artificial ageing. For the linseed oil, the average values of the P/S ratios from steps NA, UV1, UV2, and UV3, as determined using GC-MS 1 and GC-MS 2, were 1.3 and 1.5, respectively. These values align closely with the range reported in the literature (1.3–1.6) [16,54,55]. In comparison, the P/S ratios calculated for the egg binder were 1.8 (GC-MS 1) and 2.8 (GC-MS 2). In the literature, the average value of the P/S ratio is reported to be around 3 [20,54], which is similar to the GC-MS 2 results. However, the GC-MS 1 results were lower; therefore, it is not recommended to rely solely on the P/S ratio for the identification of lipid binders. Instead, other characteristic ratios and markers should be considered for accurate identification.

The A/P ratio is a useful tool for distinguishing between drying oils and egg binders, as it indicates the oxidation and drying degree, which gradually increases over time [35,53]. This is demonstrated in Table 1, where the average A/P values of all the samples rose more than twofold during the ageing experiment. The values of 0.4 and 0.2 for the non-aged linseed oil samples increased to 0.8 (GC-MS 1 and GC-MS 2), which closely aligns with the values reported in the literature [31,53]. On the other hand, the fresh egg samples exhibited a significantly lower azelaic acid content compared to the linseed oil, resulting in lower A/P values. The initial values of 0.07 and 0.02 for the fresh egg samples increased to 0.2 and 0.1 (GC-MS 1 and GC-MS 2) with ageing. According to the literature [54], the A/P values of eggs are usually below 0.3, which is consistent with the obtained values. Surprisingly, the highest A/P values were not found in the most aged samples of linseed oil (UV3) but were achieved in the UV2 and UV1 samples. This may be explained by the potential evaporation of azelaic acid from the sample surface during ageing under UV light, as described in [49].

During ageing, the O/S ratio was monitored as a crucial parameter, particularly for the oils, because it characterises their maturity level [41,56]. The results reveal a significant decrease in the O/S ratio during artificial ageing, indicating a loss of oleic acid. The average values of the egg binder dropped remarkably from 2.7 and 4.0 to 0.2 and 0.05 (GC-MS 1 and GC-MS 2), while for the linseed oil, the values came down from 1.8 and 2.2 to 0.4 and 0.1 (GC-MS 1 and GC-MS 2). These findings are consistent with earlier research that indicates O/S ratios of less than 0.5 for historical artworks [54]. The results show that artificial ageing accelerates the oxidation of oleic acid, which is highly susceptible to oxidation due to its double bond. This makes oleic acid a valuable indicator of the degree of oxidation [49,53]. The most significant decrease in the O/S values was observed between the NA and UV1 steps.

The A/Su ratio is another important parameter, which is crucial for the evaluation of the pre-heating processes that can occur in linseed oil processing, such as exposure to higher temperatures during the preparation of the so-called polymerised oil [29,53]. A substantial decrease in the average values was observed during the artificial ageing. The average values for linseed oil were 3.8 and 4.2 and decreased to 1.9 and 2.1 for the analyses using GC-MS 1 and GC-MS 2, respectively. Raw oils have values greater than 6, and heat-processed oils have values in the range of 2–3 [53]. A mixture of polymerised and purified linseed oil was used to prepare the model samples, so the obtained results are in full agreement with the literature. A similar trend was observed in the egg samples, with the average values dropping from 3.1 and 4.4 to values of 1.2 and 1.4 for GC-MS 1 and GC-MS 2, respectively. The most significant decrease occurred between the NA and UV1 steps, except for the egg samples that were measured using GC-MS 1, where the final UV3 step resulted in the lowest average value of 1.2. This confirms that the artificial ageing process was highly effective, even after the first ageing step.

The last studied parameter was the sum of the relative representation of selected dicarboxylic acids (∑D), namely suberic (C8), azelaic (C9), and sebacic (C10) acid, which are the major oxidative products of C18 unsaturated fatty acids [12,33]. During the artificial ageing process, ∑D gradually increased, particularly in the linseed oil, as it contains a larger amount of unsaturated fatty acids compared to eggs. The highest values were observed after the final ageing step, UV3. The average ∑D values were found to be 25.7% and 35.4% for the linseed oil and 9.3% and 13.5% for the egg binder (GC-MS 1 and GC-MS 2, respectively). The results, which are presented in Table 1 as the average values obtained from three repeated injections of each sample, are in accordance with the data from the literature [16,43].

The fatty acid ratios and the parameter ∑D for the egg and linseed oil binding media were also compared statistically. The normality of all the data (see Supplementary Materials—Microsoft Office Excel file) was analysed graphically, as well as using the Jarque–Bera test (J–B test) and the Kolmogorov–Smirnov test (K–S test). The test criteria χ^2_{exp} for the J–B test and D for the K–S test were compared to critical values for χ^2 and D with a significance level $\alpha = 0.05$ and degrees of freedom v ($v = n - 1$), respectively. The null hypothesis stated that the data were normally distributed (H$_0$: χ^2_{exp} (or D) $< \chi^2_{(1-\alpha)}$), and this was tested against the alternative hypothesis that the data were not normally distributed (H$_1$: χ^2_{exp} (or D) $> \chi^2_{(1-\alpha)}$). The calculated values were less than the critical values, so the null hypothesis was accepted. Then, an unpaired two-tailed Student's t-test was applied to compare the results of the fatty acid ratios and the parameter ∑D between the egg and linseed oil binders using one instrument (GC-MS 1 or GC-MS 2) for each ageing step. The average values and standard deviations from Table 1 were used for this testing. The null hypothesis H_0: $\mu_1 = \mu_2$ was tested against the alternative hypothesis H_0: $\mu_1 \neq \mu_2$. The critical values for t with a significance level $\alpha = 0.05$ and degrees of freedom ν ($\nu = n_1 + n_2 - 2$) were compared to the calculated values. Except for one sample (the A/Su ratio of the NA samples measured using GC-MS 2), the calculated values exceeded the critical values, so the null hypothesis was rejected. The t-test proved the statistically significant difference between the egg and linseed oil binders. Furthermore, an unpaired two-tailed Student's t-test (confidence level of 95%) was used for comparison of the results obtained using the GC-MS 1 and GC-MS 2 instruments. The t-values suggested a statistically significant difference between the results obtained using GC-MS 1 and GC-MS 2 at the chosen level of significance except for seven values, where there was not significant difference. These were the values of A/P, A/Su, O/S (UV1) and A/Su, O/S (NA) for the linseed oil binder and A/Su (UV1 and UV2) for the egg binding media. Such an inconsistency likely stems from the inhomogeneity of the binders used and the small number of physical replicates of each sample. The results of the statistical testing are also supported by the results from the PCA (see Section 2.6).

2.5. Other Characteristic Markers

Due to the described variations in the diagnostic parameters and the potential influence of various factors on the measured values, it is recommended to explore other specific substances to indicate the type of binder. Sterols, for instance, offer a promising marker group for the detection of egg binders. Cholesterol usually accounts for up to 5% of the total lipid content in eggs [57]; this is a large amount compared with the content of any sterol in a drying oil [12]. Thus, it is often considered an indicator of egg's presence [29,43,58,59]. However, relying solely on cholesterol may yield inaccurate results due to its autooxidation [60,61], as illustrated in Figure 1. While cholesterol was initially present in all the fresh egg binder samples, it disappeared during the initial ageing step (UV1). The subsequent investigation focused on cholesterol oxidation products (COPs) and revealed various identified COPs: Cholesta-4,6-dien-3-ol, Cholesta-4,6-dien-3-one, Cholesta-3,5-diene, Cholest-5-en-3-one, and Cholesta-3,5-dien-7-one. Cholesta-3,5-dien-7-one persists even in the most aged samples (UV3), and therefore it is used as an egg binder marker in historical samples [16,62,63]. None of these COPs or cholesterol were detected in the oil binder samples. Additionally, sterols like beta-sitosterol, stigmasterol, and isofucosterol were identified in both the egg binding media and linseed oil, although the latter two are more prevalent in linseed oil [64,65]. However, they can also appear in trace amounts in eggs [57]. Correct interpretation therefore requires a quantitative evaluation of the occurrence of these substances, preferably using a suitable statistical method, such as principal component analysis.

2.6. PCA

A combination of diagnostic ratios of fatty acids, complemented by sterols and other markers, may not invariably suffice for the accurate identification of lipid binders. Cholesterol oxidation products emerge as potential indicators of egg binders, although it could be a challenging task to identify them in real historical samples due to their frequently low quantities, often approaching or falling below the method's limit of detection [16]. As already mentioned, various factors significantly affect fatty acid ratios and contribute to considerable discrepancies in the measured values, which makes comparison with the reference values from the literature somewhat difficult. For instance, the P/S ratio in the model egg samples, subjected to the final stage of artificial ageing (UV3) and analysed using the GC-MS 1 instrument, exhibited average values of 1.8 ± 0.3. Such a P/S ratio value could be erroneously interpreted as indicative of linseed oil if evaluated without further context only through comparison with literary sources [20,31]. Therefore, the importance of including control samples to verify the reference values for analysis and distinguishing between egg and oil-based binders needs to be emphasised. However, obtaining authentic reference samples with a precisely documented origin and preparation methods is typically rare and rather challenging. The scarcity of such samples makes their routine use as reference standards impractical. The proposed approach, involving model samples with well-defined preparation protocols, ideally coupled with contemporary sophisticated artificial ageing procedures and techniques, represents an efficient and cost-effective strategy to address this limitation.

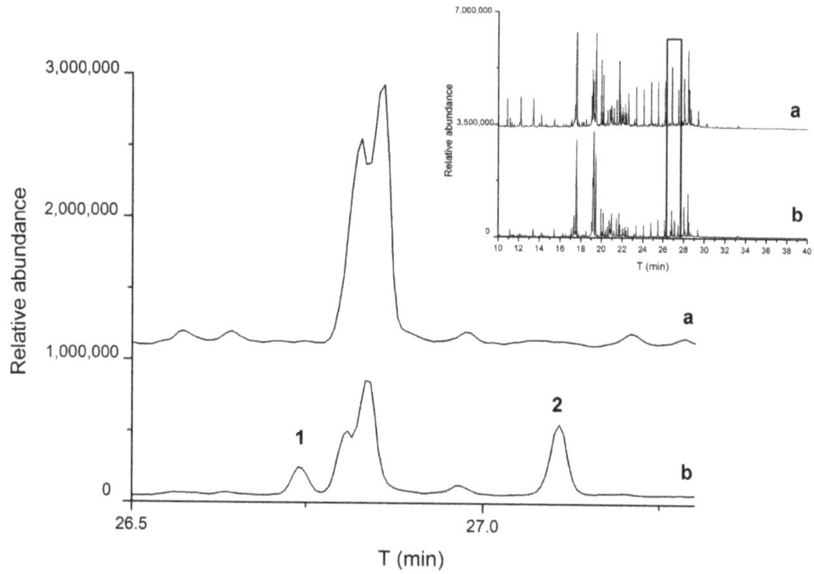

Figure 1. Zoomed-in total ion chromatograms (TICs) of egg samples without pigment after artificial ageing, step UV1 (a) and before ageing (b). 1—cholesterol methyl ether, 2—cholesterol. The inset shows the whole TICs of the sample before (b) and after artificial ageing (a) with a marked section of the expanded part.

Principal component analysis (PCA) may be an advantageous resource. This methodology is commonly used to differentiate protein binders based on the distinct representation of individual amino acids [54,66]. However, PCA has rarely been used to distinguish between lipid binders. In this study, PCA was used to evaluate the reference samples of artificially aged (UV3 step) egg and oil binders, measured using both GC-MS 1 and GC-MS 2 instruments. Six variables were considered for the PCA: the P/S, A/P, O/S, and A/Su acid ratios; the parameter $\sum D$; and the presence/absence of COPs. The first two principal components (PCs) contributed to 81.8% of the total variance in the data, as depicted in Figure 2. Principal components PC1 and PC2 accounted for 64.0% and 17.8% of the total variability, respectively. PC1 correlated mainly with the COPs, A/P, P/S, A/Su, and $\sum D$, with the strongest correlation observed between the COPs and the A/P ratio. An increase in A/P suggests an increase in A/Su and $\sum D$.

Further, a biplot was used to visually segregate the egg and oil binding media (black ellipses), with distinct clusters for the GC-MS 1 and GC-MS 2 instrumentation in the egg samples (red and green ellipses). This spatial distribution is likely attributed to the low homogeneity of the egg samples and consequently to a significant discrepancy in the P/S ratio (mean value 1.8 $\pm$ 0.3 for GC-MS 1 versus 2.9 $\pm$ 0.2 for GC-MS 2) and the O/S ratio (mean value 0.2 $\pm$ 0.1 for GC-MS 1 versus 0.05 $\pm$ 0.02 for GC-MS 2). On the contrary, the linseed oil samples exhibited an overlap of clusters, indicating a negligible difference in the variables analysed using the GC-MS 1 and GC-MS 2 instruments. In addition, the PCA biplot illustrated the relationship between the egg/oil binding media and six variables.

The P/S ratio and the presence of COPs are significantly related to the presence of eggs, indicating that cholesterol and its oxidation products are sterols generally present only in eggs. The greater proportion of palmitic acid in animal fats than in drying oils leads to the higher P/S ratio observed in eggs [13]. In contrast, the linseed oil samples were associated with variables such as the A/P, A/Su, and O/S ratios and the parameter ∑D. This reflects the higher concentration of unsaturated fatty acids (linolenic, linoleic, oleic) and their degradation products (dicarboxylic azelaic, sebacic, and suberic acids) in oils compared to eggs [33].

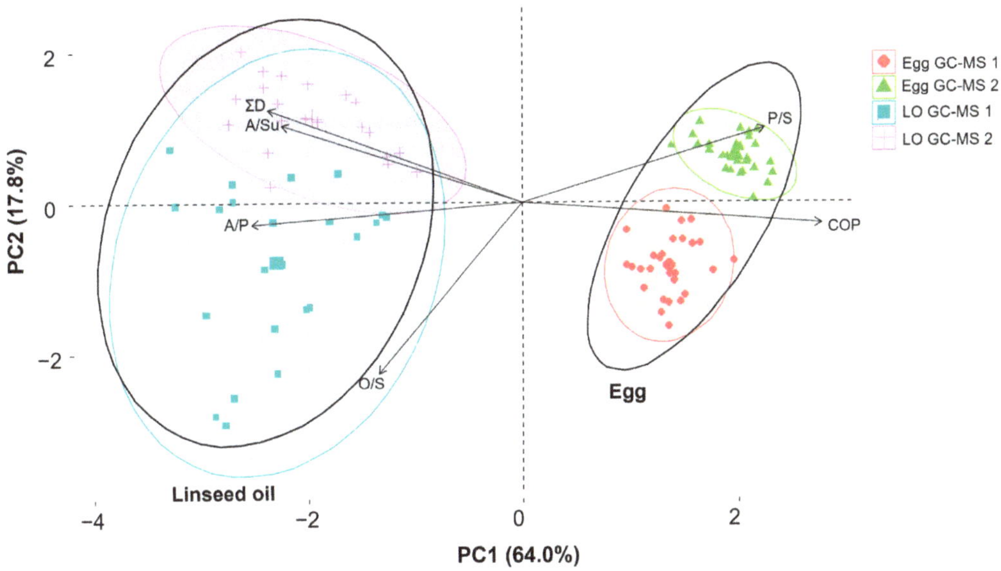

Figure 2. PCA biplot of egg and linseed oil (LO) binding media reference samples after artificial ageing (UV3 step) analysed using two GC-MS instruments (GC-MS 1 and GC-MS 2). The ellipsoids represent 95% confidence regions encompassing the centroids of each data cluster. Black vectors represent six variables' loadings: A/P—azelaic/palmitic acid ratio, A/Su—azelaic/suberic acid ratio, O/S—oleic/stearic acid ratio, P/S—palmitic/stearic acid ratio, ∑D—sum of the relative representation of azelaic, sebacic, and suberic acids, COP—presence or absence of any cholesterol oxidation product.

3. Materials and Methods

3.1. Chemicals

Methanol and *n*-hexane of analytical grade were obtained from PENTA (Praha, Czech Republic). The toluene (analytical-grade) was purchased from Sigma-Aldrich (St. Louis, MO, USA). Further, the 3-(Trifluoromethyl)phenyltrimethylammonium hydroxide (TFTMAH), 5% *w*/*v* in methanol (known as MethPrep II), was supplied by Tokyo Chemical Industry Co., Ltd. (Nihonbashi-honcho, Chuo-ku, Tokyo, Japan). The fatty acid standards—azelaic (C9), suberic (C8), sebacic (C10), palmitic (C16:0), stearic (C18:0), oleic (C18:1), linoleic (C18:2), linolenic (C18:3), and cholesta-3,5-diene-7-one—were purchased from Sigma-Aldrich. Individual stock standard solutions were prepared for fatty acids in *n*-hexane and cholesta-3,5-diene-7-one in 2-propanol at a concentration of 1 mg·mL^{-1} and stored in a refrigerator at 5 °C.

3.2. Binders and Pigments

The pigments used for the preparation of the model samples were as follows—Egyptian blue, indigo, and red earth pigment (red ochre)—obtained from Zecchi (Florence, Italy). The azurite, lead-tin yellow, and ochre were purchased from Kremer Pigmente

GmbH & Co. KG (Aichstetten, Germany). The oil binding media was prepared from a mixture of polymerised linseed oil (Umton Colours, Děčín, Czech Republic) and purified linseed oil (H. Schmincke & Co.-GmbH & Co. KG, Erkrath, Germany). A fresh hen egg was purchased from a local market. Only the egg yolk, meticulously separated from the egg white, was utilised as a binder.

3.3. Preparation of the Reference Wall Painting Samples

To simulate the historical samples and study the alteration mechanisms, lime plaster blocks were prepared according to the traditional historical recipes [67]. The base coat was made of one part white slaked lime (Vápenka Čertovy schody, Tmaň Czech Republic) and three parts white coarse silica sand (Tomi-písek s.r.o., Prague, Czech Republic). Then, the blocks were left to dry in a humid environment for four days. After the base coat matured, the surface was scratched with lines to remove any loose grains and to give a key for the finish coat. The finish coat was composed of one part slaked lime and two parts fine sharp sand (particle sizes up to 2 mm) and applied thinly on top of the base coat. The surface was smoothed with a felt float trowel and a felt brush (cat's tongue shape) and lightly dampened with water during the scouring process. Dry powdered pigments were hand-ground into the binding media on glass slides with a spatula and applied as approx. 1 cm^2 squares onto dry lime plaster (Figure 3). Samples (squares) of the lime plaster mixed with the binding media and pigment were taken from the block three days after drying and labelled as non-aged (NA). Then, the block underwent accelerated artificial ageing.

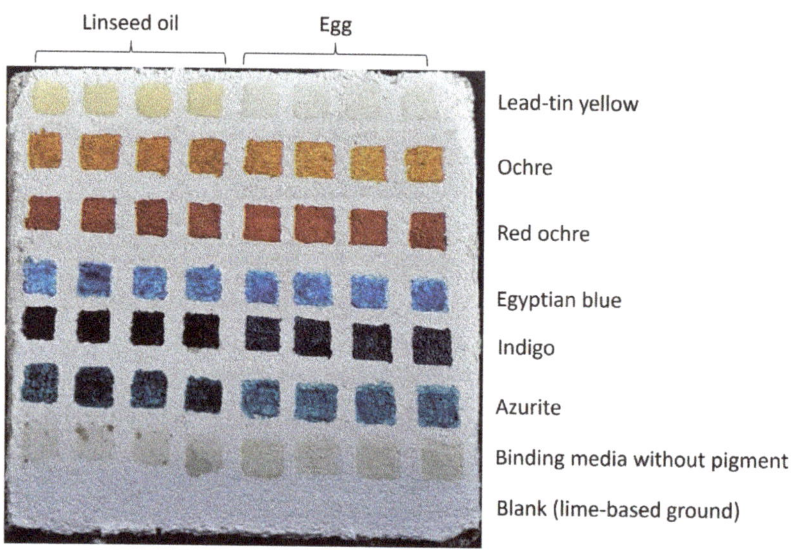

Figure 3. Lime-based plaster block with squares (1 cm^2) of linseed oil and egg binding media with six various pigments.

3.4. Accelerated Artificial Ageing

The artificial ageing was performed using a SOLARBOX 3000e RH chamber (CO.FO. ME.GRA., Milano, Italy) equipped with a xenon lamp, which replicates the total optical spectrum of the Sun. The parameters of ageing were set as follows: Xe lamp irradiation of 500 $W \cdot m^{-2}$, temperature of 30 °C, and relative humidity of ~75%. The block of lime plaster with the binders and pigments was kept under these conditions for 504 h (UV1). Afterwards, one aliquot part of the reference material was put into a glass vial and stored at −20 °C prior to the GC-MS analyses. The rest of the reference samples were placed back into the SOLARBOX chamber and aged repeatedly for 504 h twice (UV2, UV3) with a total duration of 1512 h.

3.5. Sample Preparation

A small amount of the reference material (approximately 0.1–0.2 mg) was placed into an autosampler vial with a glass conical insert and treated with 20 μL of derivatisation mixture. It consisted of one part 5% w/v methanolic solution of Meth-Prep II and two parts solvent mixture (methanol:toluene, 1:2, v/v). The resulting mixture was vortexed for 30 s, followed by sonication for 10 min, and then heated to 60 °C for 1 h. After cooling to laboratory temperature, the samples were centrifuged for 5 min at 4400 rpm. The supernatants were transferred into a new vial with conical inserts and analysed using GC-MS. A group of samples containing seven egg binder samples (one non-pigmented and six pigmented samples, including azurite, Egyptian blue, indigo, red ochre, ochre, and lead-tin yellow) and seven samples with a linseed oil binder (with the same pigments as in the egg) were analysed at each step of artificial ageing process (NA, UV1, UV2, and UV3). For each sample, there were three replicate injections, resulting in a total of 21 analyses for egg and 21 analyses for oil binder for each step of artificial ageing.

3.6. GC-MS Instrumentation and Conditions

The gas chromatographic analyses were carried out using two GC-MS instruments which were situated in Olomouc (Czech Republic) and in Vienna (Austria) and were of the same type—Agilent 6890, coupled with a mass selective detector (MSD) Agilent 5973N (Agilent, Santa Clara, CA, USA). The capillary columns OPTIMA-5 MS, 30 m × 0.25 mm × 0.25 μm (Macherey-Nagel, Düren, Germany), and J&W DB-5ms UI, 30 m × 0.25 mm × 0.25 μm (Agilent, Santa Clara, CA, USA), were operated with helium (5.5, Siad, Bergamo, Italy) as the carrier gas at a constant flow rate of 0.9 $mL \cdot min^{-1}$. The oven program was as follows: 50 °C for 1 min, 10 °C/min ramp to a final temperature of 320 °C, and held for 12 min. The total run time was 40.00 min. A 2 μL aliquot of the derivatised extract was injected in pulsed splitless mode (206.8 kPa, 18 s, 300 °C). Identification of the compounds was based on a comparison of the respective mass spectra of the analysed compounds with the spectra of analytical standards or with the spectra in NIST Library 2020 (National Institute of Standards and Technology, Gaithersburg, MD, USA).

3.7. Data Analysis

The statistical software QC.Expert 3.2 (TriloByte Ltd., Pardubice, Staré Hradiště, CZ) was used for verification of the data normality, which was checked visually (histogram and Q-Q plot) and using statistical tests: the Jarque–Bera test and the Kolmogorov–Smirnov test (K–S test). The Jarque–Bera test is based on the sample skewness and kurtosis [68,69], while the K–S test is based on the empirical distribution function (ECDF) [70,71]. The software was also used for comparison of the obtained values for the egg and linseed oil binding media using an unpaired two-tailed Student's *t*-test. This *t*-test was also used for comparison of the results obtained using the GC-MS 1 and GC-MS 2 instruments.

The GC-MS data acquired underwent analysis through unsupervised principal component analysis (PCA), incorporating univariate scaling. The resulting transformed data matrix was then transferred to the R statistical software (version 4.1.2, R Core Team 2021, Vienna, Austria), a freely available statistical computing environment. The default package "stats" was employed for conducting the PCA and facilitating visualisation. The results were evaluated by generating a biplot and employing k-means clustering to group the data into *k* clusters. The clusters are represented by 95% confidence intervals around their center (centroid), which correspond to the means of the points assigned to each cluster.

4. Conclusions

GC-MS stands as a well-established technique for lipid binder identification, relying on the diagnostic fatty acid ratios P/S, A/P, A/Su, and O/S and specific parameters ($\sum$D and COP). The P/S ratio, historically crucial for distinguishing between egg and oil binders, faces scrutiny due to potential influences from various factors. In this study, reference samples of wall paintings with oil and egg binders and six different pigments were analysed using two GC-MS instruments with minor hardware differences (type of stationary phase), aiming to assess the parameter variations. The impact of artificial ageing on the model samples was also investigated. While most of the parameters changed as expected during ageing, the P/S ratio remained nearly constant. Ageing induced a decrease in unsaturated fatty acids and the emergence of dicarboxylic acids (azelaic, sebacic, and suberic), which are the oxidation products of these unsaturated acids. Both binder types exhibited altered values in parameters such as a rise in the A/P ratio and $\sum$D and a decrease in O/S and A/Su. A notable discrepancy in the P/S values of the aged egg binder between GC-MS 1 (1.8 ± 0.3) and GC-MS 2 (2.9 ± 0.2) prompted the exploration of additional markers, leading to the identification of cholesterol oxidation products, notably cholesta-3,5-dien-7-one, specific to eggs. However, the identification of such markers is challenging concerning the amount of the samples, the binder-to-pigment ratio, and especially the initial amount of cholesterol from which these products are subsequently created. Consequently, PCA, integrating multiple parameters, proved effective in distinguishing between the oil and egg binders, even when the P/S values deviated from the literature references. In conclusion, this study emphasises the complexities of identifying wall painting binders, particularly lipidic ones, and the importance of alternative markers and innovative analytical methods for comprehensive analysis and conservation efforts. The application of PCA provides a promising avenue for effective differentiation between egg and oil binders.

Supplementary Materials: The following supporting information can be downloaded at https://www.mdpi.com/article/10.3390/molecules29071520/s1. Figure S1: Total ion chromatogram (TIC) of linseed oil sample with indigo pigment after artificial ageing (UV3). Identified fatty acid methyl esters are numbered (1-21), and degradation products of indigotin (A, B, and D) and indirubin (C) are alphabetically designated (A–D). The compounds are listed in Table S1; Table S1: Fatty acids and degradation products of indigotin and indirubin were identified in the gas chromatogram of the artificially aged (UV3) linseed oil sample with indigo pigment. The same compounds were also identified in the egg binder. The identification of compounds was based on a comparison with the corresponding mass spectra in the NIST library and with a study [72].

Author Contributions: Conceptualisation, J.N., P.B. and V.P.; methodology, J.N. and P.B.; formal analysis, J.N.; investigation, J.N. and V.P.; writing—original draft preparation, J.N. and P.B.; writing—review and editing, J.N., P.B., V.P., O.K. and L.K.; visualisation, J.N. and O.K.; supervision, P.B. and V.P.; funding acquisition, P.B. and L.K. All authors have read and agreed to the published version of the manuscript.

Funding: This research was funded by ERDF/ESF "Arteca: Advanced physical-chemical methods of research and protection of cultural and artistic heritage" (No. CZ.02.1.01/0.0/0.0/17_048/0007378), by the institutional projects of Palacký University (IGA_PrF_2023_027, IGA_PrF_2024_026), and by the Czech Science Foundation (project 23-07284K).

Institutional Review Board Statement: Not applicable.

Informed Consent Statement: Not applicable.

Data Availability Statement: Data will be made available on request.

Acknowledgments: We would like to express our gratitude to David Hrabálek, who belongs to the Academy of Fine Arts in Prague. He prepared the reference wall painting samples and shared his extensive knowledge in the field of conservation science, which proved to be valuable contributions to this study.

Conflicts of Interest: The authors declare no conflicts of interest.

References

1. Sotiropoulou, S.; Sciutto, G.; Tenorio, A.L.; Mazurek, J.; Bonaduce, I.; Prati, S.; Mazzeo, R.; Schilling, M.; Colombini, M.P. Advanced Analytical Investigation on Degradation Markers in Wall Paintings. *Microchem. J.* **2018**, *139*, 278–294. [CrossRef]

2. Geddes Da Filicaia, E.; Evershed, R.P.; Peggie, D.A. Review of Recent Advances on the Use of Mass Spectrometry Techniques for the Study of Organic Materials in Painted Artworks. *Anal. Chim. Acta* **2023**, *1246*, 340575. [CrossRef]
3. Holclajtner-Antunović, I.; Stojanović-Marić, M.; Bajuk-Bogdanović, D.; Žikić, R.; Uskoković-Marković, S. Multi-Analytical Study of Techniques and Palettes of Wall Paintings of the Monastery of Žiča, Serbia. *Spectrochim. Acta Part A Mol. Biomol. Spectrosc.* **2016**, *156*, 78–88. [CrossRef]
4. Sfarra, S.; Ibarra-Castanedo, C.; Tortora, M.; Arrizza, L.; Cerichelli, G.; Nardi, I.; Maldague, X. Diagnostics of Wall Paintings: A Smart and Reliable Approach. *J. Cult. Herit.* **2016**, *18*, 229–241. [CrossRef]
5. Daffara, C.; Ambrosini, D.; Pezzati, L.; Paoletti, D. Thermal Quasi-Reflectography: A New Imaging Tool in Art Conservation. *Opt. Express* **2012**, *20*, 14746. [CrossRef]
6. Stout, S.; Cosentino, A.; Scandurra, C. Non-Invasive Materials Analysis Using Portable X-ray Fluorescence (XRF) in the Examination of Two Mural Paintings in the Catacombs of San Giovanni, Syracuse. In *Digital Heritage. Progress in Cultural Heritage: Documentation, Preservation, and Protection*; Ioannides, M., Magnenat-Thalmann, N., Fink, E., Žarnić, R., Yen, A.-Y., Quak, E., Eds.; Lecture Notes in Computer Science; Springer International Publishing: Cham, Switzerland, 2014; Volume 8740, pp. 697–705. [CrossRef]
7. Piqué, F.; Verri, G. *Organic Materials in Wall Paintings: Project Report*; The Getty Conservation Institute: Los Angeles, CA, USA, 2015; ISBN 978-1-937433-29-1.
8. Borg, B.; Dunn, M.; Ang, A.; Villis, C. The Application of State-of-the-Art Technologies to Support Artwork Conservation: Literature Review. *J. Cult. Herit.* **2020**, *44*, 239–259. [CrossRef]
9. Mora, P.; Mora, L.; Philippot, P. *La Conservazione Delle Pitture Murali*, 2nd ed.; Compositori: Bologna, Italy, 2001.
10. Cennini, C. *Il Libro Dell'arte*, 7th ed.; Frezzato, F., Ed.; Neri Pozza: Vicenza, Italy, 2012.
11. Cennini, C.; Broecke, L. *Cennino Cennini's 'Il Libro Dell'Arte': A New English Translation and Commentary with Italian Transcription*; Archetype: London, UK, 2015.
12. Mills, J.S.; White, R. *The Organic Chemistry of Museum Objects*, 2nd ed.; Butterworth-Heinemann Ltd.: Oxford, UK, 1994.
13. Schilling, M.; Khanjian, H. Gas Chromatographic Determination of the Fatty Acid and Glycerol Content of Lipids. I. The Effects of Pigments and Aging on the Composition of Oil Paints. In Proceedings of the ICOM Committee for Conservation Preprints, 11th Triennial Meeting, Edinburgh, UK, 1–6 September 1996; pp. 220–227.
14. Vallance, S.L. Critical Review Applications of Chromatography in Art Conservation: Techniques Used for the Analysis and Identification of Proteinaceous and Gum Binding Media. *Analyst* **1997**, *122*, 75R–81R. [CrossRef]
15. Regazzoni, L.; Cavallo, G.; Biondelli, D.; Gilardi, J. Microscopic Analysis of Wall Painting Techniques: Laboratory Replicas and Romanesque Case Studies in Southern Switzerland. *Stud. Conserv.* **2018**, *63*, 326–341. [CrossRef]
16. Colombini, M.P.; Modugno, F.; Fuoco, R.; Tognazzi, A. A GC-MS Study on the Deterioration of Lipidic Paint Binders. *Microchem. J.* **2002**, *73*, 175–185. [CrossRef]
17. Chiavari, G.; Gandini, N.; Russo, P.; Fabbri, D. Characterisation of Standard Tempera Painting Layers Containing Proteinaceous Binders by Pyrolysis (/Methylation)-Gas Chromatography-Mass Spectrometry. *Chromatographia* **1998**, *47*, 420–426. [CrossRef]
18. Silva, R.H.D. The Problem of The Binding Medium Particularly in Wall Painting. *Archaeometry* **1963**, *6*, 56–64. [CrossRef]
19. Casadio, F.; Giangualano, I.; Piqué, F. Organic Materials in Wall Paintings: The Historical and Analytical Literature. *Stud. Conserv.* **2004**, *49* (Suppl. S1), 63–80. [CrossRef]
20. Kouloumpi, E.; Lawson, G.; Pavlidis, V. The Contribution of Gas Chromatography to the Resynthesis of the Post-Byzantine Artist's Technique. *Anal. Bioanal. Chem.* **2007**, *387*, 803–812. [CrossRef]
21. Mills, J.S. The Gas Chromatographic Examination of Paint Media. Part I. Fatty Acid Composition and Identification of Dried Oil Films. *Stud. Conserv.* **1966**, *11*, 92–107. [CrossRef]
22. Laurie, A.P. *The Painter's Methods*; Dover Publications, Inc.: Mineola, NY, USA, 1988.
23. Theophilus. *On Divers Arts—The Foremost Medieval Treatise on Painting, Glassmaking and Metalwork*; Dover Publications, Inc.: New York, NY, USA, 1979.
24. van den Berg, J.D.J. *Analytical Chemical Studies on Traditional Linseed Oil Paints*; Molart: Amsterdam, The Netherlands, 2002.
25. White, R.; Pilc, J. Analyses of Paint Media. *Natl. Gallery Technol. Bull.* **1996**, *17*, 91–103.
26. Bensi, P.; Danti, C.; Matteini, M.; Moles, A. La Pellicola Pittorica Nella Pittura Murale in Italia: Materiali e Tecniche Esecutive Dall'alto Medioevo al XIX Secolo. In *Le Pitture Murali: Tecniche, Problemi, Conservazione*; Centro Di: Firenze, Italy, 1990; pp. 73–102.
27. Hess, M. *Hess's Paint Film Defects, Their Causes and Cure*, 3rd ed.; Chapman and Hall: London, UK, 1979.
28. Levison, H.W. Yellowing and Bleaching of Paint Films. *J. Am. Inst. Conserv.* **1985**, *24*, 69–76. [CrossRef]
29. Mills, J.; White, R. Organic Mass-Spectrometry of Art Materials: Work in Progress. *Natl. Gallery Technol. Bull.* **1982**, *6*, 3–18.
30. Manzano, E.; Rodriguez-Simón, L.R.; Navas, N.; Checa-Moreno, R.; Romero-Gámez, M.; Capitan-Vallvey, L.F. Study of the GC–MS Determination of the Palmitic–Stearic Acid Ratio for the Characterisation of Drying Oil in Painting: La Encarnación by Alonso Cano as a Case Study. *Talanta* **2011**, *84*, 1148–1154. [CrossRef]
31. Schilling, M.; Khanjian, H.; Carson, D. Fatty Acid and Glycerol Content of Lipids; Effects of Ageing and Solvent Extraction on the Composition of Oil Paints. *J. Technol. Archit. Environ.* **1997**, *5*, 71–78.
32. Manzano, E.; Rodríguez-Simón, L.R.; Navas, N.; Capitán-Vallvey, L.F. Non-Invasive and Spectroscopic Techniques for the Study of Alonso Cano's *Visitation* from the Golden Age of Spain. *Stud. Conserv.* **2021**, *66*, 298–312. [CrossRef]

33. Colombini, M.P.; Modugno, F. (Eds.) Organic Materials in Art and Archaeology. In *Organic Mass Spectrometry in Art and Archaeology*; Wiley: Hoboken, NJ, USA, 2009; pp. 1–36. [CrossRef]
34. Colombini, M.P.; Andreotti, A.; Bonaduce, I.; Modugno, F.; Ribechini, E. Analytical Strategies for Characterizing Organic Paint Media Using Gas Chromatography/Mass Spectrometry. *Acc. Chem. Res.* **2010**, *43*, 715–727. [CrossRef]
35. Bonaduce, I.; Carlyle, L.; Colombini, M.P.; Duce, C.; Ferrari, C.; Ribechini, E.; Selleri, P.; Tiné, M.R. New Insights into the Ageing of Linseed Oil Paint Binder: A Qualitative and Quantitative Analytical Study. *PLoS ONE* **2012**, *7*, e49333. [CrossRef]
36. Schilling, M.; Carson, D.; Khanjian, H. Gas Chromatographic Determination of the Fatty Acid and Glycerol Content of Lipids. IV. Evaporation of Fatty Acids and the Formation of Ghost Images by Framed Oil Paintings. In Proceedings of the ICOM Committee for Conservation Preprints, 12th Triennial Meeting, Lyon, France, 29 August–3 September 1999; pp. 242–247.
37. Tsakalof, A.K.; Bairachtari, K.A.; Chryssoulakis, I.D. Pitfalls in Drying Oils Identification in Art Objects by Gas Chromatography. *J. Sep. Sci.* **2006**, *29*, 1642–1646. [CrossRef]
38. Rampazzi, L.; Cariati, F.; Tanda, G.; Colombini, M.P. Characterisation of Wall Paintings in the Sos Furrighesos Necropolis (Anela, Italy). *J. Cult. Herit.* **2002**, *3*, 237–240. [CrossRef]
39. Tammekivi, E.; Vahur, S.; Vilbaste, M.; Leito, I. Quantitative GC–MS Analysis of Artificially Aged Paints with Variable Pigment and Linseed Oil Ratios. *Molecules* **2021**, *26*, 2218. [CrossRef]
40. Pitthard, V.; Griesser, M.; Stanek, S. Methodology and Application of GC-MS to Study Altered Organic Binding Media from Objects of the Kunsthistorisches Museum, Vienna. *Ann. Chim.* **2006**, *96*, 561–573. [CrossRef]
41. Wei, S. A Study of Natural Organic Binding Media Used in Artworks and of Their Ageing Behaviour by GC/FID and GC/MS. Ph.D. Thesis, Vienna University of Technology, Vienna, Austria, 2007.
42. Pitthard, V.; Finch, P.; Bayerová, T. Direct Chemolysis-Gas Chromatography-Mass Spectrometry for Analysis of Paint Materials. *J. Sep. Sci.* **2004**, *27*, 200–208. [CrossRef]
43. Colombini, M.P.; Modugno, F.; Giacomelli, M.; Francesconi, S. Characterisation of Proteinaceous Binders and Drying Oils in Wall Painting Samples by Gas Chromatography–Mass Spectrometry. *J. Chromatogr. A* **1999**, *846*, 113–124. [CrossRef]
44. Schlenk, H.; Gellerman, J.L. Esterification of Fatty Acids with Diazomethane on a Small Scale. *Anal. Chem.* **1960**, *32*, 1412–1414. [CrossRef]
45. Glastrup, J. A Note on the Analysis of the Binding Medium from a Phoenician Shipwreck. *Stud. Conserv.* **1995**, *40*, 65–68. [CrossRef]
46. Blau, K.; Halket, M.J. *Handbook of Derivatives for Chromatography*, 2nd ed.; John Wiley & Sons: Chichester, UK, 1993.
47. Seppänen-Laakso, T.; Laakso, I.; Hiltunen, R. Analysis of Fatty Acids by Gas Chromatography, and Its Relevance to Research on Health and Nutrition. *Anal. Chim. Acta* **2002**, *465*, 39–62. [CrossRef]
48. Eder, K. Gas Chromatographic Analysis of Fatty Acid Methyl Esters. *J. Chromatogr. B Biomed. Sci. App.* **1995**, *671*, 113–131. [CrossRef]
49. Pitthard, V.; Stanek, S.; Griesser, M.; Muxeneder, T. Gas Chromatography—Mass Spectrometry of Binding Media from Early 20th Century Paint Samples from Arnold Schönberg's Palette. *Chromatographia* **2005**, *62*, 175–182. [CrossRef]
50. Challinor, J.M. Review: The Development and Applications of Thermally Assisted Hydrolysis and Methylation Reactions. *J. Anal. Appl. Pyrolysis* **2001**, *61*, 3–34. [CrossRef]
51. MACHEREY-NAGEL GmbH & Co., KG. Available online: https://www.mn-net.com/gc-column-nonpolar-low-polar-optima-5-ms-30-m-l-0.25-mm-id-0.25-m-df-726220.30 (accessed on 14 February 2024).
52. Agilent Technologies. Available online: https://www.agilent.com/en/product/gc-columns/low-bleed-gc-ms-columns/db-5ms-columns (accessed on 14 February 2024).
53. Izzo, F.C. 20th Century Artist's Oil Paints: A Chemical-Physical Survey. Ph.D. Thesis, Ca' Foscari University, Venice, Italy, 2011.
54. Colombini, M.P.; Modugno, F.; Menicagli, E.; Fuoco, R.; Giacomelli, A. GC-MS Characterization of Proteinaceous and Lipid Binders in UV Aged Polychrome Artifacts. *Microchem. J.* **2000**, *67*, 291–300. [CrossRef]
55. Degano, I.; La Nasa, J.; Ghelardi, E.; Modugno, F.; Colombini, M.P. Model Study of Modern Oil-Based Paint Media by Triacylglycerol Profiling in Positive and Negative Ionization Modes. *Talanta* **2016**, *161*, 62–70. [CrossRef]
56. Pitthard, V.; Griesser, M.; Stanek, S.; Bayerova, T. Study of Complex Organic Binding Media Systems on Artworks Applying GC-MS Analysis: Selected Examples from the Kunsthistorisches Museum, Vienna. *Macromol. Symp.* **2006**, *238*, 37–45. [CrossRef]
57. Antova, G.A.; Gerzilov, V.T.; Petkova, Z.Y.; Boncheva, V.N.; Bozhichkova, I.N.; St Penkov, D.; Petrov, P.B. Comparative Analysis of Nutrient Content and Energy of Eggs from Different Chicken Genotypes. *J. Sci. Food Agric.* **2019**, *99*, 5890–5898. [CrossRef]
58. Andreotti, A.; Bonaduce, I.; Colombini, M.P.; Gautier, G.; Modugno, F.; Ribechini, E. Combined GC/MS Analytical Procedure for the Characterization of Glycerolipid, Waxy, Resinous, and Proteinaceous Materials in a Unique Paint Microsample. *Anal. Chem.* **2006**, *78*, 4490–4500. [CrossRef]
59. Serefidou, M.; Bracci, S.; Tapete, D.; Andreotti, A.; Biondi, L.; Colombini, M.P.; Giannini, C.; Parenti, D. Microchemical and Microscopic Characterization of the Pictorial Quality of Egg-Tempera Polyptych, Late 14th Century, Florence, Italy. *Microchem. J.* **2016**, *127*, 187–198. [CrossRef]
60. Mills, J.S.; White, R. The Identification of Paint Media from the Analysis of Their Sterol Composition: A Critical View. *Stud. Conserv.* **1975**, *20*, 176. [CrossRef]

61. Van Den Brink, O.F.; Ferreira, E.S.B.; Van Der Horst, J.; Boon, J.J. A Direct Temperature-Resolved Tandem Mass Spectrometry Study of Cholesterol Oxidation Products in Light-Aged Egg Tempera Paints with Examples from Works of Art. *Int. J. Mass. Spectrom.* **2009**, *284*, 12–21. [CrossRef]
62. Boon, J.J.; Peulvé, S.; Van Den Brink, O.F.; Duursma, M.; Rainford, D. Molecular Aspects of Mobile and Stationary Phases in Ageing Tempera and Oil Paint Films. In *Early Italian Paintings—Techniques and Analysis*; Bakkenist, T., Hoppenbrouwers, R., Dubois, H., Eds.; Limburg Conservation Institute: Maastricht, The Netherlands, 1997; pp. 35–56.
63. Buckley, S.A.; Stott, A.W.; Evershed, R.P. Studies of Organic Residues from Ancient Egyptian Mummies Using High Temperature-Gas Chromatography-Mass Spectrometry and Sequential Thermal Desorption-Gas Chromatography-Mass Spectrometry and Pyrolysis-Gas Chromatography-Mass Spectrometry. *Analys* **1999**, *124*, 443–452. [CrossRef] [PubMed]
64. Herchi, W.; Harrabi, S.; Sebei, K.; Rochut, S.; Boukhchina, S.; Pepe, C.; Kallel, H. Phytosterols Accumulation in the Seeds of *Linum usitatissimum* L. *Plant Physiol. Biochem.* **2009**, *47*, 880–885. [CrossRef] [PubMed]
65. Rubner-Institut, M. Fatty Acid Composition, Tocopherol and Sterol Contents in Linseed (*Linum usitatissimum* L.) Varieties. *Iran. J. Chem. Chem. Eng.* **2017**, *36*, 147–152.
66. Colombini, M.P.; Modugno, F.; Giacomelli, A. Two Procedures for Suppressing Interference from Inorganic Pigments in the Analysis by Gas Chromatography–Mass Spectrometry of Proteinaceous Binders in Paintings. *J. Chromatogr. A* **1999**, *846*, 101–111. [CrossRef]
67. Cennini d'Andrea, C.; Thompson, D.V. *The Craftman's Handbook: The Italian "Il Libro Dell'Arte"*; Dover Publications, Inc.: New York, NY, USA, 1954.
68. Jarque, C.M.; Bera, A.K. Efficient Tests for Normality, Homoscedasticity and Serial Independence of Regression Residuals. *Econ. Lett.* **1980**, *6*, 255–259. [CrossRef]
69. Jarque, C.M.; Bera, A.K. A Test for Normality of Observations and Regression Residuals. *Int. Stat. Rev./Rev. Int. Stat.* **1987**, *55*, 163. [CrossRef]
70. Kolmogorov, A. Sulla Determinazione Empirica Di Una Legge Di Distribuzione. *G. Dell'ist. Ital. Degli Attuari* **1933**, *4*, 83–91.
71. Smirnov, N. Sur Les Ecarts de La Courbe de Distribution Empirique. *Recl. Mathématique (Mat. Sb.)* **1939**, *6*, 3–26.
72. Poulin, J. A New Methodology for the Characterisation of Natural Dyes on Museum Objects Using Gas Chromatography–Mass Spectrometry. *Stud. Conserv.* **2018**, *63*, 36–61. [CrossRef]

MDPI AG
Grosspeteranlage 5
4052 Basel
Switzerland
Tel.: +41 61 683 77 34

Molecules Editorial Office
E-mail: molecules@mdpi.com
www.mdpi.com/journal/molecules

www.ingramcontent.com/pod-product-compliance
Lightning Source LLC
Chambersburg PA
CBHW061547210926
51850CB00009B/2020
9783725864768